The Ecological and Societal Consequences of Biodiversity Loss

SCIENCES

Ecosystems and Environment,
Field Directors – Dominique Joly and Françoise Gaill

Biodiversity, Subject Head – Fabienne Aujard

The Ecological and Societal Consequences of Biodiversity Loss

Coordinated by
Michel Loreau
Andy Hector
Forest Isbell

iSTE

WILEY

First published 2022 in Great Britain and the United States by ISTE Ltd and John Wiley & Sons, Inc.

ISTE Ltd
27-37 St George's Road
London SW19 4EU
UK

www.iste.co.uk

John Wiley & Sons, Inc.
111 River Street
Hoboken, NJ 07030
USA

www.wiley.com

Library of Congress Control Number: 2021950400

British Library Cataloguing-in-Publication Data
A CIP record for this book is available from the British Library
ISBN 978-1-78945-072-9

ERC code:
LS8 Ecology, Evolution and Environmental Biology
LS8_1 Ecosystem and community ecology, macroecology
LS8_2 Biodiversity, conservation biology, conservation genetics

MIX
Paper from responsible sources
FSC® C013604

Contents

Introduction

The Ecological and Societal Consequences of Biodiversity Loss

Michel LOREAU[1], Andy HECTOR[2], and Forest ISBELL[3]

[1] *Theoretical and Experimental Ecology Station, CNRS, Moulis, France*

[2] *University of Oxford, UK*

[3] *University of Minnesota, St. Paul, USA*

One of the distinctive and fascinating features of ecological systems is their extraordinary complexity. An ecosystem is often composed of thousands of different species that interact in myriad ways at the scale of a single hectare. Each species is composed of many individuals that vary due to differences in their genetics and their particular experience of their local environment. These complex local systems are strongly connected to each other, and aggregate into larger and larger entities, from the landscape scale to that of the entire biosphere, where it becomes evident that they exert a major influence on the physical and chemical properties of our planet. How can such enormously complex systems be studied?

During the second half of the 20th century, two increasingly divergent approaches to ecological systems developed within ecology, which have gradually led to two largely distinct disciplines, community ecology and ecosystem ecology. A community is defined broadly as a set of species that live together in some place. The focus in community ecology has traditionally been on species diversity: what exogenous and endogenous forces lead to more or less diverse communities? How do species interactions constrain the number of species that can coexist? What patterns emerge from these interactions? An ecosystem is the entire system of biotic and abiotic components that interact in some place. The ecosystem concept is broader than the community concept because it includes a wide range of biological, physical, and chemical processes that connect organisms and their environment. But the focus in ecosystem ecology has traditionally been on the overall functioning of ecosystems as distinct entities: how is energy captured, transferred, and ultimately dissipated in different ecosystems? How are limiting nutrients recycled, thereby ensuring the renewal of the material elements necessary for growth? What factors and processes control energy and material flows, from local to global scales?

In a sense, community ecology provides a microscopic perspective on ecosystems because it analyzes their parts, while ecosystem ecology provides a macroscopic perspective on the same systems because it studies them as a whole. The distinction between micro- and macroscopic, however, does not necessarily apply to the spatial scales considered by the two disciplines. Although much of community ecology does consider species interactions at small scales, a growing fringe, known as macroecology, considers patterns of species diversity and species distributions at vast spatial scales. The focus on species – species distributions, species diversity, species interactions – is more central to the community approach than the spatial scale considered. Similarly, ecosystem ecology studies the fluxes of energy and materials at various spatial scales. What distinguishes the ecosystem approach is its focus on the system as a whole, often without considering the species that compose it.

At a time when humankind is rising to the status of a major global biogeochemical force and raising the prospect of a global ecological crisis, it is important to step back and ask whether individually studying communities and ecosystems is the best path to follow. Human environmental impacts include the destruction and fragmentation of natural habitats, pollution, climate change, overexploitation of biological resources, homogenization of biota, and biodiversity loss. These impacts affect species and ecosystems indistinctly. Moreover, they interact with each other, which may lead to synergistic effects. For instance, climate change is likely to cause massive additional biodiversity loss. Biodiversity loss in turn is likely to decrease the ability of ecosystems to resist the effects of climate change, with possible feedbacks on the climate system itself. Species, communities, and ecosystems have always been inextricably linked, but the major disruptions generated by humans in the current period make this reality plainly obvious. A synthetic approach to ecology, which integrates populations, communities, and ecosystems, is required to develop appropriate responses to the global ecological crisis we are entering.

Community ecology is a dynamic field of research in which knowledge has accumulated rapidly during the last 60 years or so based largely on a modern hypothetico-deductive approach. However, theories and hypothesis building have often outpaced empirical studies and hypothesis testing in community ecology, which hinders steady scientific progress. As a result, this subdiscipline has few "laws" or robust generalizations, except for some large-scale empirical patterns such as species–area relationships (Rosenzweig 1999). In contrast, ecosystem ecology is a subdiscipline that has traditionally had a strong empirical basis. Its theories are largely based on inductive generalizations from field measurements, with comparatively few theory-driven hypotheses and experimental tests. There is no doubt that the ecosystem approach has been instrumental in developing our understanding of the global biogeochemistry of the Earth system and of current global environmental changes. Yet, despite these successes, a number of authors have questioned its relatively static view of ecological systems and even its scientific relevance, calling for a fundamental rejuvenation of the discipline (O'Neill 2001). Strengthening theory, experimental tests and their interactions, and paying due attention to ecological dynamics and complexity are key ingredients of such a rejuvenation.

On balance, community ecology and ecosystem ecology provide two perspectives on complex ecological systems that have largely complementary strengths and weaknesses. Both disciplines have been called into question, and each would benefit from the perspective developed by the other. Developing theories about interactions between species and between these and their environment with the ultimate goal of predicting ecosystem functioning and ecosystem services would help to focus community ecology on issues that are both scientifically important and socially relevant. Incorporating the diversity, complexity, and dynamical nature of communities in its view of ecosystem functioning would help ecosystem ecology to be livelier and to provide more reliable, if probably more uncertain, predictions. It is becoming increasingly clear that merging the two perspectives is necessary both to ensure continued scientific progress and to provide society with the scientific means to face growing environmental challenges (Loreau 2010a, 2010b). A more integrative ecology also needs to include humans, not just as an external force that disrupts ecosystems, but as an integral part of the biosphere that interacts with its other components.

The new biodiversity and ecosystem functioning (BEF) research field that emerged in the 1990s and expanded over the last decades has greatly contributed to moving ecology forward in that direction. The idea that plant diversity enhances plant biomass production is arguably foundational in ecology, dating back to Darwin's Principle of Divergence (McNaughton 1993; Hector and Hooper 2002), but this idea did not catch on for more than a century because of the success of Liebig's law of the minimum and the lack of rigorous theory and experimental designs in agricultural sciences. Agriculture gradually shifted to a new model based on the industrial production of artificial fertilizers, the mechanization of agriculture, and the use of monocultures of artificially selected crop types. It is only when the detrimental ecological consequences of the modern industrial model, and in particular the threat of biodiversity loss, started to be widely recognized at the end of the last century that interest in the effects of biodiversity loss on ecosystem functioning emerged. This interest then spread rapidly, penetrated experimental and theoretical ecology, and led to the emergence of an entire new research field at the interface between community and ecosystem ecology (Loreau *et al.* 2001, 2002; Hooper *et al.* 2005; Naeem *et al.* 2009; Loreau 2010a; Cardinale *et al.* 2012; Tilman *et al.* 2014).

Interest in this issue grew largely out of practical concerns about the potential ecological consequences of current biodiversity loss caused by the increased impact of human activities on natural and managed ecosystems. There is growing recognition that the world's ecosystems provide society with a wide range of "ecosystem services" (Millennium Ecosystem Assessment 2005), or "nature's contributions to people" (Diaz *et al.* 2018), that are crucial to human well-being and sustainable development. These services or contributions are derived from the normal functioning of ecosystems, raising the important question whether impoverished ecosystems may in some way function less efficiently than the more species-rich systems from which they are derived, and hence gradually lose their ability to deliver ecosystem services to human societies. However, beyond this eminently practical motivation, the new BEF research field has had a much broader and deeper transformative role in ecology.

One of its main benefits has been to foster integration of community ecology and ecosystem ecology. Ecology has traditionally regarded, implicitly or explicitly, species diversity as an epiphenomenon driven by a combination of abiotic environmental factors (such as temperature, rainfall, and soil fertility), ecosystem processes that are themselves determined by these abiotic factors (such as productivity, biomass, and nutrient cycling), and biotic interactions within communities (such as competition and predation). This tenet was shared by both community ecology and ecosystem ecology. Community ecology was devoted historically to explaining patterns and processes of species coexistence and diversity. Ecosystem ecology often ignored species diversity as some sort of "background noise" irrelevant to ecosystem functioning. Thus, the two disciplines considered species diversity in contrasting ways, which explains their historical divergence. However, both have shared the basic assumption that species diversity is an epiphenomenon. The new BEF research field has overthrown this central tenet by considering biodiversity – in particular species and genetic diversity – as a driver of ecosystem functioning (Naeem 2002; Loreau 2010b). This simple paradigmatic change has had far deeper consequences than might appear at first sight. If biodiversity affects ecosystem functioning, ecosystem ecologists can no longer ignore the dynamics of biodiversity within ecosystems. Similarly, community ecologists can no longer ignore the potential feedback that biodiversity has on its own maintenance through ecosystem functioning. The basis for the historical separation of the two disciplines then vanishes, even though it may take some time until both sides recognize the full implications of this change.

The BEF research field also contributed to transforming ecology in other ways. Historically, ecology had an unfortunate propensity to disconnect empirical and theoretical research, with a profusion of poorly generalized empirical data and an equal profusion of poorly tested theories. As a result of this disconnect, there were few attempts at resolving controversies through consensus building within the ecological scientific community, leading to periodic shifts in fashionable research topics. In this context, the scientific process through which the BEF research field developed was quite remarkable. First, the BEF research field established a strong foothold in controlled experiments, which were still relatively rare in ecology at that time. Since the first flagship BEF experiments in the 1990s (Naeem *et al.* 1994; Tilman *et al.* 1997a; Hector *et al.* 1999), several hundred BEF experiments have been carried out to assess the repeatability of previous findings, determine their level of generality, or disentangle the mechanisms underlying the effects of biodiversity on ecosystem functioning (Cardinale *et al.* 2012; O'Connor *et al.* 2017). Second, theory quickly developed hand in hand with experiments (Tilman *et al.* 1997b; Loreau 1998, 2000) and provided a valuable approach to interpret and generalize the results of BEF experiments. Theory played a key role in resolving the early scientific controversies that arose over the interpretation of BEF experiments along with this high level of replication studies (Huston 1997).

Scientific controversies are often the result of a lack of clarity in the theoretical framework, a lack of appropriate tools, or a lack of sufficient empirical evidence to distinguish among clearly identified competing hypotheses. The controversies over BEF were no exception and combined all three problems. Theory helped to resolve these controversies in two ways. First, it helped to clarify the conceptual framework in which BEF experiments were being conceived and interpreted, yielding a consensus on the possible mechanisms and outcomes of these experiments (Loreau *et al.* 2001). Second, it helped to devise a new additive partition methodology inspired by (although conceptually different from) the Price equation in evolutionary genetics to disentangle the two main classes of mechanisms underlying the results of BEF experiments

(Loreau and Hector 2001). Application of this methodology to multiple BEF experiments revealed that the positive effects of biodiversity effects on ecosystem functioning are usually driven by a combination of both classes of mechanisms (Cardinale *et al.* 2012), but functional complementarity between species tends to play an increasingly important role as biodiversity and time after experimental manipulation increase (Loreau and Hector 2001; Cardinale *et al.* 2007; Fargione *et al.* 2007; Reich *et al.* 2012). The tight link between theory and experiments is a major legacy of the BEF research field. Experiments helped to build a more practically oriented theory, and theory helped to analyze and interpret experiments more effectively. This interactive process allowed clear but balanced conclusions to be drawn, thereby providing healthy ground for consensus building and future studies.

Like many successful new research fields, BEF research did not remain confined in its initial scientific boundaries, but expanded its scope to encompass a wide range of fundamental issues in ecology, such as the functioning of food webs (Thébault and Loreau 2003; Duffy *et al.* 2007), the spread of diseases (Keesing *et al.* 2006), and the spatial dynamics of metacommunities (Loreau *et al.* 2003), as well as re-igniting interest in the relationship between the diversity and stability of ecological systems. Progress achieved on the diversity–stability relationship has been particularly significant given the long, controversial history of this issue in ecology (Pimm 1984; McCann 2000; Ives and Carpenter 2007). Two decisive features distinguish the new theoretical and experimental approaches to the diversity–stability relationship developed within the BEF research agenda from earlier ones: first, these new approaches have explicitly differentiated, and linked, stability properties at the population level and at the aggregate community or ecosystem level; and, second, they have abandoned the traditional assumptions that the environment is constant and that populations and ecosystems reach an equilibrium in order to explicitly incorporate population dynamical responses to environmental fluctuations. As a result, we now have a much better and more complete understanding of the diversity of diversity–stability relationships that can arise at different levels of organization and of the mechanisms that generate them (Loreau and de Mazancourt 2013).

Lastly, because its initial impetus was provided by the societal relevance of the issues it was addressing, BEF research has also had a strong impact on biodiversity management and policy. It has done this by reaching out to and attracting applied ecologists as well as economists and social scientists working on environmental issues (Naeem *et al.* 2009). This has allowed the field to spread yet further and influence policy. In particular, it provided vital scientific knowledge in support of the ongoing slow but steady shift towards more sustainable practices in agriculture and forestry, and played an important role in major international assessments such as the Millennium Ecosystem Assessment (2005) and the Intergovernmental science–policy Platform on Biodiversity and Ecosystem Services.

This book bears witness to the extraordinary scientific expansion and progress of BEF research during the last three decades. It offers a comprehensive synthesis of current scientific knowledge on both the ecological and societal consequences of biodiversity loss for a wide audience of upper undergraduate students, postgraduate students, and academic and research staff. Although no book can perform all roles, we hope it will provide both an up-to-date overview of the field and serve as a textbook to support teaching in this area. In this respect, this book differs from previous books on BEF (Loreau *et al.* 2002; Naeem *et al.* 2009), the goal of which was primarily to provide research syntheses and perspectives.

The book is organized in five parts. The first part provides a general introduction to biodiversity, ecosystems, and the various concepts that are used throughout the book. Chapter 1 sets the stage for the book by providing a broad overview of the major changes in biodiversity that have occurred in the past, that are occurring in the present, and that are likely to occur in the future. Chapter 2 explains in more detail what biodiversity is and how it can be measured, which turns out to be a much more complex issue than most people imagine. Chapter 3 explains what ecosystems are, how they are shaped by abiotic and biotic factors, how they change over time and space, how they function, and how they affect and are affected by humans.

The second part of the book reviews the core principles and results of BEF research, that is, how biodiversity affects ecosystem functioning. As mentioned above, BEF research played a transformative role in ecology by tightly linking theory and experiments. The first two chapters in this part reflect this evolution. Chapter 4 provides the theoretical foundations of BEF research. It explains why biodiversity tends generally to promote ecosystem functioning from basic niche theory, and how this prediction can be extended to more complex ecosystems that have multiple trophic levels and are heterogeneous in space and time. Chapter 5 reviews the experimental evidence for how biodiversity affects ecosystem functioning. It explains why experiments are a vital tool to increase ecological knowledge, how experiments tested BEF theory and provided insights into underlying mechanisms, and how they extended classic theory. Chapter 6 shows how the basic knowledge gained from theory and experiments can be used to design and interpret observational studies. Ultimately, theory and experiments are useful to the extent that they help understand the real world. Disentangling the relationships between multiple dimensions of biodiversity and changing environmental conditions, management context, and ecosystem functioning, however, poses serious methodological challenges under natural conditions. Chapter 6 examines the benefits, challenges, and results of analyzes observational data, as well as possible ways in which the design and analysis of observational studies can be improved.

The third part of the book reviews current knowledge on how biodiversity affects ecosystem stability. As mentioned above, the relationship between the diversity and stability of ecological systems has been a controversial topic through the history of ecology. It is remarkable that BEF research was able to provide fresh perspectives on this topic, and largely resolve the historical debate, in a relatively short time. Again, the combination of theory and experiments played a key role in this progress. Therefore, this part adopts the same structure as the previous one. Chapter 7 first reviews the key theoretical advances that made progress possible. It clarifies the various meanings of the multifaceted stability concept, shows how its various components are connected, examines the mechanisms that explain the stabilizing effect of biodiversity on ecosystem stability, and discusses how diversity–stability relationships can be scaled up in space. Chapter 8 then provides an in-depth review of the available experimental evidence for these theoretical predictions. It shows that experiments have largely confirmed both the main predictions and the mechanisms identified by the new theory. Lastly, Chapter 9 assesses the balance of evidence regarding the direction of biodiversity–stability relationships and their underlying mechanisms in natural ecosystems, based on both observational studies along natural gradients of biodiversity and global change experiments assessing how environmental drivers influence temporal stability.

The fourth part of the book addresses some of the social and economic consequences of BEF research, in particular how biodiversity affects the provision of ecosystem services to human societies. Chapter 10 reviews the multiple documented benefits of crop genetic and species diversity in agriculture and forestry and argues that biodiversity has the potential to support a new green revolution in managed ecosystems. Chapter 11 examines the linkages between biodiversity and human health. It shows that the new Anthropocene era is re-organizing the great ecological and evolutionary game between microbes, hosts, non-hosts, and humans, leading to new pandemics such as the recent COVID-19 outbreak, and argues that facing future pandemics will require an acute consciousness that human health is intimately linked to biodiversity. Chapter 12 provides a roadmap of the economic valuation of biodiversity and ecosystem services. It seeks to clarify what economic valuation is (and is not) and provide a brief introduction to the existing methods of valuation, highlighting their applications and limitations.

The fifth and final part of the book zooms out to provide a global view of the future of biodiversity in our changing planet. Chapter 13 emphasizes the linkages and feedbacks that exist between biodiversity change and climate change – today's twin major global environmental crises. It describes how life on Earth responds to climate change in both terrestrial and aquatic ecosystems, how this response will affect ecosystem functioning, and how human societies can cope with these twin environmental crises. Chapter 14 shows how biodiversity and human societies are inextricably connected by multilayer feedbacks that alter decision-making, impact agriculture production, influence well-being, alter income groups, and shape ecosystems. It argues that taking these feedbacks into account is critical to devise successful biodiversity policies and management strategies. Lastly, Chapter 15 reminds us that protecting and restoring biodiversity is crucial for the sustainability of ecosystem functioning, stability, and services and discusses the merits and limitations of two widespread conservation strategies, protected areas and ecological restoration. It also describes how protecting biodiversity can simultaneously help protect ecosystem functioning and services and how increasing the levels of biodiversity in restoration schemes can increase their efficacy.

We hope the content of this book will inspire early-career students and researchers to further advance this science and apply this knowledge to conserve the world's biodiversity and ecosystems.

References

Cardinale, B.J., Wright, J.P., Cadotte, M.W. *et al.* (2007). Impacts of plant diversity on biomass production increase through time because of species complementarity. *Proceedings of the National Academy of Sciences of the USA*, 104, 18123–18128.

Cardinale, B.J., Duffy, J.E., Gonzalez, A. *et al.* (2012). Biodiversity loss and its impact on humanity. *Nature*, 486, 59–67.

Diaz, S., Pascual, U., Stenseke, M. *et al.* (2018). Assessing nature's contributions to people. *Science*, 359, 270–272.

Duffy, J.E., Cardinale, B.J., France, K.E., McIntyre, P.B., Thébault, E., Loreau, M. (2007). The functional role of biodiversity in ecosystems: Incorporating trophic complexity. *Ecology Letters*, 10, 522–538.

Fargione, J., Tilman, D., Dybzinski, R. *et al.* (2007). From selection to complementarity: Shifts in the causes of biodiversity-productivity relationships in a long-term biodiversity experiment. *Proceedings of the Royal Society B*, 274, 871–876.

Hector, A. and Hooper, R. (2002). Darwin and the first ecological experiment. *Science*, 295, 639–640.

Hector, A., Schmid, B., Beierkuhnlein, C. *et al.* (1999). Plant diversity and productivity experiments in European grasslands. *Science*, 286, 1123–1127.

Hooper, D.U., Chapin, F.S., Ewel, J.J. *et al.* (2005). Effects of biodiversity on ecosystem functioning: A consensus of current knowledge. *Ecological Monographs*, 75, 3–35.

Huston, M.A. (1997). Hidden treatments in ecological experiments: Re-evaluating the ecosystem function of biodiversity. *Oecologia*, 110, 449–460.

Ives, A.R. and Carpenter, S.R. (2007). Stability and diversity of ecosystems. *Science*, 317, 58–62.

Keesing, F., Holt, R.D., Ostfeld, R.S. (2006). Effects of species diversity on disease risk. *Ecology Letters*, 9, 485–498.

Loreau, M. (1998). Biodiversity and ecosystem functioning: A mechanistic model. *Proceedings of the National Academy of Sciences of the USA*, 95, 5632–5636.

Loreau, M. (2000). Biodiversity and ecosystem functioning: Recent theoretical advances. *Oikos*, 91, 3–17.

Loreau, M. (2010a). *From Populations to Ecosystems: Theoretical Foundations for a New Ecological Synthesis. Monographs in Population Biology*. Princeton University Press, Princeton, NJ.

Loreau, M. (2010b). Linking biodiversity and ecosystems: Towards a unifying ecological theory. *Philosophical Transactions of the Royal Society B*, 365, 49–60.

Loreau, M. and Hector, A. (2001). Partitioning selection and complementarity in biodiversity experiments. *Nature*, 412, 72–76.

Loreau, M. and de Mazancourt, C. (2013). Biodiversity and ecosystem stability: A synthesis of underlying mechanisms. *Ecology Letters*, 16(S1), 106–115.

Loreau, M., Naeem, S., Inchausti, P. *et al.* (2001). Biodiversity and ecosystem functioning: Current knowledge and future challenges. *Science*, 294, 804–808.

Loreau, M., Naeem, S., Inchausti, P. (2002). *Biodiversity and Ecosystem Functioning: Synthesis and Perspectives*. Oxford University Press, Oxford.

Loreau, M., Mouquet, N., Gonzalez, A. (2003). Biodiversity as spatial insurance in heterogeneous landscapes. *Proceedings of the National Academy of Sciences of the USA*, 100, 12765–12770.

McCann, K.S. (2000). The diversity–stability debate. *Nature*, 405, 228–233.

McNaughton, S.J. (1993). Biodiversity and stability of grazing ecosystems. In *Biodiversity and Ecosystem Function*, Schulze, E.-D. and Mooney, H.A. (eds). Springer-Verlag, Berlin.

Millennium Ecosystem Assessment (2005). *Ecosystems and Human Well-being: Biodiversity Synthesis*. Millennium Ecosystem Assessment. World Resources Institute, Washington, DC.

Naeem, S. (2002). Ecosystem consequences of biodiversity loss: The evolution of a paradigm. *Ecology*, 83, 1537–1552.

Naeem, S., Thompson, L.J., Lawler, S.P., Lawton, J.H., Woodfin, R.M. (1994). Declining biodiversity can alter the performance of ecosystems. *Nature*, 368, 734–737.

Naeem, S., Bunker, D.E., Hector, A., Loreau, M., Perrings, C. (2009). *Biodiversity, Ecosystem Functioning, and Human Wellbeing: An Ecological and Economic Perspective*. Oxford University Press, Oxford.

O'Connor, M.I., Gonzalez, A., Byrnes, J.E.K. *et al.* (2017). A general biodiversity–function relationship is mediated by trophic level. *Oikos*, 126, 18–31.

O'Neill, R.V. (2001). Is it time to bury the ecosystem concept? (With full military honors, of course!). *Ecology*, 82, 3275–3284.

Pimm, S.L. (1984). The complexity and stability of ecosystems. *Nature*, 307, 321–326.

Reich, P.B., Tilman, D., Isbell, F. *et al.* (2012). Impacts of biodiversity loss escalate through time as redundancy fades. *Science*, 336, 589–592.

Rosenzweig, M.L. (1999). *Species Diversity in Space and Time*. Cambridge University Press, Cambridge.

Thébault, E. and Loreau, M. (2003). Food-web constraints on biodiversity-ecosystem functioning relationships. *Proceedings of the National Academy of Sciences of the USA*, 100, 14949–14954.

Tilman, D., Knops, J., Wedin, D., Reich, P., Ritchie, M., Siemann, E. (1997a). The influence of functional diversity and composition on ecosystem processes. *Science*, 277, 1300–1302.

Tilman, D., Lehman, C.L., Thomson, K.T. (1997b). Plant diversity and ecosystem productivity: Theoretical considerations. *Proceedings of the National Academy of Sciences of the USA*, 94, 1857–1861.

Tilman, D., Isbell, F., Cowles, J.M. (2014). Biodiversity and ecosystem functioning. *Annual Review of Ecology, Evolution, and Systematics*, 45, 471–493.

PART 1

Biodiversity and Ecosystems: An Overview

1

Biodiversity Change: Past, Present, and Future

Andy PURVIS[1,2] and Forest ISBELL[3]

[1] *Natural History Museum, London, UK*

[2] *Department of Life Sciences, Imperial College London, Ascot, UK*

[3] *University of Minnesota, St. Paul, USA*

1.1. Setting the stage: difficulties of documenting, understanding, and communicating biodiversity change

The IPBES Global Assessment of biodiversity and ecosystem services concluded that transformative change to the global socioeconomic system is necessary if biodiversity loss is to be stopped (Díaz *et al.* 2019). As soon as it was launched, it faced concerted pushback from what might be termed "extinction denialists" (Anon 2019, Lees *et al.* 2020). These critics used a range of tactics in an effort to undermine the credibility of the evidence presented in the Global Assessment, presumably in order to maintain the (for them) comfortable status quo. They variously conflated different measures of biodiversity change, cherry-picked counterexamples to general patterns, denied that change is taking place, argued that change is always taking place and therefore there is nothing to worry about, emphasized uncertainties, argued that biodiversity loss was essentially a historical rather than current issue, argued that some of the drivers of biodiversity loss were actually solutions, and even alleged misuse of data. Although their arguments were at best disingenuous, these "doubters for hire" were able to gain some traction

The Ecological and Societal Consequences of Biodiversity Loss,
coordinated by Michel LOREAU, Andy HECTOR, and Forest ISBELL.

because of the inconvenient truth that quantifying biodiversity change and understanding the reasons for it are both surprisingly difficult. Why?

One important reason is the sheer breadth of the concept of biodiversity. As defined in the Convention on Biological Diversity (United Nations 1993), it is "the variability among living organisms from all sources including, inter alia, terrestrial, marine and other aquatic ecosystems and the ecological complexes of which they are part; this includes diversity within species, between species and of ecosystems." This broad definition encompasses:

– all measures of the variety of life in an area, within and across all taxonomic groups, including their traits and their interactions with each other and their abiotic environment; this is termed α-diversity;

– how all of these change as one moves to another area; this spatial turnover is known as spatial β-diversity;

– assessment for all sizes of area, from a single point to the entire globe; this aspect of scale is known as spatial grain.

The inclusivity of this definition of biodiversity may have helped the rapid adoption of the term by researchers in conservation and ecology – whatever we study, it is covered – but it also contributes to at least four serious problems that hamper the development of simple, clear messages about biodiversity change and their communication to policy makers or the broader public.

First, measuring biodiversity as a whole – even for the smallest area – is not practically possible. Researchers therefore routinely and necessarily take shortcuts, using subsets of biodiversity (e.g. the birds seen in daylight in a two-week period of the summer) as proxies for the whole, as illustrated in Figure 1.1. The subsets are often chosen for reasons of convenience or even habit rather than because evidence shows them to be representative of biodiversity more broadly. This assumption of representativeness is so deeply implicit that many papers do not seriously consider the possibility that it is wrong. It usually is wrong: taxonomic groups differ widely in their responses to drivers of change (Lawton *et al.* 1998) and temporal trends even within the same region (Outhwaite *et al.* 2020). Alternatively, subsets may be chosen not because they are representative but because they are unusually sensitive to environmental change. The traditional use of indicator species is to monitor for changes in the environment, rather than in biodiversity (Siddig *et al.* 2016), so a combined temporal trend for such species may suggest much more rapid biodiversity change than would be seen from a representative set of species.

Figure 1.1. *Different taxonomic or ecological subsets of biodiversity may change differently over time. In this schematic, monitoring birds gives a time series of estimated biodiversity (orange line) that differs from those that other groups of species would give (other lines). Rainforest photograph: Ben Sutherland (CC BY 2.0). For a color version of this figure, see www.iste.co.uk/loreau/biodiversity.zip*

Second, biodiversity data are even less comprehensive than this use of shortcuts implies. For example, while ~1.5 million extant species of animals have been named, estimates of the true total number range from 3–100 million (Caley *et al.* 2014). Our biased knowledge makes it hard to generalize to biodiversity as a whole. We know more about large species, terrestrial species (especially birds), and species in developed countries than we do about small species, marine species, and species in emerging economies or developing countries (Hortal *et al.* 2015; Meyer *et al.* 2015). The situation is worse still for biodiversity in the past. At the time of writing, the Global Biodiversity Information Facility database (www.gbif.org) holds nearly 1.5 billion georeferenced occurrences of species whose year of observation is known; fewer than 4% are from before 1970. Very few high-quality ecological time series cover more than a few decades (Magurran *et al.* 2010), and very few monitoring schemes began before the 1970s. Moving to fossils, most extant animals and plants are not known from the fossil record; most extinct species never fossilized; tropical rainforests are particularly bad environments for fossilization; most fossil locations have temporal gaps in the record and permit only approximate dating; and none of the hierarchical levels of classification (species, genus, etc.) mean the same thing for fossils as in the present day (Kidwell and Flessa 1996; Forey *et al.* 2004; Purvis 2008).

Third, as discussed in Chapter 2, there are uncountably many ways of quantifying the biodiversity present in a given sample of organisms, and this number rises further when spatial turnover and different spatial scales are considered (McGill *et al.* 2015). Although frameworks are emerging to organize this diversity of diversity measures (see Chapter 2) (Pereira *et al.* 2013), consolidation is far from complete. Even if all researchers agreed what taxa to sample, we would not agree on how to measure them. It follows that there are also uncountably many ways to quantify biodiversity change over time. The combination of very many possible measures and relatively little temporal data, especially over long time periods, makes it very hard to bundle all the different measurements together into a coherent tapestry showing change over time.

Fourth, most laypeople's concept of biodiversity is much narrower than the very broad Convention on Biological Diversity (CBD) definition – often the number of animal and plant species (Bermudez and Lindemann-Matthies 2020) – setting up obvious communication problems.

Despite these difficulties, there is unambiguous evidence that biodiversity has changed over time both naturally and as driven by human-caused pressures. The next three sections briefly sketch some of this evidence, followed by a summary of how biodiversity is thought likely to change in the near future. The chapter concludes with some thoughts about how biodiversity change research is itself likely to change.

1.2. Biodiversity change in Earth history

Perhaps the best-known examples of biodiversity change in geological history are mass extinctions, such as the end-Permian and end-Cretaceous events, which wiped out (among other groups) trilobites and non-avian dinosaurs, respectively. The complete and permanent disappearance from the fossil record of diverse groups of species with previously abundant fossils represents strong evidence of biodiversity change, and both the rocks laid down at these times and the ecological differences between survivors and casualties also bear testimony about the likely drivers (Erwin *et al.* 2002; Schulte *et al.* 2010; Archibald *et al.* 2010; Bond and Grasby 2017).

However, the fossil record is generally too incomplete, biased, and blurred to give an unambiguous picture of the rest of geological time. Most analyses of global biodiversity trends through deep time have been forced to study shallow-sea shelly invertebrates as a proxy, because most other species were too unlikely to leave fossils (Briggs 2003). Brachiopods and cephalopods are the only animal clades with fossil records so good that 75% of then-extant genera are recorded from a given ~5 million year period of time (Foote and Sepkoski 1999). Incompleteness can be mitigated by coarsening the analysis (Ezard and Purvis 2016), but spatio-temporal biases still complicate attempts to infer diversity dynamics. The amount and paleogeographic coverage of fossil-rich rock available for scrutiny varies widely among geological time periods, with more recent periods tending to provide more rock (Vilhena and Smith 2013) across a wider area (Close *et al.* 2020); and fossil sampling intensity varies both spatially and temporally. Methodological mitigations for these biases are ongoing (Vilhena and Smith 2013; Bokulich 2018; Close *et al.* 2018; Close *et al.* 2020). It may be premature to even attempt to estimate how global taxonomic diversity has changed over time, aside from the handful of major extinction events (Close *et al.* 2020).

One microfossil clade – macroperforate planktonic foraminifera – has such a good fossil record that its dynamics can be studied much more straightforwardly, and at the level of evolutionary species rather than more inclusive taxa such as genera. Analyzing the Cenozoic radiation showed that the per-lineage speciation rate tended to be highest when species richness was low, whereas the extinction rate tended to be highest when climate changed rapidly (Ezard *et al.* 2011). Statistical model comparisons suggested that interspecific competition imposed a finite upper limit to the clade's species richness, but this ceiling was higher in warmer periods and when more sediment packages were laid down (Ezard and Purvis 2016). Different ecologies have dominated at different times (Ezard *et al.* 2011). A similar pattern is seen among major marine invertebrate groups with the best fossil records over the last 540 million years; individual groups rise and fall while the total diversity appears relatively constant (Alroy 2010; Close *et al.* 2020). Whether zooming right in on the clade with the most precise data, or zooming right out to look at all the groups with the most data overall, the trends in diversity through time have been different for different groups of species. We should therefore remember not to conflate the set of species being studied with overall biodiversity (see Figure 1.1).

One aspect of past diversity dynamics of particular interest in the current biodiversity crisis is the "background" rate of extinction, to which present-day rates of species extinction are often compared. The background rate is conceived of as being the typical per-species rate, not including mass extinctions, and can be estimated as the inverse of the average persistence time of species (Marshall 2017).

However, comparing these rates with current (much higher) extinction rates is conceptually problematic, because the background rate is an average over a long time period during which the short-term rate of extinction apparently varied widely, with strong pulses at the ends of geological time periods even aside from mass extinction events (Foote 2005). Current short-term rates – while undoubtedly many times higher than the long-term average rate of species extinction – may therefore have many more precedents in geological history than is widely appreciated.

In summary, the fossil record provides strong evidence that the composition of life on Earth has been in continual flux as different taxonomic and ecological groups have risen and fallen; and that mass extinction events markedly reduced the known global diversity of life for a time. However, the fragmentary evidence means that precise knowledge of rates and causes of diversity change is largely restricted to case studies (particular taxa and/or regions, which might not reflect broader biodiversity), while the overarching picture remains elusive.

1.3. Pre-industrial biodiversity change

We have impacted biodiversity throughout our history as a species (Johnson *et al.* 2017; Lewis and Maslin 2015), with our footprint spreading worldwide as the human population expanded in number, range, and technology (Erlandson and Braje 2013). We have been both a destructive and a creative force.

The megafaunal extinction is the best-known early example of our destructive force. More than 10% of mammalian genera died out between 100,000 and 500 years ago, in regional waves that generally follow human establishment more closely than climate change (Johnson *et al.* 2017; Haynes 2018). The casualties tended to be species with low reproductive rates, as expected if increased mortality from hunting was a driver of their extinction, while the ecology of those slow-reproducing species that survived tended to minimize human interaction (Johnson 2002).

Widespread Holocene extinctions of island-endemic species also point to anthropogenic causes (Louys *et al.* 2021). For example, Madagascar, first colonized by people <1,500 years ago (Anderson *et al.* 2018), lost all endemic species weighing >10kg before historical records began, including multiple lineages of lemur – 17 species altogether – all larger than any extant lemur (Kistler *et al.* 2015). Butchery marks on bones attest to a very direct role of human action (Anderson *et al.* 2018).

Human actions have shaped new biodiversity too, with huge effects on the biosphere. Domesticated species have been key to transforming the planet into an easier place for us to live, reshaping ecosystems to meet our needs for food and many other materials (Zeder 2015). Although the very first animal to be domesticated – the dog (whose domestication pre-dates agriculture (Skoglund *et al.* 2015)) – is an exception, most were domesticated in the early Holocene for food (Driscoll *et al.* 2009). Mammalian livestock numbers rose steadily, such that the global biomass of (non-human) megafauna had recovered to pre-extinction levels by the Industrial Revolution (Barnosky 2008). Over 6,000 breeds of domesticated mammals and over 2,000 of domesticated birds have been produced from relatively few wild ancestors, bred for different characteristics and environmental tolerances.

Domestication of wild plants transformed both food supply and, as agriculture spread in the Neolithic from the centers of domestication, the landscapes in which people lived. Estimates of the historical spread of agriculture vary considerably, depending on when intensification started; by the Industrial Revolution, agriculture could have covered a third of the world's land (Ellis *et al.* 2013) or only 5% (Goldewijk *et al.* 2011). The spread of agriculture significantly impacted some biogeochemical cycles; for instance, paddy farming and pastoralism were driving increases in the atmospheric concentration of methane by 4,000 years ago (Fuller *et al.* 2011). By 1500, perhaps 99% of people were agriculturalists (Lewis and Maslin 2015). Hand in hand with the spread of agriculture, many species of agricultural weed and other synanthropes greatly extended their geographic ranges. Biotic exchange of crops, livestock, synanthropes, and other species was also facilitated by the growth of global trade and European colonialism (Lewis and Maslin 2015; Dyer *et al.* 2017). This colonialism led to widespread transformation of indigenous cultural landscapes, with monoculture crops and plantations replacing natural forest (and other ecosystems), accelerating a long-term decline in global forest cover (Pongratz *et al.* 2008). Many of these human impacts were so significant that they have been discussed as possible markers of the formal start of a new period of geological time, the Anthropocene (Lewis and Maslin 2015). However, the impacts since the Industrial Revolution have been still more pervasive and intense.

1.4. Biodiversity change in the "Anthropocene"

The Industrial Revolution facilitated and drove sophisticated and intensive management of ecosystems to meet the needs of the growing population. The demand for power drove rapid deforestation until new technologies unlocked fossil fuel deposits. Successive agricultural innovations massively increased crop and livestock yields. New mass-transit technologies accelerated long-distance trade, weakening – in rich countries anyway – local ecological constraints on human population growth and resource consumption. The global urban population rose from about 70 million in 1800 (7% of the world's total) to over 260 million by 1900 (16% of the total) and around 2.9 billion (47% of the total) by 2000 (Klein Goldewijk *et al.* 2010).

Although human actions have been the dominant force driving biodiversity change in some places for thousands of years, anthropogenic drivers accelerated markedly from the mid-20th century (Steffen *et al.* 2015). We live in a human-dominated world. This section draws on the IPBES Global Assessment (Díaz *et al.* 2019), particularly the chapter on the status and trends of nature (Purvis *et al.* 2019).

Much of nature has already been used by people for croplands, livestock, forestry, or fishing. As little as 13% of land and 23% of the ocean is sufficiently free of obvious human impact to still be classed as "wilderness". Croplands cover about 1.6 billion ha of land, and land grazed by livestock covers more than twice that area (Hurtt *et al.* 2020). Croplands, pastures, and forestry have shifted location to some extent over time, leaving few fertile places untouched by agriculture's sprawling impacts. Only about a quarter of the 4.1 billion ha of forests worldwide are relatively undisturbed primary forests (FAO 2020). Urban areas only cover about 1% of land (Hurtt *et al.* 2020) but draw resources from a much wider footprint.

Agricultural and urban land use substantially reduce local biodiversity, with the worst-affected habitats having 75% fewer species and 40% fewer organisms (Newbold *et al.* 2015). Most areas intensely used for plantations, croplands, pastures, or urban areas have lost more than 20% of their species locally (Newbold *et al.* 2015). The Biodiversity Intactness Index, which quantifies the intactness of community composition, averages only about 79% across terrestrial ecosystems (Hill *et al.* 2018). Some local communities have gained more species than they have lost in recent decades (Dornelas *et al.* 2014; Vellend *et al.* 2013), especially where ecosystems are recovering after past disturbances (Gonzalez *et al.* 2016; Isbell *et al.* 2019).

The abundances of many wild species have greatly declined in recent centuries and decades. The biomass of humans is nearly an order of magnitude higher than that of all wild animals combined and the biomass of livestock is nearly twice that of humans (Bar-On *et al.* 2018). Wild animal abundances are substantially reduced in areas intensely used for croplands, pastures, or urban areas (Newbold *et al.* 2015). Abundances of some insects and birds appear to be rapidly decreasing in some locations (van Klink *et al.* 2020; Rosenberg *et al.* 2019). Harvest of fish and timber have both increased by nearly 50% since 1970 (Díaz *et al.* 2019).

Although average local biodiversity and species abundances have declined globally, some places have gained more species in recent history than they have lost.

For example, New Zealand now has more than 4,000 plant species, roughly half of them introduced from elsewhere (Sax *et al.* 2002). As few of the native plants have gone extinct, New Zealand's plant richness has risen markedly in recent decades. Similar trends have been observed in some temperate ocean ecosystems, where tropical fish species are expanding into temperate latitudes, presumably due to warming oceans (Hyndes *et al.* 2016). It is perhaps unsurprising that species richness can increase in a large region over relatively short periods of time, given that increases in richness are relatively rapid because they require the addition of but one individual, whereas decreases in richness are much slower because they require the loss of all individuals. It remains unclear whether these regions will maintain their higher numbers of species or whether many more extinctions will eventually occur.

Other regions have lost more species in recent decades than they have gained. For example, a recent survey of Iowa prairies found only about 55% of previously known native grassland plants (Wilsey *et al.* 2005): far more native species were no longer found than introduced species were newly found. Similarly, a resampling of prairies in Wisconsin also found that 8–60% of the original plant species were lost from individual remnants over a 32- to 55-year period (Leach and Givnish 1996).

At the global scale, over 500 vertebrate and 500 flowering plant species have gone extinct in historical times, corresponding to rates of species extinction that are at least tens to hundreds of times background rates (Vos *et al.* 2014; Ceballos *et al.* 2015; Humphreys *et al.* 2019). More alarmingly still, roughly a million animal and plant species are currently threatened with extinction (Díaz *et al.* 2019). This figure is arrived at by combining estimates of numbers of species and percentage of species threatened in different animal and plant groups. In well-studied groups of animals, approximately 25% of species are currently threatened with extinction, while around 39% of vascular plants are currently threatened with extinction (Nic Lughadha *et al.* 2020). For understudied and hyperdiverse groups of animals, notably insects, the prevalence of extinction risk may be lower but available evidence suggests it is unlikely to be <10% (Purvis *et al.* 2019). A million threatened species may be an underestimate if insects are as threatened as the best-known taxa, or if their diversity is nearer to the high end of the wide range of published estimates (Purvis *et al.* 2019). Without further action, many of these threatened species could become extinct in the coming decades and centuries, potentially dwarfing today's extinction rates.

1.5. Future of biodiversity change

How might biodiversity continue to change in the future in response to human pressures on the natural world? Answering this deceptively simple question well is absolutely crucial to our chances of a "Good Anthropocene" in which both nature and human well-being can be sustained. However, the usual way of constructing biodiversity indicators – by compiling empirical time-series – limits their usefulness for questions about the future. Consequently, Tittensor *et al.*'s (2014) assessment of whether the Aichi biodiversity targets would be met could do no better than carefully extrapolating indicator trends forward to the year 2020.

Doing better requires an understanding of how the indicator responds to anthropogenic pressures, that is, a model. Every indicator's construction involves at least an implicit statistical model, but indicators vary widely in how much they explicitly model effects of anthropogenic drivers. At one extreme, "direct-compilation" indicators such as the Red List Index (Butchart *et al.* 2007) and the Living Planet Index (McRae *et al.* 2017) are produced directly by compiling empirical data without explicitly considering drivers. Making these indicators forward-looking requires specification of how they will change in response to drivers (Visconti *et al.* 2016); there may be little evidence on which to base these choices, and model mis-specification will mean that past trends and future projections are not comparable as they do not represent exactly the same aspect of biodiversity. At the other extreme, "model-derived" indicators such as the Biodiversity Intactness Index (Hill *et al.* 2018), MSA (Schipper *et al.* 2020), and the Biodiversity Habitat Index (Hoskins *et al.* 2020) depend on models of how biodiversity responds to drivers, sometimes derived from very large global data sets. For such model-derived indicators, the transition between past trends and future projection is seamless; but model mis-specification will cause systematic bias in indicator estimates. Note that, when making future projections, both approaches assume that the effects of unmodeled drivers are negligible (Titeux *et al.* 2016). Both approaches, then, have serious problems to overcome; but the potential benefits of getting them right are enormous.

Most obviously, models can let us compare the likely consequences for biodiversity (more specifically, the biodiversity variables being modeled) of plausible possible futures (Wearn *et al.* 2012), (Schipper *et al.* 2020). Such exploratory scenarios help to frame the policy agenda and enable decision-makers to consider biodiversity alongside other relevant variables when making their choices (IPBES 2016) – no guarantee of a sustainable future, but surely a pre-requisite.

These plausible futures are developed by applying biodiversity models to scenarios of how drivers will change over coming decades. These scenarios, mostly developed to support climate-change assessments, consider a range of levels of radiative climate forcing (van Vuuren *et al.* 2011) against a variety of plausible socioeconomic and geopolitical backdrops that pose different challenges for climate adaptation and mitigation (O'Neill *et al.* 2014).

Projections under "business-as-usual" scenarios (in which sociopolitical and climate trajectories continue current trends) point consistently to ongoing significant reduction of terrestrial, freshwater, and marine biodiversity at all levels from genetic diversity to major ecosystems and biomes (Díaz *et al.* 2019, Shin *et al.* 2019). The number of species threatened with extinction is predicted to rise, and widespread disruption of species interactions is expected to ramify through ecological communities with effects that are hard to predict. Ecological communities without natural analogues are expected to become more numerous and widespread. Even under the most optimistic of the plausible futures – with the lowest levels of climate change and the most environmentally-minded sociopolitical changes – many aspects of biodiversity are projected to decline; for example, ocean net primary production and fish production are both projected to fall by around 3% (Shin *et al.* 2019).

As well as exploring plausible futures for biodiversity under different storylines, models have the potential to develop target-seeking, or "normative", scenarios – plotting possible routes to an agreed-upon future state (IPBES 2016). "Bending the curve" – stopping and beginning to reverse biodiversity loss (Mace *et al.* 2018) – is one possible target that has been used in this way (Figure 1.2). An ensemble of eight biodiversity models, linked to the land-use component of four integrated assessment models, was interrogated to find strategies able to bend the curve of terrestrial biodiversity (Leclère *et al.* 2020). A strategy integrating reduced agricultural demand (through reduced food waste and more plant-based diets), increased agricultural efficiency (through, for example, sustainable intensification), habitat restoration, and effective conservation was able to bend the curve in all models, whether they were focusing on abundance, community composition, or species persistence. Most projected trends were positive by 2050, though models differed in the speed and degree of bending.

Combining multiple kinds of biodiversity models in this way is valuable given the wide quantitative variation in their predictions (Leclère *et al.* 2020; Bellard *et al.* 2012). Even so, because few if any biodiversity models yet incorporate multiple drivers in a truly comparable way, these scenarios do not incorporate the effects of climate change on biodiversity (Leclère *et al.* 2020).

1.6. Future of biodiversity change research

Biodiversity change research will itself have to continue to change, given the challenges involved in navigating to a sustainable future. Quantifying biodiversity loss will continue to be important, but the more pressing need is to project how alternative policies and decisions will change biodiversity and ecosystem services. "No net loss" and "net gain" are increasingly likely to be criteria that planned policies and activities will have to meet, possibly in combination with net-zero or net-negative greenhouse-gas emissions. Only models can assess plans against such criteria.

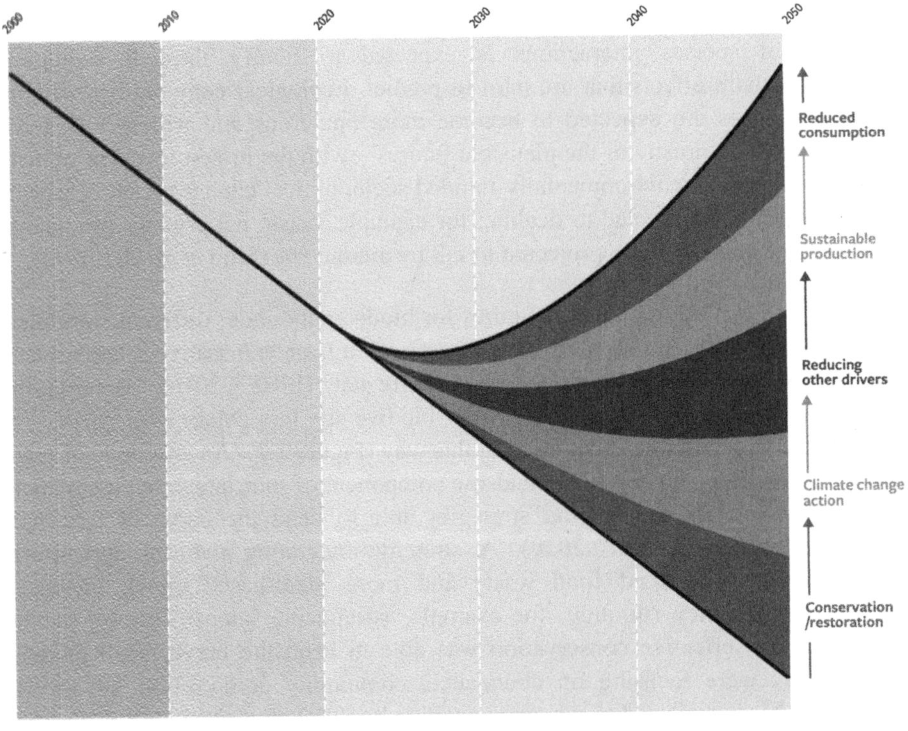

Figure 1.2. *Trends in biodiversity (various metrics) will continue to decline under business as usual (lower black line), but may be slowed or reversed (bending the curve, upper black line) by ambitious coordinated actions (right) (CBD 2020). For a color version of this figure, see www.iste.co.uk/loreau/biodiversity.zip*

Are the models yet up to the job? Yes and no. Yes, because models are simply the only way to do this, so we should start using them; and no, because the existing biodiversity models have not yet been refined and improved through multiple rounds of model–model and model–data comparison as, for example, climate models have (Flato *et al.* 2013). Furthermore, model–data comparison – comparing model outputs with observational data to estimate model skill (Olsen *et al.* 2016) – is far from straightforward for models that aim to represent broad-sense biodiversity, rather than just a narrow subset: one motivation for the development of these models has been the lack of time-series data covering a broad range of taxa (Scholes and Biggs 2005). Robust evaluation of biodiversity models and the projections they make is clearly a crucial challenge for future research (Fulton *et al.* 2019; Watermeyer *et al.* 2021).

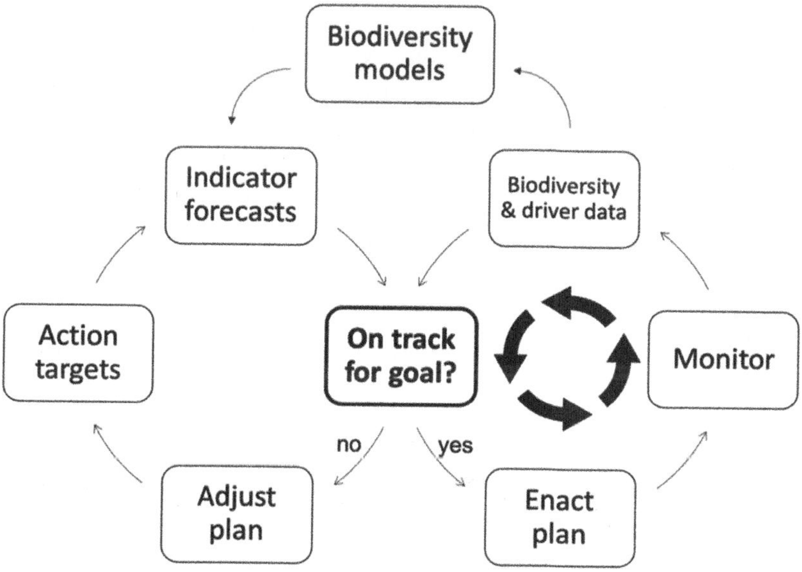

Figure 1.3. *Integrating models and monitoring to reach biodiversity targets efficiently. To assess whether a given planned action or policy will achieve an agreed biodiversity goal, the plan is combined with biodiversity models (fitted using existing data) to forecast the expected outcomes. If the projection misses the goal, the plan must be adjusted (e.g. through additional mitigation). Enacting the plan includes monitoring biodiversity and relevant drivers, with the new data being compared to the previous projections, and the model is then updated. If updated projections miss the goal, the plan is adjusted. The tight circle of arrows in the right-hand loop is to indicate the need to greatly speed up the monitoring/modeling loop*

The future – starting now – needs biodiversity models and relevant data streams to be much more closely integrated than at present. Figure 1.3 sketches how integration might ensure that biodiversity targets are achieved, while also driving ongoing model improvement through model–data comparisons and the accumulation of data. This framework could be applied at any scale, from ensuring a housing development delivers net biodiversity gain to setting global action targets to bend the curve by 2050. Monitoring data – on pertinent, not just convenient, taxa – should ideally flow seamlessly into global repositories of ready-to-use data (as climate data already do), which will therefore have to be much better resourced than at present (as climate data repositories already are) and supported by stronger incentives to share data openly. This accumulation of high-quality time-series data will help future models to overcome some widespread structural limitations of current ones, such as a narrow range of drivers and simple assumptions about how drivers combine and how responses unfold over time, and how other factors influence these aspects. It will also allow us to establish how remotely-sensed data can most cost-effectively be combined with different sources of field data to infer biodiversity change. Models must be as open as the data. Biodiversity observation and modeling both simply have to become much more cumulative disciplines than they are today. They must also both do a better job of reflecting broad-sense biodiversity rather than just the best-known groups. A well-resourced "Biodiversity Model Intercomparison Project", analogous to the Coupled and Atmospheric Model Intercomparison Projects for climate models, would greatly accelerate model improvement.

The future also needs much better information about the past. Climate models benefited from extending time series back before the instrumental record (through use of data from such sources as tree rings and ice cores) and even into deep time (using paleoenvironmental proxies). In the biodiversity domain, the main opportunities for extending the record backwards in an analogous way are the digitization of specimens in natural history collections (Lister 2011) and text-mining relevant old literature (Rivera-Quiroz *et al.* 2020), provided that models can account for the opportunistic nature of the samples (Jönsson *et al.* 2021). Such historical data, despite their undoubted limitations, will also be invaluable in the context of habitat restoration.

If we are to achieve a sustainable future, we need biodiversity to be at the heart of scenarios, rather than being bolted to scenarios developed for climate assessments. The overlapping causes and solutions of the climate and biodiversity crises mean that existing scenario frameworks will continue to be useful, but new frameworks, with biodiversity fully involved in social-ecological feedbacks, can better capture the range of positive human–nature relationships (Pereira *et al.* 2020; Bennett *et al.* 2016).

Lastly, we urgently need to improve in how we communicate biodiversity change. Despite over 30 years of common usage within life sciences, the word "biodiversity" is still jargon to most outsiders and the general public. Its complexity and difficulty of measurement hamper clear messaging: any simple quantitative statements about *biodiversity* change over time are unlikely to be strictly true. Coalescing instead around a small set of idealized indicators of important but complementary aspects of biodiversity as a whole would bring real advantages. For example, Mace *et al.* (2018) proposed global extinction risk, wild species abundance, and the average "health" of local ecosystems as a possible minimal set of axes. The narrower definition of such indicators would allow them to have informative – even resonant – jargon-free names that could be used in strongly worded quantitative messaging, with "biodiversity" used only as the umbrella term we know it to be. The science would also benefit: modeling and monitoring would focus on estimating these indicators, simplifying model intercomparisons, and driving ongoing improvements in both modeling and monitoring. Such improvement is hampered by the current conflation of indicator definitions with particular implementations of them, while the widespread conflation of biodiversity as a whole with individual indicators hampers communication.

With a million animal and plant species already threatened with extinction, with some iconic ecosystems approaching catastrophic thresholds, and with extinction denialists claiming there is no problem, biodiversity's future is far from secure. The science of biodiversity change is in the spotlight like never before, with a crucial role to play if society is to navigate a pathway to a "Good Anthropocene". The next decade will be an exciting time for biodiversity science, but the stakes could not be higher.

1.7. Acknowledgements

This work was supported by the Natural Environment Research Council (grant NE/M014533/1) and by the Prince Albert II of Monaco Foundation (*Plants Under Pressure 2*).

1.8. References

Alroy, J. (2010). The shifting balance of diversity among major marine animal groups. *Science*, 329(5996), 1191–1194.

Anderson, A., Clark, G., Haberle, S. *et al.* (2018). New evidence of megafaunal bone damage indicates late colonization of Madagascar. *PLoS ONE*, 13(10), e0204368.

Anon (2019). Biodiversity centre stage. *Nature Ecology and Evolution*, 3(6), 861–861.

Archibald, J.D., Clemens, W.A., Padian, K., *et al.* (2010). Cretaceous extinctions: Multiple causes. *Science*, 328(5981), 973–973.

Bar-On, Y.M., Phillips, R., Milo, R. (2018). The biomass distribution on Earth. *Proceedings of the National Academy of Sciences*, 115(25), 6506–6511.

Barnosky, A.D. (2008). Megafauna biomass tradeoff as a driver of Quaternary and future extinctions. *Proceedings of the National Academy of Sciences*, 105(Supplement 1), 11543–11548.

Bellard, C., Bertelsmeier, C., Leadley, P., Thuiller, W., Courchamp, F. (2012). Impacts of climate change on the future of biodiversity. *Ecology Letters*, 15(4), 365–377.

Bennett, E.M., Solan, M., Biggs, R. *et al.* (2016). Bright spots: Seeds of a good Anthropocene. *Frontiers in Ecology and the Environment*, 14(8), 441–448.

Bermudez, G.M.A. and Lindemann-Matthies, P. (2020). "What matters is species richness" – High school students' understanding of the components of biodiversity. *Research in Science Education*, 50, 2158–2187.

Bokulich, A. (2018). Using models to correct data: Paleodiversity and the fossil record. *Synthese* [Online]. Available at: https://doi.org/10.1007/s11229-018-1820-x.

Bond, D.P.G. and Grasby, S.E. (2017). On the causes of mass extinctions. *Palaeogeography, Palaeoclimatology, Palaeoecology*, 478, 3–29.

Briggs, D.E.G. (2003). The role of decay and mineralization in the preservation of soft-bodied fossils. *Annual Review of Earth and Planetary Sciences*, 31(1), 275–301.

Butchart, S.H.M., Akçakaya, H.R., Chanson, J. *et al.* (2007). Improvements to the Red List Index. *PLoS ONE*, 2(1), e140.

Caley, M.J., Fisher, R., Mengersen, K. (2014). Global species richness estimates have not converged. *Trends in Ecology and Evolution*, 29(4), 187–188.

CBD (2020). Global Biodiversity Outlook 5. Report, CBD, Montreal.

Ceballos, G., Ehrlich, P.R., Barnosky, A.D. *et al.* (2015). Accelerated modern human–induced species losses: Entering the sixth mass extinction. *Science Advances*, 1, e1400253.

Close, R.A., Evers, S.W., Alroy, J., Butler, R.J. (2018). How should we estimate diversity in the fossil record? Testing richness estimators using sampling-standardised discovery curves. *Methods in Ecology and Evolution*, 9(6), 1386–1400.

Close, R.A., Benson, R.B.J., Saupe, E.E., Clapham, M.E., Butler, R.J. (2020). The spatial structure of Phanerozoic marine animal diversity. *Science*, 368(6489), 420–424.

Díaz, S., Settele, J., Brondízio, E.S. *et al.* (2019). Pervasive human-driven decline of life on Earth points to the need for transformative change. *Science*, 366(6471), eaax3100.

Dornelas, M., Gotelli, N.J., Mcgill, B. *et al.* (2014). Assemblage time series reveal biodiversity change but not systematic loss. *Science*, 344(6181), 296–299.

Driscoll, C.A., Macdonald, D.W., O'Brien, S.J. (2009). From wild animals to domestic pets, an evolutionary view of domestication. *Proceedings of the National Academy of Sciences*, 106(Supplement 1), 9971–9978.

Dyer, E.E., Cassey, P., Redding, D.W. *et al.* (2017). The global distribution and drivers of alien bird species richness. *PLOS Biology*, 15(1), e2000942.

Ellis, E.C., Kaplan, J.O., Fuller, D.Q. *et al.* (2013). Used planet: A global history. *Proceedings of the National Academy of Sciences of the United States of America*, 110(20), 7978–7985.

Erlandson, J. and Braje, T. (2013). Archeology and the Anthropocene. *Anthropocene*, 4, 1–7.

Erwin, D., Bowring, S., Yugan, J. (2002). End-Permian mass extinctions: A review. In *Catastrophic Events and Mass Extinctions: Impacts and Beyond*, Koeberl, C. and MacLeod, K.G. (eds). Geological Society of America, Boulder.

Ezard, T. and Purvis, A. (2016). Environmental changes define ecological limits to species richness and reveal the mode of macroevolutionary competition. *Ecology Letters*, 19.

Ezard, T., Aze, T., Pearson, P.N., Purvis, A. (2011). Interplay between changing climate and species' ecology drives macroevolutionary dynamics. *Science*, 332(6027), 349–351.

FAOSTAT (2020). *FAOSTAT*. Food and Agriculture Organization of the United Nations [Online]. Available at: www.fao.org/faostat [Accessed December 2020].

Flato, G., Marotzke, J., Abiodun, B. *et al.* (2013). Evaluation of climate models. In *Climate Change 2013: The Physical Science Basis. Contribution of Working Group I to the Fifth Assessment Report of the Intergovernmental Panel on Climate Change*. Cambridge University Press, Cambridge.

Foote, M. (2005). Pulsed origination and extinction in the marine realm. *Paleobiology*, 31, 6–20.

Foote, M. and Sepkoski, J.J. (1999). Absolute measures of the completeness of the fossil record. *Nature*, 398(6726), 415–417.

Forey, P.L., Fortey, R.A., Kenrick, P., Smith, A.B. (2004). Taxonomy and fossils: A critical appraisal. *Philosophical Transactions of the Royal Society of London. Series B: Biological Sciences*, 359(1444), 639–653.

Fuller, D., van Etten, J., Manning, K. *et al.* (2011). The contribution of rice agriculture and livestock pastoralism to prehistoric methane levels: An archaeological assessment. *Holocene*, 21, 743–759.

Fulton, E.A., Blanchard, J.L., Melbourne-Thomas, J., Plagányi, É.E., Tulloch, V.J.D. (2019). Where the ecological gaps remain, a modelers' perspective. *Frontiers in Ecology and Evolution*, 7, 424.

Goldewijk, K.K., Beusen, A., Drecht, G.V., Vos, M.D. (2011). The HYDE 3.1 spatially explicit database of human-induced global land-use change over the past 12,000 years. *Global Ecology and Biogeography*, 20(1), 73–86.

Gonzalez, A., Cardinale, B.J., Allington, G.R.H. *et al.* (2016). Estimating local biodiversity change: A critique of papers claiming no net loss of local diversity. *Ecology*, 97(8), 1949–1960.

Haynes, G. (2018). The evidence for human agency in the Late Pleistocene megafaunal extinctions. In *Encyclopedia of the Anthropocene*, DellaSala, D. and Goldstein, M. (eds). Elsevier, Amsterdam.

Hill, S.L.L., Gonzalez, R., Sanchez-Ortiz, K. *et al.* (2018). Worldwide impacts of past and projected future land-use change on local species richness and the Biodiversity Intactness Index. *bioRxiv*, 311787.

Hortal, J., de Bello, F., Diniz-Filho, J.A.F. *et al.* (2015). Seven shortfalls that beset large-scale knowledge of biodiversity. *Annual Review of Ecology, Evolution, and Systematics*, 46(1), 523–549.

Hoskins, A.J., Harwood, T.D., Ware, C. *et al.* (2020). BILBI: Supporting global biodiversity assessment through high-resolution macroecological modeling. *Environmental Modeling and Software*, 132, 104806.

Humphreys, A.M., Govaerts, R., Ficinski, S.Z., Nic Lughadha, E., Vorontsova, M.S. (2019). Global dataset shows geography and life form predict modern plant extinction and rediscovery. *Nature Ecology and Evolution*, 3(7), 1043–1047.

Hurtt, G.C., Chini, L., Sahajpal, R. *et al.* (2020). Harmonization of global land use change and management for the period 850–2100 (LUH2) for CMIP6. *Geosci. Model Dev.*, 13(11), 5425–5464.

Hyndes, G.A., Heck Jr., K.L., Vergés, A. *et al.* (2016). Accelerating tropicalization and the transformation of temperate seagrass meadows. *BioScience*, 66(11), 938–948.

IPBES (2016). Summary for policymakers of the methodological assessment of scenarios and models of biodiversity and ecosystem services of the Intergovernmental Science-Policy Platform on Biodiversity and Ecosystem Services. Report, Secretariat of the Intergovernmental Science-Policy Platform on Biodiversity and Ecosystem Services, Bonn, Germany.

Isbell, F., Tilman, D., Reich, P.B., Clark, A.T. (2019). Deficits of biodiversity and productivity linger a century after agricultural abandonment. *Nature Ecology and Evolution*, 3(11), 1533–1538.

Johnson, C.N. (2002). Determinants of loss of mammal species during the Late Quaternary "megafauna" extinctions: Life history and ecology, but not body size. *Proceedings of the Royal Society of London. Series B: Biological Sciences*, 269(1506), 2221–2227.

Johnson, C.N., Balmford, A., Brook, B.W. *et al.* (2017). Biodiversity losses and conservation responses in the Anthropocene. *Science*, 356(6335), 270–275.

Jönsson, G.M., Broad, G.R., Sumner, S., Isaac, N.J.B. (2021). A century of social wasp occupancy trends from natural history collections: Spatiotemporal resolutions have little effect on model performance. *Insect Conservation and Diversity* [Online]. Available at: https://doi.org/10.1111/icad.12494.

Kidwell, S.M. and Flessa, K.W. (1996). The quality of fossil record: Populations, species, and communities. *Annual Review of Earth and Planetary Sciences*, 24(1), 433–464.

Kistler, L., Ratan, A., Godfrey, L.R. *et al.* (2015). Comparative and population mitogenomic analyses of Madagascar's extinct, giant "subfossil" lemurs. *Journal of Human Evolution*, 79, 45–54.

Klein Goldewijk, K., Beusen, A., Janssen, P. (2010). Long-term dynamic modeling of global population and built-up area in a spatially explicit way: HYDE 3.1. *The Holocene*, 20(4), 565–573.

van Klink, R., Bowler, D.E., Gongalsky, K.B. *et al.* (2020). Meta-analysis reveals declines in terrestrial but increases in freshwater insect abundances. *Science*, 368(6489), 417.

Lawton, J.H., Bignell, D.E., Bolton, B. *et al.* (1998). Biodiversity inventories, indicator taxa and effects of habitat modification in tropical forest. *Nature*, 391(6662), 72–76.

Leach, M.K. and Givnish, T.J. (1996). Ecological determinants of species loss in remnant prairies. *Science*, 273(5281), 1555–1558.

Leclère, D., Obersteiner, M., Barrett, M. *et al.* (2020). Bending the curve of terrestrial biodiversity needs an integrated strategy. *Nature*, 585, 551–556.

Lees, A.C., Attwood, S., Barlow, J., Phalan, B. (2020). Biodiversity scientists must fight the creeping rise of extinction denial. *Nature Ecology and Evolution*, 4(11), 1440–1443.

Lewis, S.L. and Maslin, M.A. (2015). Defining the Anthropocene. *Nature*, 519(7542), 171–180.

Lister, A.M. (2011). Natural history collections as sources of long-term datasets. *Trends in Ecology and Evolution*, 26(4), 153–154.

Louys, J., Braje, T.J., Chang, C.-H. *et al.* (2021). No evidence for widespread island extinctions after Pleistocene hominin arrival. *Proceedings of the National Academy of Sciences*, 118(20), e2023005118.

Mace, G.M., Barrett, M., Burgess, N.D. *et al.* (2018). Aiming higher to bend the curve of biodiversity loss. *Nature Sustainability*, 1(9), 448–451.

Magurran, A.E., Baillie, S.R., Buckland, S.T. *et al.* (2010). Long-term datasets in biodiversity research and monitoring: Assessing change in ecological communities through time. *Trends in Ecology and Evolution*, 25(10), 574–582.

Marshall, C.R. (2017). Five palaeobiological laws needed to understand the evolution of the living biota. *Nature Ecology and Evolution*, 1(0165).

McGill, B.J., Dornelas, M., Gotelli, N.J., Magurran, A.E. (2015). Fifteen forms of biodiversity trend in the Anthropocene. *Trends in Ecology and Evolution*, 30(2), 104–113.

McRae, L., Deinet, S., Freeman, R. (2017). The diversity-weighted living planet index: Controlling for taxonomic bias in a global biodiversity indicator. *PLoS ONE*, 12(1), e0169156.

Meyer, C., Kreft, H., Guralnick, R., Jetz, W. (2015). Global priorities for an effective information basis of biodiversity distributions. *Nature Communications*, 6(1), 8221.

Newbold, T., Hudson, L.N., Hill, S.L.L. *et al.* (2015). Global effects of land use on local terrestrial biodiversity. *Nature*, 520(7545), 45–50.

Nic Lughadha, E., Bachman, S.P., Leão, T.C.C. *et al.* (2020). Extinction risk and threats to plants and fungi. *Plants, People, Planet*, 2(5), 389–408.

O'Neill, B.C., Kriegler, E., Riahi, K. *et al.* (2014). A new scenario framework for climate change research: The concept of shared socioeconomic pathways. *Climatic Change*, 122(3), 387–400.

Olsen, E., Fay, G., Gaichas, S. *et al.* (2016). Ecosystem Model Skill Assessment. Yes We Can! *PLoS ONE*, 11(1), e0146467.

Outhwaite, C.L., Gregory, R.D., Chandler, R.E., Collen, B., Isaac, N.J.B. (2020). Complex long-term biodiversity change among invertebrates, bryophytes and lichens. *Nature Ecology and Evolution*, 4(3), 384–392.

Pereira, H.M., Ferrier, S., Walters, M. *et al.* (2013). Essential biodiversity variables. *Science*, 339(6117), 277–278.

Pereira, L.M., Davies, K.K., Den Belder, E. *et al.* (2020). Developing multiscale and integrative nature–people scenarios using the Nature Futures Framework. *People and Nature*, 2(4), 1172–1195.

Pongratz, J., Reick, C., Raddatz, T., Claussen, M. (2008). A reconstruction of global agricultural areas and land cover for the last millennium. *Global Biogeochemical Cycles*, 22, 3.

Purvis, A. (2008). Phylogenetic approaches to the study of extinction. *Annual Review of Ecology, Evolution, and Systematics*, 39(1), 301–319.

Purvis, A., Molnar, Z., Obura, D. *et al.* (2019). Status and trends – Nature. In *Global Assessment Report on Biodiversity and Ecosystem Services of the Intergovernmental Science-Policy Platform on Biodiversity and Ecosystem Services*, Brondiízio, E.S., Settele, J., Díaz, S., Ngo, H.T. (eds). IPBES Secretariat, Bonn, Germany.

Rivera-Quiroz, F.A., Petcharad, B., Miller, J.A. (2020). Mining data from legacy taxonomic literature and application for sampling spiders of the Teutamus group (Araneae; Liocranidae) in Southeast Asia. *Scientific Reports*, 10(1), 15787.

Rosenberg, K.V., Dokter, A.M., Blancher, P.J. *et al.* (2019). Decline of the North American avifauna. *Science*, 366(6461), 120.

Sax, D.F., Gaines, S.D., Brown, J.H. (2002). Species invasions exceed extinctions on islands worldwide: A comparative study of plants and birds. *American Naturalist*, 160(6), 766–783.

Schipper, A.M., Hilbers, J.P., Meijer, J.R. *et al.* (2020). Projecting terrestrial biodiversity intactness with GLOBIO 4. *Global Change Biology*, 26(2), 760–771.

Scholes, R.J. and Biggs, R. (2005). A biodiversity intactness index. *Nature*, 434(7029), 45–49.

Schulte, P., Alegret, L., Arenillas, I. *et al.* (2010). The Chicxulub Asteroid impact and mass extinction at the Cretaceous-Paleogene boundary. *Science*, 327(5970), 1214–1218.

Shin, Y.J., Arneth, A., Roy Chowdhury, R. *et al.* (2019). Plausible futures of nature, its contributions to people and their good quality of life. In *Global Assessment Report on Biodiversity and Ecosystem Services of the Intergovernmental Science-Policy Platform on Biodiversity and Ecosystem Services*, Brondiízio, E.S., Settele, J., Díaz, S., Ngo, H.T. (eds). IPBES Secretariat, Bonn, Germany.

Siddig, A.A.H., Ellison, A.M., Ochs, A., Villar-Leeman, C., Lau, M.K. (2016). How do ecologists select and use indicator species to monitor ecological change? Insights from 14 years of publication in Ecological Indicators. *Ecological Indicators*, 60, 223–230.

Steffen, W., Richardson, K., Rockström, J. *et al.* (2015). Planetary boundaries: Guiding human development on a changing planet. *Science*, 347(6223), 1259855.

Titeux, N., Henle, K., Mihoub, J.-B. *et al.* (2016). Biodiversity scenarios neglect future landuse changes. *Global Change Biology*, 22(7), 2505–2515.

Tittensor, D.P., Walpole, M., Hill, S.L.L. *et al.* (2014). A mid-term analysis of progress toward international biodiversity targets. *Science*, 346(6206), 241–244.

United Nations (1993). United Nations convention on biological diversity. Report, UNCED, Rio de Janeiro.

Vellend, M., Baeten, L., Myers-Smith, I.H. *et al.* (2013). Global meta-analysis reveals no net change in local-scale plant biodiversity over time. *Proceedings of the National Academy of Sciences of the United States of America*, 110(48), 19456–19459.

Vilhena, D. and Smith, A. (2013). Spatial bias in the marine fossil record. *PLoS ONE*, 8, e74470.

Visconti, P., Bakkenes, M., Baisero, D. *et al.* (2016). Projecting global biodiversity indicators under future development scenarios. *Conservation Letters*, 9(1), 5–13.

Vos, J., Joppa, L., Gittleman, J., Stephens, P., Pimm, S. (2014). Estimating the normal background rate of species extinction. *Conservation Biology: The Journal of the Society for Conservation Biology*, 29.

van Vuuren, D.P., Edmonds, J., Kainuma, M. *et al.* (2011). The representative concentration pathways: An overview. *Climatic Change*, 109(1), 5.

Watermeyer, K.E., Guillera-Arroita, G., Bal, P. *et al.* (2021). Using decision science to evaluate global biodiversity indices. *Conservation Biology*, 35(2), 492–501.

Wearn, O.R., Reuman, D.C., Ewers, R.M. (2012). Extinction debt and windows of conservation opportunity in the Brazilian Amazon. *Science*, 337(6091), 228–232.

Wilsey, B.J., Martin, L.M., Polley, H.W. (2005). Predicting plant extinction based on species–area curves in prairie fragments with high beta richness. *Conservation Biology*, 19(6), 1835–1841.

Zeder, M.A. (2015). Core questions in domestication research. *Proceedings of the National Academy of Sciences*, 112(11), 3191–3198.

2

Biodiversity: Concepts, Dimensions, and Measures

Anne CHAO[1] and Robert K. COLWELL[2,3,4]
[1] *Institute of Statistics, National Tsing Hua University, Hsin-Chu, Taiwan*
[2] *University of Connecticut, Storrs, USA*
[3] *University of Colorado Museum of Natural History, Boulder, Colorado, USA*
[4] *Federal University of Goiás, Goiânia, Brazil*

2.1. Introduction

Biodiversity (biological diversity), in a broad sense, encompasses the variety and variability among living and extinct organisms from all sources and the ecological complexes in which they occur. Although the term biodiversity was first used in the literature in the 1980s (Wilson and Peter 1988), the concept of biological/natural diversity had been developing since the 19th century and continues to acquire much of its modern meaning in adapting to technological advances. As indicated in the recent Intergovernmental Science-Policy Platform on Biodiversity and Ecosystem Services (IPBES) Conceptual Framework, biodiversity includes "variation in genetic, phenotypic, phylogenetic, and functional attributes, as well as changes in abundance and distribution over time and space, within and among species, biological communities and ecosystems" (Díaz *et al.* 2015, p. 8). Biodiversity is a hierarchical concept covering different levels of organization, from genes to ecosystems, but also the scope and variety of interactions between entities within and between these levels.

The Ecological and Societal Consequences of Biodiversity Loss,
coordinated by Michel LOREAU, Andy HECTOR, and Forest ISBELL.

Global warming, environmental degradation, and human activities/disturbances have led to an unprecedented threat to Earth's ecosystems and widespread concern about the loss of biodiversity. It is essential to understand the causes and processes of biodiversity, assess the effects of human disturbance on biodiversity, and predict the impacts/consequences of loss in biodiversity on the functional properties of ecosystems. Both biodiversity and ecosystem function are multifaceted concepts covering many dimensions across multiple scales; neither can be characterized by one or a few measures. Indeed, "we are all men (and women) trying to describe a monstrous elephant of ecological and evolutionary diversity" (Nanny 2004, p. 721). Further establishing the relationship between biodiversity and ecosystem functioning and stability (BEFS) is inevitably complicated and challenging, as reflected by extensive research and pertinent debates (e.g. Hooper *et al.* 2005; Naeem *et al.* 2009; Loreau *et al.* 2010, and other chapters in this volume). This chapter focuses primarily on quantifying biodiversity and its spatial/temporal change under a unified framework (Hill numbers and their generalizations). This framework covers the most widely used taxonomic measures and their generalized versions that incorporate evolutionary history among species and species traits. The measures derived from this framework possess mathematical properties that directly support biological reasoning about diversity. Based on our very limited background in BEFS, we also attempt to relate some of the measures to those used in BEFS studies.

An enormous number of diversity measures from different perspectives have been proposed (Magurran 2004), not only in ecological sciences but also in other sciences studying complex systems of any kind (e.g. economics, genetics, information science, physics, statistics, etc.). In section 2.2, we briefly review progress in quantifying taxonomic diversity, from the classic species richness and conventional complexity indices toward a unified framework based on the notion of *effective number of species* (*ENS*) – a concept originally developed in two pioneering papers (MacArthur 1965; Hill 1973). We also elaborate, intuitively, to explain how the ENS approach provides a robust and meaningful framework in support of biological reasoning about diversity.

In section 2.3.1, we review a continuum of species/taxonomic diversity (TD) measures (*Hill numbers*) based on the concept of the ENS (Hill 1973). Taxonomic diversity includes two components – species richness and evenness among species abundances. Species richness features prominently in BEFS studies and is often used as a surrogate for biodiversity (Isbell *et al.* 2017), but it ignores species abundances. The evenness of species abundances has been demonstrated to also affect ecosystem productivity. Evenness may play a more important role in biotic changes, because evenness usually responds more rapidly to human activities than changes in species richness (Hillebrand *et al.* 2008). In section 2.3.2, we introduce a class of evenness measures that can be connected to BEFS research.

In TD, all species are treated as equally distinct from one another, and thus species relatedness/differences are not considered. Remarkable progress in our understanding of phylogenies and the ongoing, intensive documentation of species traits have made the quantification of biodiversity an increasingly complex topic in recent years. Pielou (1975, p. 17) was the first to propose that the concept of diversity could be broadened to incorporate species differences in the classification levels of species, genus, family, etc. Since the 1990s, many researchers have recognized the importance of incorporating relatedness/differences among species based on their evolutionary histories, as phylogenetic diversity (PD), and/or their functional traits, as functional diversity (FD), into the expanded concept of diversity. The responses of species to environmental drivers and/or the effect of species on ecosystem processes and functioning are likely to be affected by the relative roles of their abundances, phylogenies, and trait values. Disentangling and assessing changes across the three dimensions of biodiversity (TD, PD, and FD) is imperative to understanding biodiversity change.

One advantage of using the ENS framework is that Hill numbers can be generalized from TD to PD (Chao *et al.* 2010) and FD (Chao *et al.* 2019). In section 2.4, we review a unified framework that integrates three dimensions of diversity. Under the unified framework, TD quantifies the effective number of species, PD quantifies the effective total branch length, mean-PD (PD divided by tree depth) quantifies the effective number of lineages, and FD quantifies the effective number of virtual functional groups (or functional "species"). Thus, TD, mean-PD, and FD are all in the same units of species/lineage equivalents and can be meaningfully compared.

In addition to within-community diversity, the assessment of among-community beta diversity and dissimilarity measures is critical for quantifying biodiversity change and compositional differentiation over a geographical area, across time, or along an environmental gradient. In section 2.5, we briefly present the partitioning scheme, which multiplicatively decomposes the diversity of a pooled community (gamma diversity) into its within-community component (alpha diversity) and among-community component (beta diversity) as well as related dissimilarity measures. In section 2.6, two data sets are used for illustrating the unified ENS-based diversity measures presented in this chapter. Due to space constraints, references are limited; refer to the review papers mentioned in each section for relevant papers in the primary research literature.

2.2. Progress in measuring taxonomic diversity

The number of species, or species richness, in a community is the simplest and most frequently used diversity measure in BEFS studies. Species richness possesses intuitive mathematical properties; its decomposition into multiple scales is direct. In many biodiversity analyses, only species presence (or detection) information for each locality is available, practical, or most relevant. For biodiversity assessments based on sampling data, some species generally remain undetected, so that true species richness is very difficult to infer from samples of limited size, especially in hyper-diverse communities. In such cases, asymptotic estimation of true species richness via statistical inference or non-asymptotic standardization of samples via rarefaction (interpolation) and extrapolation (iNEXT) is required for fair comparisons of diversity among multiple communities (Colwell *et al.* 2012, Chao *et al.* 2014, Hsieh *et al.* 2016). In a majority of BEFS analyses, species richness is experimentally manipulated, thus obviating issues of statistical estimation and standardization for richness. As most BEFS analyses are made conditional on species richness, this feature simplifies the quantification of diversity and its change, but further assumptions and methods may be required to make inferences and evaluate the variances/uncertainties of the resulting data and patterns.

Historically, the earliest quantitative diversity index can be traced to Fisher *et al.*'s 1943 paper for Malayan butterfly data. In Fisher *et al.*'s statistical model, species rank abundances follow a log-series distribution. One of the parameters characterizing the log series is the well-known Fisher's "alpha", which has been used as a diversity index for analyzing ecological data. However, Fisher's alpha uses only the information of observed species richness and the total number of individuals, regardless of the sampling scheme; species abundances are ignored.

Species abundances can be incorporated into diversity analyses in two ways: the first approach (the parametric approach) models species abundances with a statistical distribution with one or two parameters. Parameters of the underlying model are then used to characterize diversity. Many models have been developed to fit species abundance data. These include the log-normal, negative binomial, geometric, Zipf–Mandelbrot, and Poisson-compound distributions; see Magurran (2004) and Colwell (2009) for reviews. Under parametric models, diversity analysis can be simplified to just one or two parameters involved in the model. However, simulations show that these parametric models work well only when the correct form of the species abundance distribution is known, but such information is generally not available. Furthermore, it may be difficult to select an appropriate parametric model. These constraints limit the applications of parametric models to diversity analysis.

The second approach (the non-parametric approach) makes no assumptions about the mathematical form of the underlying abundance distributions. Scientists in other disciplines developed "complexity" measures (which consider both evenness and richness) much earlier than ecologists. For example, a popular complexity measure, the *Gini–Simpson index,* was originally developed by the Italian statistician and sociologist Corrado Gini around 1912. The Gini–Simpson index represents the probability that two randomly chosen individuals belong to different species. Simpson, in 1949, introduced the complement measure (i.e. Simpson concentration index) to the ecology community. Another popular measure of complexity is *Shannon entropy* developed by Claude Shannon in 1948 in information science, which measures the uncertainty in the species identity of a randomly chosen individual in the community. Ramón Margalef introduced Shannon entropy to ecology as a diversity index around 1957. Shannon entropy and the Gini–Simpson index are special cases of a family of generalized entropy, called "Tsallis entropy", in physics.

Shannon entropy (in units of information) and the Gini–Simpson index (a probability) use different units, so they cannot be compared with each other. Although the two indices have been useful in various disciplines, they do not fulfill an essential requirement for any diversity measure, that is, the *replication principle,* or equivalently, the *doubling property.* The doubling property states that if two communities are of equal diversity with equal numbers of individuals, but share no species in common (species abundance distributions are allowed to be different), then the diversity of the pooled community must be double the diversity of either individual community.

Why the doubling property is an essential requirement for a diversity measure can be intuitively understood as follows. Consider a community with S equally abundant species. Suppose we build this community by adding one equally abundant species at a time, and, after each addition, observe the increase in Shannon entropy and the Gini–Simpson index. For the successive order of additional species 1, 2, ..., S, Shannon entropy values are 0, 0.693, 1.099, 1.386, 1.609, ..., $\log S$; the corresponding Gini–Simpson values are 0, 0.5, 0.667, 0.75, 0.8, ..., $1-1/S$. The incremental contribution of each species for the two measures is not constant, which it ought to be, for consistency. Jost (2007) further showed that failure to satisfy the doubling property would cause inconsistent or counter-intuitive interpretations, especially in conservation analysis.

The ecologist Robert MacArthur was the first to resolve these problems by converting the complexity measures to the "effective number of species" (*ENS*), that is, the number of equally abundant species that are needed to give the same value of the diversity measure (MacArthur 1965). To do so, Shannon entropy is converted to its exponential, and the Gini–Simpson index to the inverse of the Simpson concentration index. The two converted effective numbers satisfy the doubling property and are linear with the addition of equally abundant species, so that they behave like real species when all species are equally abundant. Thus, the diversity of a community with two species is double the diversity of the community with only one species; the diversity of a community with three species is triple the diversity of the community with only one species, etc. This property not only makes the magnitude of changes in these measures interpretable and the order of addition of species immaterial, but also ensures logical consistency in conservation analyses. When the focus is on beta diversity, this property also ensures that the ratio of gamma to alpha diversity reflects species compositional differentiation.

2.3. Taxonomic diversity and evenness measures

Assume that there are S species in a community, indexed by $i = 1, 2, \ldots, S$. Let z_i represent the raw abundance (number of individuals) of species i, but it could be any other metric such as the biomass, coverage of plants or corals, or basal area of plants.

2.3.1. *Taxonomic diversity: effective number of species*

Hill (1973) integrated the two converted Shannon and Gini–Simpson measures (section 2.2) and species richness into a continuum of diversity measures, referred to as Hill numbers, defined for a diversity order $q \geq 0, q \neq 1$, as

$$^{q}TD = \left[\sum_{i=1}^{S} (z_i / z_+)^q\right]^{1/(1-q)} = \left(\sum_{i=1}^{S} p_i^q\right)^{1/(1-q)} \qquad [2.1]$$

where $z_+ = \sum_{i=1}^{S} z_i$ denotes the total number of individuals in the community, and $p_i = z_i / z_+$ denotes the relative abundance of species i. Hill numbers with negative orders can also be calculated, but they place so much weight on rare species that they have poor sampling properties.

The parameter q (which need not be an integer) determines how much the measure discounts rare species. When $q = 0$, ${}^{0}TD$ is simply species richness, which counts *species* equally without regard to their abundances and thus is disproportionally sensitive to individuals of rare species. The Hill number of order $q = 1$ (${}^{1}TD$) is defined as the limit of ${}^{q}TD$ as q tends to 1; this limit exists and gives the exponential of Shannon entropy. The measure for $q = 1$ counts *individuals* equally and weighs each species in proportion to its abundance, without favoring either rare or abundant species. Thus, ${}^{1}TD$ can be interpreted as the effective number of abundant species in the community. When $q = 2$, the Hill number reduces to the inverse of the Simpson concentration index; each species is weighted by its squared abundance and thus disproportionately sensitive to dominant species. The measure ${}^{2}TD$ can be interpreted as the effective number of highly abundant or dominant species in the community.

Hill numbers for all orders $q \geq 0$ fulfill the doubling property, and all have the same units as species richness. It is possible to plot them on a single graph as a function of the parameter q. This *diversity profile/curve* conveys all the information in the species abundance distribution. In the comparison of two communities, when their profiles do not cross, one community is unambiguously more diverse than the other. If their profiles cross, then their ranking can only be made conditional on diversity order q. In most applications, the diversity profiles beyond $q = 3$ stabilize and contain little information on species abundance. A diversity profile/curve for all values (including non-integers) of q from 0 to 2 or 3 would be sufficient to characterize TD in a community. Since the publication of articles in a multi-author *Ecology* forum (Ellison 2010 and the papers following it), a consensus seems to have emerged that Hill numbers should be used for quantifying TD.

In BEFS, Shannon entropy was adopted to quantify community diversity in studies of invasibility and compositional stability (e.g. Foster *et al.* 2002). Recently, Wang and Loreau (2016) applied the Hill number of order $q = 2$ (i.e. the inverse of Simpson's concentration index) to assess the relation between biodiversity and temporal invariability across scales. We suggest that, in future work, it would be more informative to simultaneously consider TD of $q = 0$, 1, and 2 (or a profile) to quantify diversity in analyzing BEFS data, so that the effects of rare ($q = 0$), abundant ($q = 1$), and dominant ($q = 2$) species on temporal stability can be evaluated and disentangled. A debate in BEFS has focused on the relative importance of species richness versus species composition (Hector *et al.* 2011). Since species proportional abundance generally varies with species composition, TD of $q > 0$ responds to differences in species composition; the relative importance of richness versus species composition may be assessed via TD profiles.

2.3.2. *Evenness measures*

Compared to diversity, measuring evenness among species relative abundances in a community is an even more complicated issue. The seemingly endless list of evenness measures suggests that a consensus on the concept and measures of evenness is a daunting goal. Chao and Ricotta (2019) proposed a unified methodology that encompasses many of the most useful evenness measures. In their approach, unevenness (or its opposite, evenness) among species relative abundances is measured by a normalized divergence (or its opposite, closeness) between the vector of species relative abundances and those of a completely even community. Among the several classes of evenness measures developed in Chao and Ricotta (2019), we focus on only one class of measures, which can be usefully connected to BEFS research.

The class of evenness measures is expressed ${}^qE = [1-({}^qTD)^{(1-q)}]/(1-S^{(1-q)})$, where ${}^qTD$ denotes Hill number of order $q > 0$. (For $q = 0$, species abundances are disregarded, so it is not meaningful to evaluate evenness.) All measures lie in a fixed interval [0, 1], with the minimum value of 0 for a maximally uneven S-species community (when one species is super-dominant, while all the others are vanishingly scarce), and the maximum value of 1 for a completely even community (all species are equally abundant), regardless of species richness. The corresponding unevenness measure is simply $1-{}^qE$, which also has a fixed range of [0, 1].

Let CV denote the coefficient of variation of species abundances, that is, $CV = \sigma/\mu$, where μ and σ^2 denote, respectively, the mean and variance based on the species abundance set $\{z_1, z_2, ..., z_S\}$. For $q = 2$, the evenness measure reduces to an index proposed by Smith and Wilson (1996). The corresponding unevenness measure for $q = 2$ can be linked to CV by the relationship $1-{}^2E = CV^2/S$. Here the divisor S in the formula is used to *adjust for the difference in species richness,* because the maximum value that CV^2 can attain, for S species, is S. For example, when $S = 2$, the maximum value of CV^2 is 2 (one species very common, the other with negligible abundance); when S = 10, the maximum value of CV^2 is 10 (one species very common, the other nine with negligible abundance). A magnitude of $CV^2 = 2$ represents maximal unevenness for the two-species case, whereas the same value of 2 for ten species represents a low degree of unevenness. The divisor S is thus required for CV^2 (i.e. using CV^2/S as an unevenness measure) when comparing communities with differing S. Since CV^2/S (a measure of q = 2) is dominated by highly abundant species, one can apply a square-root transformation to avoid this dominance, that is, use $CV/\sqrt{S}$ as an unevenness measure; see Kvålseth (2015) for additional properties.

In the special case of $q = 1$, the limit of ${}^qE$ when q tends to 1 reduces to the widely used Pielou's J' evenness index, i.e., ${}^1E = J' = H/\log(S)$ (Pielou 1966), where H denotes Shannon entropy. Here the divisor log(S) is also required when comparing communities with differing richness, because the maximum value that Shannon entropy can attain, for S species, is log(S). One can construct a continuous profile that depicts ${}^qE$ as a function of diversity order $q > 0$. This profile can be compared among communities even if species richness is not fixed across communities.

In BEFS studies, temporal variability of ecosystems is often measured by CV (or CV^2) based on a time series of community-aggregate biomass, productivity, or other pertinent variables; temporal invariability/stability is measured by the inverse of CV (or the inverse of CV^2) (e.g. Pimm 1984; Tilman *et al.* 2006). The above evenness theory in biodiversity can be readily applied to generalize these popular temporal variability measures to account for the inherent effect on CV of time-series length (the number of time points). When variables are measured over K time points, the maximum value that CV^2 can attain is K. As illustrated above, a magnitude of $CV^2 = 2$ represents maximal temporal variability based on two time points, whereas the same value of 2 for ten time points represents a relatively low degree of variability. Therefore, to compare temporal variability across studies with different numbers of time points, the above theory suggests that CV (or CV^2) should be generalized to $CV/\sqrt{K}$ (or CV^2/K), which adjusts for differences in time series length. The corresponding measure of temporal stability is $1 - CV/\sqrt{K}$ (or $1 - CV^2/K$). Such generalizations assure that the same magnitude of CV (or CV^2) quantifies the same degree of variability, even when time series length is different. Unlike the two original stability measures ($1/CV$ or $1/CV^2$), which take a value of infinity when $CV = 0$, the generalized measures of temporal stability and variability are one-complements of each other, so both are in the range of [0, 1] and the infinity issue does not arise.

In addition to the $q = 2$ measure, one can also consider Shannon entropy H based on biomass/productivity time series and obtain the information-based temporal variability measure $1 - H/\log K$ and the corresponding stability measure $H/\log K$. MacArthur (1955) was the first to use Shannon entropy to quantify community stability in a trophic web. A normalized form of Shannon entropy was also applied in Hairston *et al.* (1968) to assess community stability in an experimental study with protozoa and bacteria. As Shannon entropy can be partitioned across hierarchical levels, it may prove worthwhile to apply information-based measures to decompose stability across spatial or temporal scales in future stability analyses.

2.4. A unified framework integrating diversities (TD, PD, and FD)

As indicated in the Introduction, biodiversity includes "variation in genetic, phenotypic, phylogenetic, and functional attributes". Chao *et al.* (2021) proposed a unified *attribute-diversity* approach that integrates ordinary Hill numbers for TD, the PD measures derived in Chao *et al.* (2010), and the FD measures developed by Chao *et al.* (2019). In the unified approach, all measures can be formulated as Hill numbers of a *hypothetical* community that can be decomposed into M sub-communities: the i-th sub-community consists of v_i attributes, each with raw abundance a_i, $i = 1, 2, \ldots, M$. Here a *taxonomic attribute* means a species in TD, a *phylogenetic attribute* means a unit-length branch segment in a phylogeny in PD, and a *functional attribute* means a virtual functional group in FD. The total abundance among all attributes is denoted as $V = \sum_{i=1}^{M} v_i a_i$. Thus, the relative abundance of any attribute in the i-th sub-community is a_i / V. The *attribute diversity* (AD) or *Hill–Chao numbers* of order q are defined as the Hill number of order q for the hypothetical community:

$$ {}^{q}AD = \left\{ \sum_{i=1}^{M} v_i \left(\frac{a_i}{\sum_{j=1}^{M} v_j a_j} \right)^q \right\}^{1/(1-q)} = \left\{ \sum_{i=1}^{M} v_i \left(\frac{a_i}{V} \right)^q \right\}^{1/(1-q)}, \quad q \geq 0,\ q \neq 1. \qquad [2.2] $$

For $q = 0$, AD is interpreted as the total effective number of attributes (or attribute richness). For $q = 1$ and $q = 2$, AD can be interpreted, respectively, as the effective number of abundant and dominant attributes.

By design, Hill numbers (equation [2.1]) are a special case of Hill–Chao numbers. Suppose that some hypothetical community can be decomposed into S sub-communities, each comprised of only one species. Then equation [2.2] reduces to the ordinary Hill number of order q. In the following two sections, we show that both PD and FD can be fit into the AD framework as special cases; Table 2.1 summarizes the formulas of TD, PD, and FD and their corresponding values of M, v_i, and abundance a_i, $i = 1, 2, \ldots, M$, as special cases of Hill–Chao numbers.

AD: $^{q}AD=\left\{\sum_{i=1}^{M} v_i \left(\frac{a_i}{V}\right)^q\right\}^{1/(1-q)}$; $^{1}AD=\lim_{q\to 1} {}^{q}AD=\exp\left(-\sum_{i=1}^{M} v_i \frac{a_i}{V}\log\frac{a_i}{V}\right)$ AD = Hill numbers of a hypothetical community that is decomposed into M sub-communities: the i-th sub-community consists of v_i attributes, each with raw abundance a_i, $i = 1, 2, \ldots, M$, and $V=\sum_{i=1}^{M} v_i a_i$ (total abundance).
TD: $^{q}TD=\left\{\sum_{i=1}^{S} \left(\frac{z_i}{z_+}\right)^q\right\}^{1/(1-q)}$ (Ordinary Hill numbers)   As a special case of AD: $M = S$ (# species), $v_i = 1$ (species), $a_i = z_i$ (abundance of species i), and $V = z_+$.
PD: $^{q}PD=\left\{\sum_{i=1}^{B} L_i \left(\frac{z_i^*}{z_+T}\right)^q\right\}^{1/(1-q)}$; $mean\,^{q}PD=\frac{^{q}PD}{T}$; T = tree depth   As a special case of AD: $M = B$ (# branches), $v_i = L_i$ (branch length) and $a_i = z_i^*$ (total abundance descended from node/branch i), and $V = z_+T$.
FD: $^{q}FD=\left\{\sum_{i=1}^{S} v_i(\tau) \left(\frac{a_i(\tau)}{z_+}\right)^q\right\}^{1/(1-q)}$; τ = threshold for functional distinctness   As a special case of AD: $M = S$ (# species), $a_i = a_i(\tau)=\sum_{j=1}^{S}[1-d_{ij}(\tau)/\tau]z_j$ (abundance of functionally indistinct set of species i), $v_i = v_i(\tau)=z_i/a_i(\tau)$ (functional groups contributed by species i), and $V = z_+$, where d_{ij} denotes the trait-based distance between species i and species j, and $d_{ij}(\tau) = \min(\tau, d_{ij})$.

Table 2.1. *Taxonomic, phylogenetic, and functional diversities are integrated under a unified framework based on attribute diversity (AD), or Hill–Chao numbers. TD, PD, and FD for any order q and q = 1 can be obtained by substituting the appropriate v_i and a_i values listed in the table into the AD formulas. Here an attribute means a species in TD, a unit-length branch segment in a phylogenetic tree in PD, or a virtual functional group in FD*

2.4.1. *Phylogenetic diversity as a special case of attribute diversity*

Consider two communities with identical species richness and evenness but with no shared species. The species in the first community are all closely related (i.e. they share a recent common ancestor). The species in the second community are distantly related (i.e. they share a much older common ancestor). Traditional TD values would be identical for both communities, but we can easily see that the second community is more phylogenetically diverse. To broaden the concept of TD to cope with the evolutionary history or phylogeny among species, we assume that all species of a focal community are connected by a rooted ultrametric phylogenetic tree, with the S species as tip nodes. Chao *et al.* (2010) proposed that one must specify the time period that PD refers to; they thus introduced an additional parameter T (a time reference point) for quantifying PD. In our applications, the reference time is chosen to be the age of the root of the phylogenetic tree spanned by all species in the community.

Most previously proposed phylogenetic measures are descended from TD measures (see section 2.2). A widely used measure is the total branch length in the phylogenetic tree (Faith 1992). As will be explained later, Faith's PD can be regarded as a phylogenetic generalization of species richness. However, species/node abundance is disregarded in Faith's PD. The Gini–Simpson index was generalized to Rao's quadratic entropy, which measures the mean phylogenetic distance between any two individuals randomly selected from the community. The phylogenetic entropy is a generalization of Shannon entropy to incorporate phylogenetic distances. Shannon entropy and the Gini–Simpson index (in their original forms) do not obey the replication principle; neither do their phylogenetic generalizations. The two generalized phylogenetic measures thus cause the same kinds of inconsistent or counter-intuitive interpretations as their parent measures (see section 2.2). This motivated Chao *et al.* (2010) to extend Hill numbers to take into account phylogeny among species, and also generalize Faith's PD to include species abundance.

Chao *et al.*'s approach can be fit into the unified framework by treating each unit-length branch segment as a phylogenetic attribute. Within the phylogenetic tree with depth T, assume that there are B branches/nodes. Let L_i denote the length of the i-th branch, $i = 1, 2,..., B$. Consider a *hypothetical* community that is decomposed into B sub-communities: the i-th sub-community consists of L_i attributes (unit-length branch segments), each with node abundance z_i^* (the total abundance descended from the i-th node/branch). Substituting these into Hill–Chao numbers (equation [2.2]) leads to the ${}^qPD$ measure derived in Chao *et al.* (2010); see Table 2.1 for formulas. PD is interpreted as the effective total phylogenetic attributes or effective total branch length (because each phylogenetic attribute is unit-length). For $q = 0$, it reduces to Faith's PD, that is, a phylogenetic generalization of species richness.

Dividing ${}^qPD$ by tree depth, we obtain the *mean* ${}^qPD$ (mean-PD of order q), which quantifies the effective number of equally divergent lineages, each with lineage length T. Thus, mean-PD has the same units as TD. When there are no interior nodes in the phylogenetic tree, *mean* ^{q}PD reduces to TD of order q. The mean-PD measures of orders $q = 0$, 1, and 2 unify three well-known phylogenetic indices: mean-PD of $q = 0$ reduces to Faith's PD divided by tree depth; mean-PD of $q = 1$ and $q = 2$ are, respectively, simple transformations of phylogenetic entropy and Rao's quadratic entropy. These transformations are similar to those we discussed for transforming Shannon entropy and the Gini–Simpson index to Hill numbers. The mean-PD measures convert the two phylogenetic generalizations of classic complexity indices to measures in units of species/lineages equivalents that satisfy the replication principle, and thus resolve the interpretation problems mentioned above.

Cadotte *et al.* (2008) demonstrated, for plant productivity, that Faith's PD explained significantly more variation than species richness or functional groups. Their results show that evolutionary history among species can provide critical information that taxonomic diversity cannot. Incorporating the additional abundance-sensitive PD measures of $q = 1$ and 2 or a profile should provide additional insights into BEFS experiments and related research.

2.4.2. *Functional diversity as a special case of attribute diversity*

Tilman (2001) defined functional diversity (FD) as "the range and value of those species and organismal traits that influence ecosystem functioning". Since its introduction, FD has revolutionized BEFS research. FD is a key to understanding ecosystem processes and response to environmental degradation and disturbance and thus has been used extensively to assess their effects on ecosystem function (Hooper *et al.* 2005). When each species is characterized by one or more traits that are relevant to ecosystem function, a wide range of approaches have been proposed to quantify FD. Different perspectives have led to the development of measures that vary in how they quantify particular aspects of trait space; see Chao *et al.* (2019) for a review.

A classic FD index is the number of functional groups (e.g. grass, forb, legume, woody plants, etc.) in a community. Each group comprises species that share similar traits and similar effects on ecosystem functioning. In practice, defining functional-group diversity may be difficult (Hooper *et al.* 2005). Another classic method is the dendrogram-based approach, in which a functional dendrogram is constructed by applying a clustering algorithm to the species pairwise distance matrix calculated from species traits (Petchey and Gaston 2002). Then FD is defined as the total branch length of the dendrogram. Virtual functional groups can be determined by "cutting off" the dendrogram at a threshold distance level. In these classic approaches, species abundances are not considered.

The generalization of Hill numbers to FD is a distance-based approach to classify species into virtual functional groups for any $q \geq 0$. Species pairwise distance is typically calculated by the Gower distance, which can deal with both categorical traits and continuous traits. In this approach, all FD measures are sensitive to both species abundance and trait-based species-pairwise distances. Just as a "cut-off" level was used in Petchey and Gaston's approach, Chao *et al.* (2019) introduced an essential parameter for FD, τ (tau), which defines the threshold for functional distinctness between any two species; τ can be chosen to be *any* positive value. Any two species with functional distance $\geq \tau$ between them are treated as functionally equally-distinct at the level of τ. The basic motivation is from statistical clustering: imagine that all species are placed in a functional space with specified pairwise distances, and we cluster them into virtual functional groups. As in most clustering algorithms, one must first determine a threshold level such that any two species with distances greater than or equal to the specified threshold level are in different clusters and vice versa. If the threshold is set to be very low, then each species forms a functional group, all species are equally distinct from one another, and FD reduces to TD. In contrast, if we use a very high threshold, then nearly all species are classified into a single group and the diversity measure becomes weakly sensitive to species abundances and traits.

Since two or more species may be functionally similar to some extent, or even functionally identical, Chao *et al.* (2019) first expanded the set of individuals belonging to species i to a *functionally indistinct* set of species i, for any specified level τ. The abundance of the functionally indistinct set of species i is denoted as $a_i(\tau)$; see Table 2.1 for formulas. When the "abundance" of species i is expanded to $a_i(\tau)$, species i no longer contributes to a single functional group, but a proportion $v_i(\tau) = z_i \,/\, a_i(\tau)$ of the group, so each individual is only counted only once in the formulation. This proportion represents the contribution of species i after its functional redundancy with other species is accounted for.

Chao *et al.* (2019) considered a hypothetical community that is decomposed into S sub-communities with the i-th sub-community consisting of $v_i(\tau)$ functional groups, each with abundance $a_i(\tau)$. Then equation [2.2] reduces to the ${}^qFD$ measure derived in Chao *et al.* (2019); see Table 2.1 for formulas. When the threshold level τ is low, each species forms a virtual functional group and ${}^qFD$ reduces to TD of order q. The measure ${}^qFD$ quantifies the effective number of equally distinct functional groups (or functional attributes or functional "species") at a given threshold level τ. Since Gower distance is between 0 and 1, we can consider all plausible thresholds in the interval [0, 1] and depict ${}^qFD$ as a function of τ. In our applications, we compute the area under the τ-profile in [0, 1] and obtain an overall, integrated functional measure.

2.5. Diversity in space and time

Diversity varies with time and space. Spatial and temporal variation in species composition (turnover) is one of the most fundamental features of the natural world. Beta diversity measures the extent of species compositional differentiation among communities. Ever since the pioneering work by Whittaker (1960, 1972), there have been many definitions, concepts, and measures of beta diversity and the closely-related (dis)similarities among communities. Chao and Chiu (2016) bridged two major approaches (the variance framework and diversity decomposition) by showing that the two approaches lead to the same (dis)similarity measures. We present here only the diversity decomposition approach for TD; it can be directly extended to PD and FD by replacing "species", below, with phylogenetic or functional attributes.

The diversity decomposition approach leads to an intuitive interpretation: gamma diversity refers to the effective number of species in the pooled community, whereas alpha diversity refers to the average effective number of species per community if communities are equally weighted. Beta diversity is gamma diversity divided by alpha diversity. That is, we apply Whittaker's original multiplicative definition of beta diversity but use Hill numbers for any diversity order $q \geq 0$. Beta diversity is then expressed in units of "community equivalents" or the effective number of communities. Beta diversity always attains a *fixed minimum* value of unity if all N communities are identical in terms of class identity and abundance, and beta diversity attains a *fixed maximum* value of N (the number of communities) when no classes are shared among communities (i.e. complete turnover). As beta diversity lies in the range of $[1, N]$, it can be compared across studies only for a fixed value of N. To illustrate with a simple example, a beta value of 2 would signify maximal differentiation between two communities ($N = 2$), whereas the same beta value for a ten-communty case ($N = 10$) signifies low differentiation. When N varies across studies, proper normalizations are needed to transform beta diversity to dissimilarity measures in a fixed range of [0, 1] by means of monotonic transformations; see Chao *et al.* (2019) for four types of transformations. The resulting similarity measures (including the multiple-community generalizations of the classic Sørensen, Jaccard, Horn, and Morisita–Horn measures) can be compared even if N varies across studies.

In comparison with alpha diversity, beta diversity has received relatively less attention in BEFS studies, primarily because most BEFS experiments have been focused on local/small scales. Mori *et al.* (2018) listed several studies that measure the relationship between beta diversity and ecosystem function. To extend BEFS across spatial scales, Wang and Loreau (2014, 2016) developed a framework to decompose gamma (regional) diversity and variability into alpha (local) and spatial beta components. Their decomposition was based on TD of order $q = 2$, which is dominated by highly abundant species. We emphasize that TD of order $q = 1$ weighs species precisely by their abundance and possesses good decomposition properties across hierarchical levels and thus can also be recommended for future BEFS research.

2.6. Examples

2.6.1. *Coral data*

We use the coral data collected by Warwick and Clarke (1990) to illustrate changes in taxonomic gamma, alpha, and beta diversity of coral communities at the South Tikus Islands, Indonesia, in three surveys (1981, 1983, and 1985) bracketing a coral bleaching event caused by El Niño in 1982. These data were also analyzed by Anderson *et al.* (2011), and Chao and Chiu (2016). Within each year, the percentage cover (as a spatial measure of species abundance) for each of the observed coral species for $N = 10$ line transects was recorded. If we designate each transect a "community", then beta diversity represents community differentiation in percentage cover among the ten transects, within each year. Here we treat percentage cover as true species abundances, without considering sampling errors/uncertainties.

Figure 2.1 shows the plots for gamma, alpha, and beta diversity profiles within each year. The diversity profiles reveal that, in 1983, following the El Niño event in 1982, both alpha and gamma diversities had decreased to some extent from their 1981 values. A large decline in rare species was reflected by the substantial drop in species richness ($q = 0$), a moderate decline for abundant species ($q = 1$), and a mild decline in dominant species ($q = 2$). From 1983 to 1985, gamma diversity of orders $q = 1$ and $q = 2$ continued to decline, but gamma diversity of $q = 0$ and alpha diversity for all q values increased.

In contrast to the findings in Anderson *et al.* (2011), Figure 2.1(c) shows that beta diversity consistently exhibited an increasing trend from 1981 to 1983, followed by a large decreasing trend from 1983 to 1985. Beta diversity declined to the 1981 level (for $q < 0.5$ measures) or even lower (for $q > 0.5$ measures), but peaked in 1983. As the number of communities is the same for each year, all dissimilarity measures (Chao and Chiu 2016) exhibit a pattern similar to beta diversity.

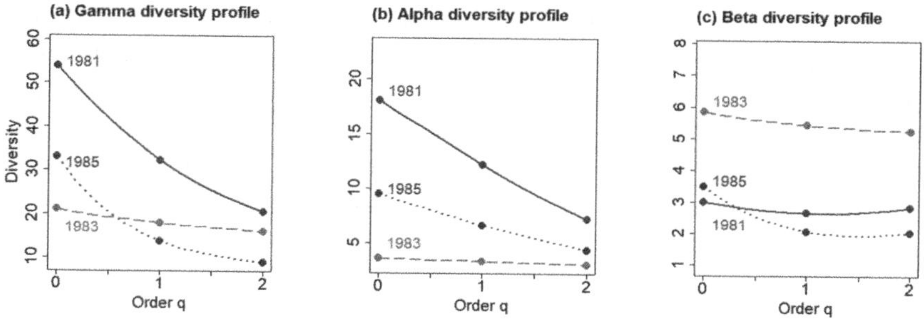

Figure 2.1. *(a) Taxonomic gamma, (b) alpha, and (c) beta diversity profiles as functions of q for the Tikus Island coral dataset (Warwick and Clarke 1990) for 1981 (black solid line), 1983 (red dashed line), and 1985 (blue dotted line). This figure is an expanded version of Figure 1 in Chao and Chiu (2016). The solid dots on each curve denote the corresponding values of q = 0, 1, and 2. For a color version of this figure, see www.iste.co.uk/loreau/biodiversity.zip*

2.6.2. *Saproxylic beetle data*

We use data on saproxylic beetles collected in a temperate mountain forest ecosystem to illustrate TD, PD, FD, and their decompositions. The data were previously analyzed by Thorn *et al.* (2016) and Chao *et al.* (2019). Here we focus on comparing beetle diversity for two bark treatments: debarked and (untreated) control. Species abundance data were collected from 12 plots. In each plot, the bark of one tree was completely removed (the "debarked" treatment), and another, untreated tree served as a control. The abundance data in the analysis were pooled from the 12 plots.

In all, 84 beetle species (11,346 individuals) were trapped in the control trees, and 61 species (3,201 individuals) in the debarked trees. Taxonomic data are from Thorn *et al.* (2016); the phylogenetic tree and species traits for all species found in the two treatments are based on Seibold *et al.* (2015). The age of the root was 76 Mys. All species were characterized for a set of ten functional traits; see Chao *et al.* (2019) for details on traits and species pairwise distances computed from the Gower distance.

For each of the two treatments, Figure 2.2(a) shows for $0 \leq q \leq 2$ a general ordering TD > mean-PD > FD among the three dimensions. Regardless of diversity type, the diversity profile of the control group lies above that of the debarked group, revealing that beetle diversity for control trees is greater than that of debarked trees; the largest difference between the two bark treatments occurs at $q = 0$, and the magnitude of the difference diminishes when the diversity order q increases. That is, rare species contribute disproportionally to high TD, mean-PD, and FD diversity in the control trees. The main differences in the diversities were likely due to the loss of many phloem-feeding and other rare species associated with bark, and these rare species are associated with rare lineages. Also, the control trees host species with a wide range of different functions, whereas debarked trees host primarily saproxylic species, which mainly bore into heartwood (Thorn *et al.* 2016).

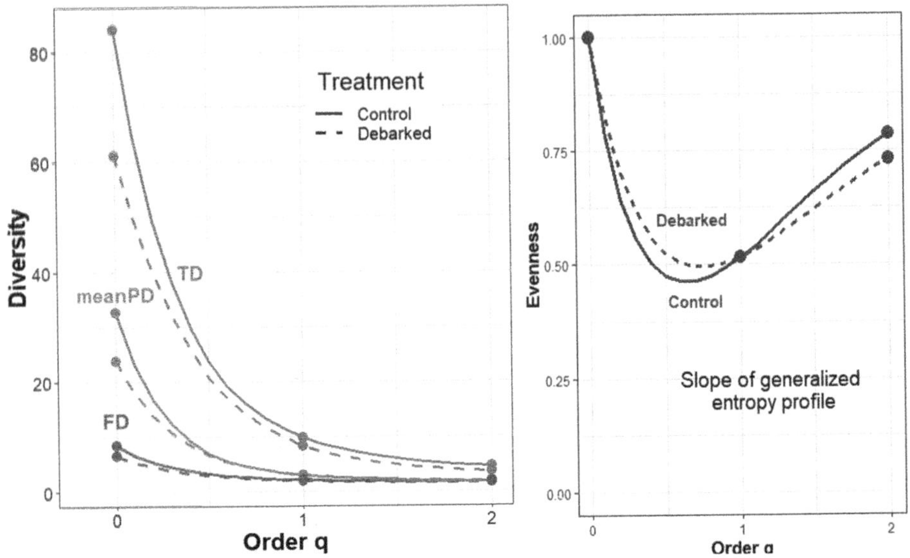

(a) Within-treatment diversity profile (b) Evenness profile

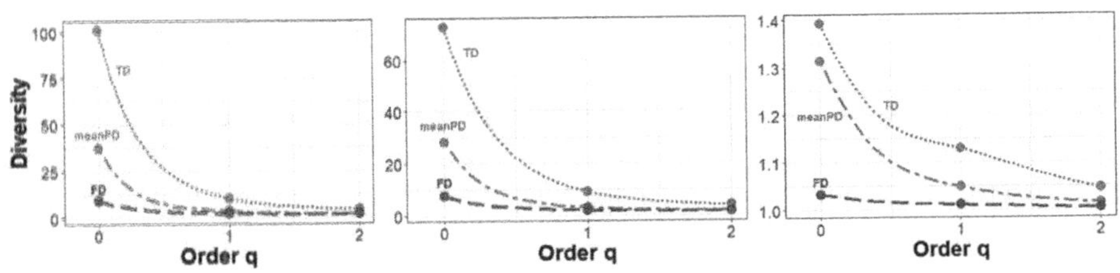

(c) Gamma diversity profile (d) Alpha diversity profile (e) Beta diversity profile

Figure 2.2. *(a) The within-treatment TD, mean-PD, and FD diversity profiles; (b) the evenness profile based on slopes of generalized entropy; and (c) the corresponding gamma based on pooling data of the two treatments, (d) alpha and (e) between-treatment beta diversity profiles. All profiles are plotted as a function of diversity order q, 0 ≤ q ≤ 2, based on beetle species relative abundance data for two treatments (control and debarked). The solid dots on each curve denote the corresponding values of q = 0, 1, and 2. For a color version of this figure, see www.iste.co.uk/loreau/biodiversity.zip*

Figure 2.2(b) shows the profiles of the two groups for the evenness measure ${}^qE$ (defined in section 2.3). The two profiles cross at $q = 1$. For $q < 1$ (rare species are in focus), the evenness profile reveals that the control group is slightly less even, mainly because of the existence of more relatively rare species than in the debarked trees. When the emphasis is on common species ($q = 1$), the two groups have roughly the same degree of evenness. When the emphasis is shifted to dominant species ($q > 1$), the control group is more even; but the two groups differ to a limited extent.

Figures 2.2(c)–(e) show the TD, mean-PD, and FD profiles of the gamma, alpha, and between-treatment beta diversity for $0 \leq q \leq 2$. As with the within-treatment profiles in Figure 2.1(a), all gamma, alpha, and beta profiles exhibit consistent ordering: TD > mean-PD > FD among the three dimensions; moreover, all decline as q is increased. This consistent pattern implies that the compositional differences in beetle species, lineages, and functions between the two bark treatments are all mainly attributable to rare species/lineages/functions. The plots of dissimilarity measures are omitted because their patterns are similar to the pattern of beta diversity.

All computations and graphics are based on the software iNEXT (specifically for TD) and iNEXT.3D (for TD, PD, and FD). R packages are available on CRAN and Shiny web interfaces are available at https://chao.shinyapps.io/iNEXTOnline/ and https://chao.shinyapps.io/iNEXT_3D/.

2.7. Conclusion

This chapter introduces a unified framework based on Hill numbers and their generalizations to quantify within-community TD, PD, and FD; see equation [2.2] and Table 2.1. The framework also features interpretable measures to assess evenness of species abundance and to compare among-community beta diversity changes in time and space. Figures 2.1 and 2.2 show how these measures can be applied to real-world examples. Some measures and their decompositions have been applied to BEFS studies. Other measures (especially the generalized stability measures and abundance-sensitive PD and FD measures) are recommended for future BEFS research.

2.8. Acknowledgements

We thank Simon Thorn for providing the beetle data used in section 2.6 and for his insightful comments on interpreting the analysis results. We also thank Shaopeng Wang and Amelia A. Wolf for helpful comments on an earlier version.

2.9. References

Anderson, M.J., Crist, T.O., Chase, J.M. *et al.* (2011). Navigating the multiple meanings of β diversity: A roadmap for the practicing ecologist. *Ecology Letters*, 14, 19–28.

Cadotte, M.W., Cardinale, B.J., Oakley, T.H. (2008). Evolutionary history and the effect of biodiversity on plant productivity. *Proceedings of the National Academy of Sciences*, 105, 17012–17017.

Chao, A. and Chiu, C.H. (2016). Bridging two major approaches (the variance framework and diversity decomposition) to beta diversity and related similarity and differentiation measures. *Methods in Ecology and Evolution*, 7, 919–928.

Chao, A. and Ricotta, C. (2019). Quantifying evenness and linking it to diversity, beta diversity, and similarity. *Ecology*, 100, e02852.

Chao, A., Chiu, C.H., Jost, L. (2010). Phylogenetic diversity measures based on Hill numbers. *Philosophical Transactions of the Royal Society B: Biological Sciences*, 365, 3599–3609.

Chao, A., Gotelli, N.G., Hsieh, T.C. *et al.* (2014). Rarefaction and extrapolation with Hill numbers: A framework for sampling and estimation in species biodiversity studies. *Ecological Monographs*, 84, 45–67.

Chao, A., Chiu, C.H., Villéger, S. *et al.* (2019). An attribute-diversity approach to functional diversity, functional beta diversity, and related (dis)similarity measures. *Ecological Monographs*, 89, e01343.

Chao, A., Henderson, P.A., Chiu, C.H. *et al.* (2021). Measuring temporal change in alpha diversity: A framework integrating taxonomic, phylogenetic and functional diversity and the iNEXT.3D standardization. *Methods in Ecology and Evolution* [Online]. Available at: https:// besjournals.onlinelibrary.wiley.com/doi/10.1111/2041-210X.13682.

Colwell, R.K. (2009). Biodiversity: Concepts, patterns, and measurement. *The Princeton Guide to Ecology*, 663, 257–263.

Colwell, R.K., Chao, A., Gotelli, N.J. *et al.* (2012). Models and estimators linking individual-based and sample-based rarefaction, extrapolation and comparison of assemblages. *Journal of Plant Ecology*, 5, 3–21.

Díaz, S., Demissew, S., Carabias, J. *et al.* (2015). The IPBES Conceptual Framework – Connecting nature and people. *Current Opinion in Environmental Sustainability*, 14, 1–16.

Ellison, A.M. (2010). Partitioning diversity. *Ecology*, 91, 1962–1963.

Faith, D.P. (1992). Conservation evaluation and phylogenetic diversity. *Biological Conservation*, 61, 1–10.

Foster, B.L., Smith, V.H., Dickson, T.L., Hildebrand, T. (2002). Invasibility and compositional stability in a grassland community: Relationships to diversity and extrinsic factors. *Oikos*, 99, 300–307.

Hairston, N., Allan, J., Colwell, R.K. *et al.* (1968). The relationship between species diversity and stability: An experimental approach with protozoa and bacteria. *Ecology*, 49, 1091–1101.

Hector, A., Bell, T., Hautier, Y. *et al.* (2011). BUGS in the analysis of biodiversity experiments: Species richness and composition are of similar importance for grassland productivity. *PLoS ONE*, 6, e17434.

Hill, M.O. (1973). Diversity and evenness: A unifying notation and its consequences. *Ecology*, 54, 427–432.

Hillebrand, H., Bennett, D.M., Cadotte, M.W. (2008). Consequences of dominance: A review of evenness effects on local and regional ecosystem processes. *Ecology*, 89, 1510–1520.

Hooper, D.U., Chapin III, F.S., Ewel, J.J. *et al.* (2005). Effects of biodiversity on ecosystem functioning: A consensus of current knowledge. *Ecological Monographs*, 75, 3–35.

Hsieh, T.C., Ma, K.H., Chao, A. (2016). iNEXT: An R package for rarefaction and extrapolation of species diversity (Hill numbers). *Methods in Ecology and Evolution*, 7, 1451–1456.

Isbell, F., Gonzalez, A., Loreau, M. *et al.* (2017). Linking the influence and dependence of people on biodiversity across scales. *Nature*, 546, 65–72.

Jost, L. (2007). Partitioning diversity into independent alpha and beta components. *Ecology*, 88, 2427–2439.

Kvålseth, T.O. (2015). Evenness indices once again: Critical analysis of properties. *SpringerPlus*, 4, 1–12.

Loreau, M. (2010). *From Populations to Ecosystems: Theoretical Foundations for a New Ecological Synthesis (MPB-46)*. Princeton University Press, Princeton.

MacArthur, R.H. (1955). Fluctuations of animal populations, and a measure of community stability. *Ecology*, 36, 533–536.

MacArthur, R.H. (1965). Patterns of species diversity. *Biological Reviews*, 40, 510–533.

Magurran, A. (2004). *Measuring Biological Diversity.* Blackwell Publishing, Oxford.

Mori, A.S., Isbell, F., Seidl, R. (2018). β-diversity, community assembly, and ecosystem functioning. *Trends in Ecology and Evolution*, 33, 549–564.

Naeem, S., Bunker, D.E., Hector, A., Loreau, M., Perrings, C. (2009). *Biodiversity, Ecosystem Functioning, and Human Wellbeing: An Ecological and Economic Perspective*. Oxford University Press, Oxford.

Nanney, D.L. (2004). No trivial pursuit. *BioScience*, 54, 720–721.

Petchey, O.L. and Gaston, K.J. (2002). Functional diversity (FD), species richness and community composition. *Ecology Letters*, 5, 402–411.

Pielou, E.C. (1966). The measurement of diversity in different types of biological collections. *Journal of Theoretical Biology*, 13, 131–144.

Pielou, E.C. (1975). *Ecological Diversity*. John Wiley, New York.

Pimm, S.L. (1984). The complexity and stability of ecosystems. *Nature*, 307, 321–326.

Seibold, S., Brandl, R., Buse, J. *et al.* (2015). Association of extinction risk of saproxylic beetles with ecological degradation of forests in Europe. *Conservation Biology*, 29, 382–390.

Smith, B. and Wilson, J.B. (1996). A consumer's guide to evenness indices. *Oikos*, 76, 70–82.

Thorn, S., Bässler, C., Bußler, H. *et al.* (2016). Bark-scratching of storm-felled trees preserves biodiversity at lower economic costs compared to debarking. *Forest Ecology and Management*, 364, 10–16.

Tilman, D. (2001). Functional diversity. In *Encyclopedia of Biodiversity*, Levin, S.A. (ed.). Elsevier, New York.

Tilman, D., Reich, P.B., Knops, J.M.H. (2006). Biodiversity and ecosystem stability in a decade-long grassland experiment. *Nature*, 441, 629–632.

Wang, S. and Loreau, M. (2014). Ecosystem stability in space: α, β and γ variability. *Ecology Letters*, 17, 891–901.

Wang, S. and Loreau, M. (2016). Biodiversity and ecosystem stability across scales in metacommunities. *Ecology Letters*, 19, 510–518.

Warwick, R. and Clarke, K. (1990). A statistical analysis of coral community responses to the 1982–83 El Niño in the Thousand Islands, Indonesia. *Coral Reefs*, 8, 171–179.

Whittaker, R.H. (1960). Vegetation of the Siskiyou mountains, Oregon and California. *Ecological Monographs*, 30, 279–338.

Whittaker, R.H. (1972). Evolution and measurement of species diversity. *Taxon*, 21, 213–251.

Wilson, E.O. and Peter, F.M. (eds) (1988). *Biodiversity*. National Academy Press, Washington, DC.

3

Ecosystems: An Overview

Amelia A. WOLF, Sarah K. ORTIZ, and Chase J. RAKOWSKI
University of Texas at Austin, USA

3.1. An introduction to ecosystems

In order to understand the effects of biodiversity loss on ecosystems, it is helpful to review some key aspects of ecosystems themselves. While ecosystems have been recognized as an important level of biological organization for well over a century, our understanding of what shapes ecosystems and ecosystem processes has been continuously evolving. In the past several decades, ecologists have made huge strides in illuminating the roles that the diversity of organisms, and the interactions between those organisms, play in structuring ecosystem processes and the benefits that humans receive from ecosystems around the globe.

An *ecosystem* is defined as the biotic and abiotic components of an area and the ways in which those component parts interact to form a dynamic system. The concept of an ecosystem explicitly focuses on understanding integrated interactions rather than focusing on individual species or abiotic processes. *Ecosystem ecology*, the subdiscipline of ecology that takes an ecosystems approach to studying the biosphere, seeks to understand how energy and resources flow through an integrated system, from abiotic to biotic components and back again. The cyclic nature of these processes leads to complex feedbacks among many of the component parts. Ecosystems are affected by characteristic *disturbance regimes*: events that are discrete in time and space that cause a disruption to resource availability or the physical environment (White and Pickett 1985). Due to these complexities,

The Ecological and Societal Consequences of Biodiversity Loss,
coordinated by Michel LOREAU, Andy HECTOR, and Forest ISBELL.

ecosystems are generally recognized as being dynamic, nonequilibrium systems rather than having a single stable equilibrium.

Most definitions of ecosystems do not explicitly denote spatial scale – rather, the spatial scale of an ecosystem is defined by the size of the organisms and the extent of the processes in question. In theory, this could lead to ecosystems within a drop of lake water, or a single Earth-sized ecosystem, but in practice, ecosystems generally fall in spatial scale between the more-encompassing *biome* – defined as a characteristic community type found in a geographic area with a certain climate – and the less-encompassing *habitat* – the area defined by the organisms and abiotic conditions in which a particular species is found. The related concept of *community* encompasses the biotic factors and interactions but does not integrate the abiotic factors, including biogeochemical processes, that are essential for defining an ecosystem.

Why are different ecosystems found in different areas across the Earth? Ecosystem extent, or the boundary between one ecosystem and another, in terrestrial systems is determined by five *state factors* that include the abiotic factors of climate, bedrock parent material, topography, and time, along with the biotic community (Jenny 1941). In aquatic communities, ecosystems are defined by a variety of abiotic factors including flow rate, water depth, water chemistry, incoming solar radiation, and substrate type.

3.1.1. *Ecosystem extent: abiotic factors in terrestrial systems*

Climate

Climate is a fundamental determinant of ecosystems and in many ways dictates where and how organisms live. *Climate* is the long-term average of temperature and precipitation in a region. These long-term averages are primarily shaped by differential incoming solar radiation across latitudes, which in turn shape global-scale atmospheric circulation patterns.

The equator receives the largest and most consistent proportion of solar radiation. As a result, the climate at the equator is warm and wet year-round, a prerequisite for most tropical ecosystems. The polar regions receive the lowest amount of solar radiation due to the tangential angle to the sun, giving rise to the polar ecosystems. Temperate ecosystems are classified by the large variation in climate due to the annual fluctuations in solar radiation.

Topography

Regional topography and proximity to oceans create local climatic differences. For example, mountainsides that face the prevailing winds experience higher precipitation due to the rising, cooling, and release of moisture from the air. As a result, a rain shadow effect occurs on mountainsides opposite the prevailing winds, where precipitation is sparse due to the cool, dry air descending and absorbing moisture as it warms.

Bedrock

Bedrock, or *parent material*, refers to the substructure of hard rock under the Earth's surface. *Chemical weathering*, the erosion caused by chemical reactions, and *physical weathering*, the erosion caused by physical forces, of bedrock determine the type of soil in a region. Variation in the chemical and mineralogical composition of bedrock strongly influences the texture of soils underlain by different types of bedrock as well as nutrient and water availability to plants living in those soils. The physical and chemical weathering of bedrock breaks down the subsurface material and alters the chemical composition such that essential nutrients – primarily phosphorous (P) but also nitrogen (N) and other plant-essential elements including calcium, magnesium, and potassium – are released and become available to biota. Additionally, variation in bedrock properties can influence the availability of water stored within bedrock, a potentially large reservoir for many perennial plants during times of extended drought.

Ecosystem age

Certain major disturbances, including glaciation, volcanic activity, and tectonic activity, can strip away most of the biota and reset an ecosystem back to bare bedrock. In areas with recent volcanic activity, for instance, newly cooled lava forms a blank surface of rock that microorganisms, plants, and animals eventually colonize; a soil layer slowly develops. The length of time since this type of major disturbance determines an ecosystem's age, a factor that plays a major role in nutrient availability.

In many ecosystems, a single nutrient is in the shortest supply, and thus becomes the limiting factor for biotic growth. (Nutrients can also be co-limiting, such that two or more nutrients are in low supply; water and sunlight are also common limiting factors for biotic growth.) Globally, either nitrogen or phosphorus is most often the limiting nutrient in ecosystems. As nitrogen and phosphorus derive from different sources within an ecosystem, ecosystem age has a strong influence on which nutrient is limiting; climate influences the relative rates at which ecosystems age due to the influence of climate on rock weathering rates (Chapin *et al.* 2011).

The main source of new nitrogen to ecosystems is nitrogen fixation, a process by which bacteria, either on their own or in partnership with certain plants, turn atmospheric N_2 into a plant-usable form of nitrogen. The main source of new phosphorus is bedrock weathering. Ecosystems go through predictable stages based on age, where the youngest ecosystems are deficient in both nitrogen and phosphorus, eventually becoming nitrogen-limited as rock-derived phosphorus becomes available via weathering, and finally reaching phosphorus limitation as rock-derived phosphorus is depleted but nitrogen fixation supplies sufficient nitrogen (Vitousek and Farrington 1997). As a result, vegetation composition will shift across ecosystem age in response to the shift in nutrient availability.

Disturbance

Large-scale abiotic disturbances impact entire ecosystems and can range from rare and intense to frequent and mild. These types of disturbances influence many abiotic factors, including water and nutrient availability, as well as the types of biotic communities that thrive in an area. Fires and hurricanes are examples of relatively rare (usually occurring less than once a year) but intense disturbances. In grasslands, fires are essential for maintaining the ecosystem as a means of preventing shrub encroachment and forest development. In rocky intertidal regions, ecosystems experience recurrent (return interval of less than a minute) mild disturbances through waves crashing onto rocks. This level of disturbance helps maintain the diversity of the intertidal regions by disrupting dominant species that would otherwise exclude other species (Sousa 1979).

Ecosystem extent: abiotic factors in aquatic ecosystems

Ecosystem age and disturbance also have important influences on aquatic ecosystems. For example, young lakes tend to be oligotrophic and deep, and then gradually fill in and accumulate nutrients over time, eventually becoming shallower and eutrophic. However, climate and bedrock have smaller influences on aquatic compared to terrestrial ecosystems. Instead, several other abiotic factors play more important roles, some of which relate to topography. Water flow rate has a large influence on biotic and abiotic factors, as evidenced by limnologists' classification of freshwater ecosystems as either *lotic* (high flow) ecosystems, such as streams, or *lentic* (low flow) ecosystems, such as lakes. Light availability is another key abiotic factor, relating both to water depth and to incoming solar radiation. For example, high light penetration in shallow waters is important for coral reef formation and drives high productivity in such ecosystems.

Water chemistry is another key determinant of aquatic ecosystems, encompassing an array of properties including pH, salinity, dissolved oxygen, dissolved organic matter, nutrient levels, pollutants, and other factors. This diversity of chemical properties is determined by a similarly diverse assortment of properties of both the immediate area and the broader catchment, from soil properties to terrestrial detritus inputs to anthropogenic pollution. Finally, the substrate type often plays an important role in determining aquatic ecosystem extent. For example, intertidal ecosystems with sandy substrates harbor a distinct array of organisms and are influenced by distinct processes such as sand deposition and removal, in contrast to rocky intertidal ecosystems.

3.1.2. *Ecosystem extent: biotic factors*

Species composition and traits

Ecosystems are shaped not only by abiotic factors but also by the species present in and around them. In turn, the presence of particular species in a given area at a given time depends not only on abiotic factors, but also on biotic factors such as dispersal (HilleRisLambers *et al.* 2012). For example, the presence of nitrogen-fixing plants in an area leads to increased levels of nitrogen in the soil, which in turn has knock-on effects on other species (Vitousek *et al.* 1987).

Measurable characteristics of an organism, such as dispersal ability, nitrogen-fixing ability, leaf shape, body size, growth rate, or metabolic rate, are termed *traits*. Species traits play a role both in how species respond to variability and changes in the environment, as well as in the effect that species have on ecosystem processes; because of this, determining how traits mediate the effects of biodiversity loss on ecosystem functioning has been the subject of many experiments. In general, higher trait diversity (see Chapter 2) allows for more complete filling of the niche space within an ecosystem and leads to maximizing a variety of ecosystem functions (multifunctionality; see Part 2 of this book).

Many ecosystems have characteristic dominant species compositions (and therefore trait compositions), such as temperate forests named for their most dominant tree species (e.g. oak-hickory forest). These dominant species maintain a relatively stable presence and therefore exert a large influence on the characteristics of the ecosystem. This influence creates a feedback wherein conditions favor the dominant taxa, cementing their stable dominance (at least within a certain range of conditions). For example, grasses burn more readily than trees and are quicker to colonize and grow freshly burned ground. Therefore, grasses tend to maintain their own dominance with the help of relatively frequent fires, allowing relatively stable grasslands to exist in climates where trees would otherwise also be able to thrive and form forests (e.g. savannahs of central Africa; see Part 3 of this book on stability) (Staver *et al.* 2011).

The long-term presence of particular plant species and traits leads to differences in soil properties which can often benefit the resident species (Ehrenfeld *et al.* 2005). Similarly, foundation species have an outsized influence on ecosystems, such that the presence of many other species and abiotic characteristics depends on the presence of the foundation species. For example, the presence of kelp maintains many other species that use the kelp as their habitat or food source as well as increasing primary production and detritus and decreasing light, collectively forming a kelp forest. Similar areas lacking kelp are dramatically different in biotic and abiotic characteristics (Graham 2004). However, many other ecosystems do not have a single species that can be easily pointed to as the dominant influence. Even in these ecosystems, though, the overall distribution of species traits has important ecosystem effects. For example, the various plant species typical of more productive environments lead to faster decomposition rates as compared to plants found in less productive environments (Chapin 2003).

Species interactions

In addition to species composition, species interactions play an important role in shaping ecosystems. Competition is a ubiquitous, albeit often unseen, force underlying the presence, abundance, and distribution of species in a given ecosystem (and therefore the species composition effects discussed above). Rather than leading to static species abundance distributions, competition can often lead to temporal dynamics in species dominance as different competing species gain advantages at different times (Albrecht and Gotelli 2001). In addition, competition among primary producers leads to the depletion of abiotic resources and changes in their relative availability (Goldberg 1990). The strength of competition among primary producers, in turn, is modulated by the availability and ratio of abiotic resources in the area as well as the strength of disturbance, facilitation, and top-down forces (herbivory and predation), which can serve to release species from competition when strong (Paine 1979; Chase *et al.* 2002; Wright *et al.* 2014).

Mutualisms, or interactions that benefit both interacting species, underlie key processes in many ecosystems. Plant–pollinator mutualisms, in which pollinators receive nourishment from plants in exchange for their pollination services, provide the foundation of many terrestrial ecosystems and of much of their diversity (Mitchell *et al.* 2009). Meanwhile, hidden underground in and around most plants' roots lies an equally important mutualism: the symbiosis between plants and mycorrhizal fungi, in which fungi provide plants with additional water and nutrients in exchange for carbohydrates (van der Heijden *et al.* 2015). More diffuse mutualisms may be much more common than previously recognized and may be an important determinant of ecosystem structure and a mechanism of biodiversity maintenance (Hay *et al.* 2004).

Trophic interactions, in which one organism benefits at the expense of another, can also play an important role in shaping ecosystems. Trophic interactions encompass predation, herbivory, detritivory, parasitism, and disease. Trophic cascades can be important determinants of the abundance of organisms across an ecosystem: these are a form of top-down control (that is, control by consumers) where the abundance of species in lower trophic levels, including primary producers, is controlled by the activities of consumers (Carpenter and Kitchell 1996). More generally, trophic interactions influence species composition and diversity (Schoener and Spiller 1996). When a trophic level is consumed or subject to disease, more-resistant taxa are selected for, and diversity is boosted as competitively dominant taxa are suppressed (Paine 1974). On the other end of the trophic interaction, the diversity and composition of the consumers is influenced by the composition and diversity of consumed species. Furthermore, trophic interactions play important roles in nutrient cycling, as consumers act as vectors to transport nutrients from their food sources to other parts of the ecosystem or to other ecosystems altogether (Schmitz *et al.* 2010).

Niches

An organism's niche is the way in which that organism interacts with other components of the ecosystem, including both its dependence on various abiotic and biotic factors and its effects on abiotic and biotic factors. Thus, the concept of the niche is closely intertwined with the concept of an ecosystem. There is an interplay between abiotic and biotic ecosystem components, as the heterogeneity of abiotic components (e.g. substrate, essential resources) affects the number of niches in an ecosystem, the number of niches affects the number of species able to coexist, and the number of species in turn affects the number of niches, as species are important components of niches themselves (Chase and Leibold 2003). The niche concept and its relationship to biodiversity effects is discussed in more detail in Chapter 4.

3.1.3. ***Major ecosystem types***

Aquatic ecosystems

Most habitats on Earth are found in the oceans. There are several ecosystem types that harbor distinct marine habitats. Most of the open ocean is shaped by severe nutrient limitation, resulting in a low density of organisms (Field 1998;

Moore *et al.* 2013). All photosynthesis takes place within the photic zone in the upper layer of the ocean, but the vast majority of the marine habitat lies below in the aphotic zone, where organisms depend largely on sinking detritus for nutrient input. In contrast, coastal areas have much greater nutrient input from terrestrial habitats and upwelling, and their shallow depths allow sunlight to reach the substrate. Thus, most marine life occurs in the shallows near coasts, including coral reefs and kelp forests. The intertidal zone represents a unique ecosystem, part marine and part terrestrial, whose denizens depend strongly on substrate type (Raffaelli and Hawkins 2012). Estuaries are similarly defined by transition – from freshwater to marine – and are some of the most productive ecosystems on Earth, although they are often dominated by one or a few plant species. Rivers and streams are freshwater lotic, ecosystems that transfer large amounts of water, sediments, nutrients, detritus, and other materials, as well as some organisms, downstream to other ecosystems. Lakes and ponds are lentic, landlocked aquatic ecosystems strongly influenced by inflow and outflow. Lakes with little or no outflow become salt lakes (Wetzel 2001). Finally, wetlands are shallow lentic ecosystems where waterlogged conditions lead to low soil oxygen levels; these systems often alternate between periods of inundation and partial drying (Kayranli *et al.* 2010).

Terrestrial ecosystems

Differences in state factors lead to a huge variety of terrestrial ecosystems around the globe. Broadly, the combination of annual rainfall and average temperature predicts where we find desert, grassland, forest, and tundra/polar ecosystems. Deserts are found in regions with very low rainfall, and the biota consists of organisms that have adaptations to long dry spells and often to very high heat. Differences in bedrock and topography, as well as climate, can lead to differentiation among different desert regions. Forests, on the other hand, tend to develop in places where rainfall is high but temperatures do not get too hot; slightly drier, slightly warmer areas are often occupied by grasslands. As described above, fire can play a pivotal role along with climate in determining the delineation between grassland and forest. Some ecosystem boundaries are delineated by changes in bedrock that underlies an area – serpentine ecosystems, for example, are often distinctive ecosystems with a highly *endemic* (specific to that area; not found elsewhere) biota that is adapted to high concentrations of heavy metals found in the bedrock.

Terrestrial versus aquatic ecosystems

The difference between water and air as a medium in which to live leads to several important differences between aquatic and terrestrial ecosystems. First, water temperatures are more stable than air temperatures. Second, light does not penetrate nearly as far through water, meaning most aquatic life is concentrated in shallow water and resulting in interesting dynamics between the photic and aphotic zones in deeper water bodies (e.g. diel vertical migrations of plankton to hide from predators during the day and access food in the shallows at night) (Hays 2003). Pelagic zones, which form the majority of aquatic habitats, are characterized by relatively little abiotic heterogeneity and therefore putatively fewer niches; nonetheless, as Hutchinson (1961) pointed out in his "Paradox of the Plankton", these ecosystems can be surprisingly biodiverse.

Water flow is intimately coupled with nutrient cycling in aquatic ecosystems, whether caused by gravity (downstream flow) or density differences due to chemistry and temperature differences (e.g. seasonal lake mixing and ocean currents). In contrast to terrestrial ecosystems, primary production is mostly performed by phytoplankton, but also by benthic microalgae, macroalgae, and aquatic plants (Wetzel 2001). Microalgae have faster dynamics than terrestrial plants, including fast growth rates, generation times, and mortality rates (Rip and McCann 2011). This often leads to dramatic sudden blooms and crashes of algal biomass, as well as marked seasonal succession patterns (Anneville *et al.* 2002). Compared to terrestrial ecosystems, freshwater ecosystems are more often P-limited, and open ocean ecosystems are often iron-limited; N-fixing cyanobacteria can dominate both freshwater and marine phytoplankton (Rabalais 2002). Finally, aquatic ecosystems are generally characterized by stronger trophic interactions, which can lead to inverted biomass pyramids in which predators outweigh herbivores which outweigh primary producers (most terrestrial ecosystems have the opposite pattern) (McCauley *et al.* 2018).

3.1.4. *Meta-ecosystems*

Ecosystems are linked together into meta-ecosystems via movement of organisms, or dispersal, as well as the movement of abiotic components (Loreau *et al.* 2003). Therefore, there are no true ecosystem boundaries, and some phenomena are difficult to understand unless considered on a larger scale. For instance, strong, sustained winds blow dust rich in iron and phosphorus from the Sahara Desert to the Atlantic Ocean and Amazonia, fertilizing both areas (Bristow *et al.* 2010). Even on smaller scales, movement of species and materials plays an important role in shaping ecosystems. For instance, source–sink dynamics, in which dispersal from patches where organisms have high fitness maintains populations in patches where they have low fitness, can be a mechanism for biodiversity maintenance, as well as a means of nutrient transport via the dispersing organisms (Gravel *et al.* 2010).

3.1.5. *Ecosystem dynamics and change over time and space*

Succession

Succession is the change in species composition and abundance over time driven by biotic and abiotic factors. Agents of change are classified into *disturbances*, events that injure or kill individuals, and *stressors*, which reduce the growth, reproduction, and survival of individuals. Biotic agents of change are often termed *ecosystem engineers*, species that modify the physical environment.

The effect of succession on ecosystem dynamics depends on the agent of change that is promoting succession. Primary succession occurs after catastrophic disturbances that either create new bedrock substrate (e.g. volcanic eruptions) or destroy all previous biotic inhabitants and strip the substrate down to bedrock (e.g. glacial retreat), as described in section 3.1.1. Secondary succession occurs after disturbances that disrupt the biota but do not remove it (or the soil substrate) entirely. This allows for a biotic community to recolonize quickly during secondary succession. Especially in forests, earlier-successional ecosystems often have high primary productivity as the canopy develops, increasing carbon sequestration rates. As succession proceeds, the canopy closes and growth becomes limited by nutrient and light availability and by biotic interactions; the rate of carbon sequestration begins to slow (Lorenz and Lal 2010). Stressors, such as nitrogen limitation for plants, can alter the community structure by altering fitness in the context of the environment, which drives species turnover and promotes succession. Nitrogen-fixers can act as ecosystem engineers when N availability is low by gradually increasing N availability to other plant species (Chapman 1935).

Climate variability

Short-term variability

Seasonal variability causes changes in the availability of resources such as light, water, nutrients. In general, ecosystems with dramatic seasonal changes are more productive in the warmer months and are primarily dormant in the colder months. Drought can also vary seasonally, affecting the average water availability within an ecosystem that further impacts the plant community structure and productivity over time. In the tropics, seasonal variation is described with respect to wet and dry seasons, as the temperature varies relatively little on average.

In aquatic systems, seasonality is defined by *seasonal stratification*, the differences in water temperature with depth in temperate and polar regions due to the differential heating of surface waters. Stratification determines the movement and availability of oxygen and nutrients to aquatic organisms seasonally, affecting productivity through time.

Long-term variability

Trends in long-term climate variability can also affect ecosystem dynamics over time. The El Niño Southern Oscillation occurs every three to eight years and is associated with unusual climatic conditions in the tropical Pacific. These climatic conditions include drier than average conditions in Southeast Asia and Australia, increasing the likelihood of fires. In the southern USA, there is often more rain than average, typically increasing primary productivity. This increase in precipitation is often followed by a drought that can increase the likelihood of fires in the area.

Over even longer periods of time are the *Milankovitch cycles* which describe the warming and cooling cycles that occur about every 100,000 years due to changes in shape of the Earth's orbit. Smaller climatic fluctuations arise because of the change in the tilt of Earth's axis of rotation and the orientation of the Earth's axis in relation to other celestial bodies. These climatic shifts can drastically change the global distributions of biomes.

3.2. Ecosystem functioning

Nearly every ecosystem on Earth is powered by the sun. Primary producers use energy from the sun and resources gained from soil, rock, and water to grow; energy and resources are transferred to other biotic and abiotic components of the ecosystem via trophic interactions, senescence, disturbance, and decomposition. Each of these steps is composed of many individual processes – pools of resources change size, fluxes of resources denote exchange among pools, and energy is transferred among ecosystem components. Each of the individual processes within an ecosystem that describes a pool or flux of energy or resources is termed an *ecosystem function*; collectively, these processes are referred to as *ecosystem functioning* (Hooper *et al.* 2005). One of the major findings of research on the impacts of biodiversity loss has been that as biodiversity decreases, ecosystem functioning decreases (see Part 2 of this book). Ecosystem goods and services often quantify the same processes as one or more ecosystem functions, but assign a human-centric value to the processes that comprise a dynamic ecosystem (see Part 4 of this book).

Productivity

Productivity is the most commonly quantified ecosystem function in biodiversity research because it forms the entry point for solar energy into the biotic system and it represents the foundation of food webs; it also has clear tie-ins with human-centric metrics including agricultural yield and carbon sequestration. Net primary productivity (hereafter productivity) is measured as the amount of new primary producer biomass (dried, after the water contained in the plant material has been removed) that grows over a defined time period and spatial scale. The productivity of an ecosystem differs greatly depending on climate, nutrient availability, soil characteristics, and time (including seasonality, successional stage, and ecosystem age).

Measuring productivity can be relatively straightforward in some ecosystems and extremely difficult in others. In ecosystems dominated by small-statured herbaceous plants, annual productivity is often quantified as the dry mass of live plants harvested at peak seasonal biomass – all live plants in a defined plot are clipped at the ground level, dried, and weighed as a measure of biomass per area (m^2, km^2, etc.) per time (day, year, etc.). Often, plots are fenced or caged for a time prior to biomass harvest to remove the effects of herbivory from productivity measurements. It is important to note that this measurement does not quantify all plant productivity, but only *aboveground productivity*. *Belowground productivity* can be quantified in a similar manner, but the process of collecting, identifying, and sorting belowground root samples is much more difficult. On the other end of the spectrum, productivity is quite challenging and time consuming to measure in ecosystems dominated by large-statured woody species (i.e. forests). In these systems, the change in size of trees over time is quantified, generally by taking repeat measurements of tree size across several years; measurements are then converted to productivity using allometric equations. Accounting for confounding variables such as herbivory can be quite a bit more difficult in ecosystems dominated by large-statured plants, and measurements are generally less accurate than in herbaceous-dominated ecosystems. Again, significant issues with quantifying belowground productivity exist in forest ecosystems as well.

Measuring productivity in aquatic ecosystems brings its own set of challenges. There is often no way to quantify the temporal component of phytoplankton productivity by measuring the amount of new phytoplankton biomass produced per unit time. Instead, phytoplankton biomass is measured and the results may or may not be assumed to relate to productivity. Even so, there is no straightforward or accurate way to estimate phytoplankton biomass either, so aquatic ecologists are normally forced to use proxies. The most commonly used proxy for phytoplankton biomass is chlorophyll-*a* concentration, which can be estimated by measuring the pigment in water samples using spectrophotometry. Another common proxy is biovolume, or the volume of cells per unit volume of water, which is calculated by measuring phytoplankton cell concentrations and cell volumes. Productivity of attached algae and macrophytes (aquatic vascular plants) can be measured in a simple manner analogous to measurement of terrestrial aboveground biomass: an area of substrate is cleared or clean tiles are placed in the water, and then the attached producers that grow are harvested and weighed after a known period of time.

Productivity: terrestrial versus aquatic ecosystems

While water and temperature limit productivity in many terrestrial ecosystems, productivity in aquatic ecosystems is mostly limited by light or nutrients. In aquatic ecosystems with limited light, such as forest streams or the depths of lakes and oceans, most life is supported by detritus rather than primary production (i.e. brown food webs). As light penetrates only a limited distance through water, especially through turbid water or water covered by ice, light tends to limit primary production in lakes and oceans except very near the surface (Wetzel 2001). However, nutrients also play a key role in regulating primary production in aquatic ecosystems. Nutrients are transported in aquatic ecosystems by currents, meaning water currents have a large influence on productivity. For example, ecosystems fed by streams (e.g. lakes or estuaries) often receive much of their nutrient load from those streams, and the amount of nutrient loading depends strongly on the stream discharge, meaning rainfall has a large indirect influence (Wetzel 2001). On the other hand, productivity in temperate lakes tends to follow seasonal cycles generated by temperature change-induced mixing: spring mixing brings nutrient-rich cold water from the lake depths to the surface, causing a spike in productivity (Anneville *et al.* 2002). Differences in productivity across aquatic systems are also heavily impacted by anthropogenic nutrient input, especially from agricultural fertilizers and urban wastewaters. Productivity in oceans is generally highest near the coasts due to upwelling and to stream discharge, especially from rivers with high anthropogenic nutrient-loading (Rabalais 2002).

Trophic transfers

Another important class of ecosystem functions which has been the subject of much BEF research is trophic transfers. *Trophic transfers* are transfers of energy and materials passed from one trophic level to the next during trophic interactions, including predation, herbivory, detritivory, parasitism, and disease. Some of the same principles apply to trophic transfers as to primary production: while producers compete for and partition abiotic resources such as water and nitrogen, consumers compete for and partition biotic resources (the organisms they consume). When organisms consume other living or dead organisms, they obtain organic compounds rich in chemical potential energy, along with nutrients (essential elements) and indigestible materials. The ratio of energy to other materials obtained from the food depends on factors such as the food quality, which is lowest for primary producers and generally increases with trophic level. Therefore, herbivory tends to involve much greater transfers of non-digestible materials that pass through the herbivore and become detritus, and the loss of most of the energy contained in the plant material as heat from the herbivore. Primary producers can have much more variable elemental ratios, and generally have much higher ratios of carbon to essential nutrients such as nitrogen and phosphorus, as compared to animals. The C:N ratio of predators more closely matches that of their prey, so the needs of a predator are better met by consuming a prey item than those of an herbivore of equivalent mass consuming the same mass of food. However, it is generally more difficult and energy-intensive to find and catch prey than to eat plants, so predator populations are often limited by their ability to obtain sufficient food (Denno and Fagan 2003).

Differences across terrestrial and aquatic ecosystems

A similar difference exists between terrestrial and aquatic ecosystems, where trophic transfers in aquatic ecosystems tend to be more efficient, with more coupling between energy and material transfers and less loss of energy and materials from the consumers to the environment (Hall *et al.* 2007). In addition, the rates of trophic transfers relative to primary production are higher in aquatic ecosystems (Rip and McCann 2011). Hypotheses explaining this difference include that algae tends to be a better quality food than plants, algae are less well-defended against herbivory, algae have a faster turnover (higher production to biomass ratio), aquatic habitats or food webs are more simple, and aquatic consumers tend to be much larger than their prey, among others (Shurin and Seabloom 2005). As a result, aquatic ecosystems are more likely to be regulated by top-down (consumptive) factors and can often exhibit inverted biomass pyramids despite high rates of primary production (McCauley *et al.* 2018).

Coupled green/brown food webs

While food webs have often been oversimplified as deriving all their energy and materials from plants (i.e. the "green food web"), it has long been established that many consumers directly or indirectly derive their energy and materials from detritus (i.e. the "brown food web"). In reality, these food webs are commonly tangled together by interactions, as predators consume both detritivores and herbivores, and omnivores (even putative herbivores) consume both plant material and detritus (Wolkovich *et al.* 2014). Similar principles apply to detritivory as to herbivory, with a major difference being food quality. Detritus tends to be lower in nutrients and easily digestible organic compounds, meaning detritivory tends to be a slower and less efficient trophic transfer (Evans-White and Halvorson 2017). This slowness often gives detritivory a stabilizing effect on food webs (Moore *et al.* 2004).

Decomposition

Decomposition is the process by which *detritus* (dead organic material) is broken down into simple organic or inorganic matter. Litter, animal carcasses, and wood are first physically fragmented by animals like earthworms and termites known as *detritivores* (organisms that consume detritus). Further breakdown by bacteria and fungi releases soluble nutrients accessible to plants.

Decomposition is a key step in nutrient cycling (see Figure 3.2). Nutrients, including nitrogen and phosphorous, are released during the process of decomposition and made available for plant and animal uptake. As a result, decomposition is an essential function of ecosystems with respect to nutrient availability and directly affects productivity. The rate of decomposition is highly sensitive to species (plant and microbial) diversity and abiotic factors such as temperature and soil moisture (Paul and Clark 1996).

Nutrient cycling

Nutrient cycling refers to the biological, chemical, and physical transformations of nutrients as they continuously move through various nutrient pools (see Figure 3.2). Major pools of nutrients include the atmosphere, bedrock, soil, the ocean, and biomass. The rate of cycling between these pools depends on the nutrient under consideration, as well as abiotic and biotic variables. The rate at which nutrients are cycled influences productivity because it determines nutrient availability within an ecosystem and as such has been a focus for BEF research. Higher species diversity generally leads to more complete use of available nutrients in an ecosystem, which in turn supports higher productivity.

Figure 3.1. *Examples of ecosystems around the world. a) Temperate grassland, South Dakota, USA (photo credit: A.A. Wolf); b) temperate forest, California, USA (photo credit: A.A. Wolf); c) tropical grassland, Amboseli, Kenya (photo credit: A.A. Wolf); d) tropical river, Tiputini River, Ecuador (photo credit: C.J. Rakowski); e) tropical forest, Costa Rica (photo credit: S.K. Ortiz); f) subtropical marsh, Everglades, Florida, USA (photo credit: C.J. Rakowski)*

Nutrient limitation is a major inhibitor to primary productivity in almost every type of ecosystem. Countless field experiments across terrestrial and aquatic ecosystems have found that adding nutrients, primarily nitrogen or phosphorous, increases productivity substantially, indicating that nutrient availability is a limiting factor (Elser *et al.* 2007). Nitrogen and phosphorous are essential for organismal function. In primary producers, nitrogen is a key component of the photosynthetic machinery that captures and transforms energy from the sun. Phosphorous is necessary for plant metabolism, structure, and reproduction.

Nutrient limitation varies widely within and across ecosystems. In general, phosphorous is the limiting nutrient in tropical rainforests, while temperate and boreal ecosystems are nitrogen-limited. In lakes and streams, nutrients are scarce, often leading to co-limitation by nitrogen and phosphorous. In marine systems, iron is often a limiting nutrient.

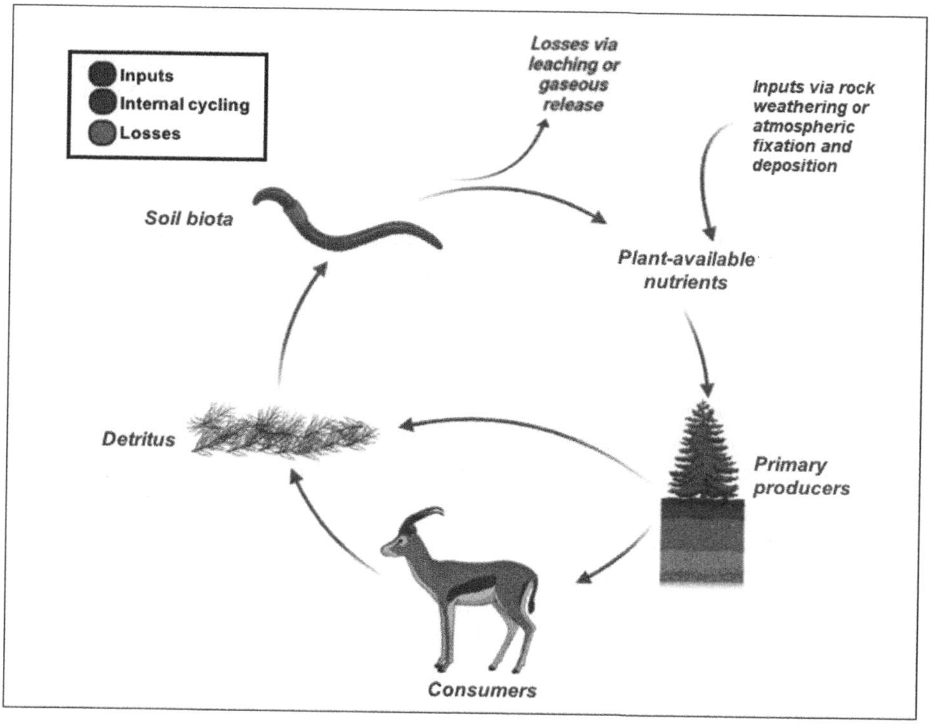

Figure 3.2. *A simplified diagram of a terrestrial nutrient cycle. For a color version of this figure, see www.iste.co.uk/loreau/biodiversity.zip*

The nitrogen cycle

Atmospheric N_2 is by far the largest pool of nitrogen on Earth, but it is largely inaccessible to organisms. Nitrogen in the atmospheric pool becomes accessible to organisms via *nitrogen fixation*, the process that transforms N_2 gas into a plant-accessible form of nitrogen. Nitrogen fixation is an energetically costly process that requires breaking the triple bond in the N_2 molecule. Abiotic forces (i.e. lightning) can fix nitrogen, but biological nitrogen fixation is the dominant pathway by which newly fixed nitrogen enters ecosystems. Biological nitrogen fixation is conducted by diazotrophs, bacteria that can convert N_2 into a plant-accessible form (Schlesinger and Bernhardt 2020). These bacteria can be free-living in the soil or water; alternatively, diazotrophic bacteria can be symbiotically associated with plants (often, though not solely, plants in the Legume family), fixing nitrogen in exchange for carbohydrates. Following fixation, organisms are able to uptake nitrogen and assimilate it into tissue; nitrogen from primary producers is incorporated into higher trophic levels via trophic transfers. When an organism dies, the organic nitrogen in its tissue becomes detritus and is generally broken down via decomposition, entering the soil nitrogen pool. Nitrate (NO_3^-), a common inorganic form of nitrogen in the soil pool, is relatively labile and thus can be exported from ecosystems via leaching losses at high rates.

The phosphorous cycle

The phosphorous cycle often moves much more slowly than the nitrogen cycle. The primary source of phosphorous to an ecosystem is bedrock; the mineral apatite has a particularly high concentration of phosphorous. Bedrock and the minerals that comprise it are broken down via physical and chemical weathering, freeing up phosphorous that is available to organisms. Primary producers use a variety of strategies to obtain or extract phosphorus from soils, including excreting organic acids and engaging in symbioses with mycorrhizal fungi. Phosphorus is incorporated into primary producer tissue, which may be consumed by another organism or senesce. Once an organism dies, the phosphorus in its tissue is eventually returned to the soil phosphorus pool via decomposition. Phosphate (PO_4^{3-}), the main inorganic form of phosphorus in soils, is relatively immobile and leaching losses of phosphorus from an ecosystem are relatively low. Given enough time, however, even slow losses of phosphorus leaching lead to decreases in plant-available phosphorus and ecosystem phosphorus limitation.

Cycles of other elements

Nitrogen and phosphorous are not the only elements necessary for organismal function. For example, potassium is needed to regulate pH and maintain osmotic balance in plants. Calcium is necessary for cell wall structure and sulfur is often used for building amino acids. Iron is essential for proteins in plants as well. The cycles of these other elements are often coupled with the nitrogen or phosphorous cycle. Most other essential elements enter an ecosystem via weathering of bedrock. In certain places, these other elements can limit productivity rather than nitrogen or phosphorous. Calcium and potassium, for instance, can co-limit productivity in tropical rainforests. In marine systems, iron is often a limiting element, particularly in the open ocean (Martin *et al.* 1994).

Invasion and disease resistance

As increasingly more invasive species are spread across the globe by human activity, ecologists have become interested in both predicting and preventing invasions. Most non-native species do not become invasive, and even the most dominating invasive species do not invade every ecosystem in the region they enter. This is due in part to abiotic factors: like all species, invasive species require a certain range of abiotic conditions, such as temperature and moisture, to survive and reproduce. However, biotic factors can also play a role in preventing or allowing invasions. When a native species with a very similar niche is present, it can compete strongly with the invader and reduce the chances of establishment. Additionally, the presence of predators that consume an invader can prevent invasion, especially in freshwater ecosystems (Alofs and Jackson 2014). Analogous to invasive species, novel pathogens are sometimes but not always successful in starting an outbreak upon entering an ecosystem; pathogens can be limited by both abiotic and biotic factors, including temperature and host population density (Joseph *et al.* 2013). The relative abilities of ecosystems to resist invasion and disease are considered ecosystem functions, with higher-diversity systems generally having greater capability to resist novel introductions.

3.3. Ecosystem stability

Ecosystems are not composed of static, constant pools of biota and materials connected by constant fluxes. Rather, they are dynamic, with all components varying over time. Some of this variation is due to changing abiotic factors: for example, as seasons change, demographic rates of species respond and populations wax and wane (Anneville *et al.* 2002). Some variation is also caused by biotic factors. Classic predator–prey cycles involve large fluctuations in predator and prey populations solely caused by the predator–prey interaction itself (Volterra 1926). Besides fluctuating abundances, ecosystems can also shift into drastically different states, where community composition or standing stocks do not resemble the previous state, and the ecosystem may or may not be easily pushed back into the previous state, a phenomenon known as alternative stable states (Schröder *et al.* 2005). *Ecosystem stability* is the general ability of an ecosystem to maintain a relatively consistent state. Many different definitions for ecosystem stability have been used, dating back at least to work by Elton (1958), and new definitions are still being proposed (e.g. Arnoldi *et al.* 2016). The majority of ecosystem stability research performed in a BEF context has focused on one of two metrics: the temporal variability (CV) of community biomass (or sometimes its inverse, "invariability"), and resistance, or the change in an ecosystem state in response to a perturbation. Other important definitions include resilience, or the time following a perturbation for an ecosystem to return to its pre-perturbated state, and persistence, which describes the tendency of a community to return to its original state (or not) following a perturbation. All of these and many other definitions of ecosystem stability have been investigated in terms of their dependence on biodiversity (Kefi *et al.* 2019). See Part 3 of this book for a full explanation of this topic, including theoretical foundations (Chapter 7), experimental evidence (Chapter 8), and observational evidence (Chapter 9).

Cycles and feedbacks

Factors influencing the populations and pools of materials that constitute ecosystems undergo natural variation, from regular cycles to more chaotic changes. Regular cycles can follow phenomena such as predator-prey oscillations, weather patterns, seasonal changes, and longer-term climatic changes. Ecosystems and their components vary in the degree to which they create or track such cycles, leading to differences in the degree of temporal variation across ecosystems (Rip and McCann 2011). However, beyond simple linear responses, ecosystems also exhibit nonlinear responses to their changing environments. *Positive feedbacks*, in which a small change builds on itself in a self-reinforcing pattern, can lead to drastic changes and alternative stable states. An example is the introduction of an invasive species: a few individuals reproduce, leading to exponential population growth (at least for a time). On the other hand, *negative feedbacks*, in which a change from equilibrium tends to be reversed, encourage stability. For example, as a plant population increases, its pathogens may build in the soil, creating a check on the population (Kardol *et al.* 2013). As all ecosystems find themselves in a finite world, resource limitation and competition are unavoidable and create checks and balances which confer a degree of stability on even the most complex ecosystems. As well as feedbacks between components of ecosystems, which influence ecosystem stability, there are feedbacks between ecosystem stability itself and characteristics of ecosystems such as species composition. A relatively stable state can allow species that are vulnerable to fluctuations in their environment to flourish and can also allow higher biodiversity, although if conditions are too stable the strongest competitors may reduce biodiversity. An unstable ecosystem may contain only the most resilient species (Collins 2000). The composition and diversity of species present in an ecosystem influence ecosystem stability through the various responses of species to abiotic conditions and through biotic interactions such as competition and predation (see Chapter 7). Therefore, characteristics of ecosystems such as species composition and diversity are influenced by, and influencers of, ecosystem stability.

3.4. Ecosystem services

Ecosystem services are defined as any of the biophysical necessities and benefits that humans obtain from ecosystems. This concept dates back to the 1960s as a means to provide valuation to the beneficial services ecosystems provide humans in an attempt to promote conservation and protection against anthropogenic forces. Much BEF research has been dedicated to connecting biodiversity to ecosystem services to determine whether and how biodiversity loss might threaten the well-being of humans (see Part 4 of this book for further discussion).

There are four main categories of ecosystem services: supporting, provisioning, regulating, and cultural services (Daily *et al.* 2003). *Supporting services* allow for life to persist. These services include nutrient cycling, water cycling, soil formation, and photosynthesis. *Provisioning services* provide goods that can be directly extracted from nature such as food, fiber for clothes and buildings, fuel, fresh water, and medicinal resources. *Regulating services* moderate natural processes that help humans thrive, and include air purification, erosion control, natural hazard mitigation, and pollination. Finally, *cultural services* provide non-material benefits that contribute to the development of peoples, cultures, and art. Ethical values, existence values, education, recreation, and ecotourism also fall under cultural services.

Services across ecosystems

Ecosystems vary in the degree and efficiency to which they provide ecosystem services. For example, grassland ecosystems are essential for provisioning services like crop production and fiber for clothes. In addition, grasslands provide carbon sequestration in the soil, maintain pollinator populations, and ensure water quality by retaining nutrients in the system, thereby reducing leakage into waterways (Sollenberger *et al.* 2019).

Forest ecosystems provide provisioning services such as timber production, used for building material and fuel. Moreover, tropical forests provide a large stock of genetic information due to the high diversity that can be exploited for medicinal resources or crop breeding. Forests also provide a suite of regulatory and supporting services such as water supply and regulation, erosion control, and climate regulation via carbon sequestration, primarily in vegetation biomass (Pearce 2001).

Freshwater ecosystems provide food, water, and energy in the form of hydroelectric power. These ecosystems also provide regulating and supporting services such as flood and drought mitigation, water purification, carbon sequestration, and nutrient and water cycling (Hering *et al.* 2015). Marine ecosystems supply provisioning services through fisheries and raw materials, regulating services that assist climate and water management, and have an integral role in nutrient cycling, an important supporting service (Remoundou *et al.* 2009).

While many ecosystems provide material services that are necessary for life to persist, they also provide cultural services that are fundamental to human health, growth, and well-being. These services include aesthetic value, cultural connection, and places to learn, create, and foster a sense of community.

3.5. Human alterations to ecosystems

Due to the pervasive and increasing pressures that humans are putting on ecosystems worldwide, some researchers have suggested that human activity rivals state factors such as climate and natural disturbance in its influence on ecosystem extent and functioning. While humans have been a part of ecosystems for hundreds of thousands of years, the magnitude of human impacts on ecosystems since the industrial revolution has ballooned. Indeed, our influence has led to unique combinations of biotic and abiotic components in many areas across the globe. *Novel ecosystems* are ecosystems that have no pre-Anthropocene analog and result from extensive human modification of the state factors of an ecosystem, especially to biotic communities and climate. *Agroecosystems* cover about 50% of Earth's habitable land (Ritchie and Roser 2013) and are characterized by a high homogeneity of biotic and abiotic components, often with a single dominant species and very high nutrient and other chemical inputs. Even where human modifications have not entirely altered ecosystems, human impacts are still extensive. Natural processes are being disrupted by anthropogenic factors that influence ecosystem extent, function, stability, and services. Our influence has drastically altered fundamental aspects of nutrient cycling, greatly increasing the pool of CO_2 in the atmosphere and contributing vast amounts of nitrogen and phosphorus to terrestrial and aquatic ecosystems. Land conversion for agriculture and urbanization has led to fragmentation of ecosystems; apex predators and other keystone species have been hunted extensively and extirpated; introduced species have changed the functioning of ecosystems across the globe. Notably, human activities have drastically altered the Earth's climate, changing the fundamental state factors that control ecosystems worldwide. While this book focuses on the consequences of biodiversity loss, the other concurrent effects of humans must be considered simultaneously for a full understanding of the impact that humans are causing.

3.6. References

Albrecht, M. and Gotelli, N.J. (2001). Spatial and temporal niche partitioning in grassland ants. *Oecologia*, 126, 134–141.

Alofs, K.M. and Jackson, D.A. (2014). Meta-analysis suggests biotic resistance in freshwater environments is driven by consumption rather than competition. *Ecology*, 95, 3259–3270.

Anneville, O., Souissi, S., Ibanez, F., Ginot, V., Druart, J.C., Angeli, N. (2002). Temporal mapping of phytoplankton assemblages in Lake Geneva: Annual and interannual changes in their patterns of succession. *Limnology and Oceanography*, 47, 1355–1366.

Arnoldi, J.-F., Loreau, M., Haegeman, B. (2016). Resilience, reactivity and variability: A mathematical comparison of ecological stability measures. *Journal of Theoretical Biology*, 389, 47–59.

Borer, E.T., Seabloom, E.W., Shurin, J. B. *et al.* (2005). What determines the strength of a trophic cascade? *Ecology*, 86, 528–537.

Bristow, C.S., Hudson-Edwards, K.A., Chappell, A. (2010). Fertilizing the Amazon and equatorial Atlantic with West African dust. *Geophysical Research Letters*, 37(14).

Carpenter, S.R. and Kitchell, J.F. (eds) (1996). *The Trophic Cascade in Lakes*. Cambridge University Press, Cambridge.

Chapin III, F.S. (2003). Effects of plant traits on ecosystem and regional processes: A conceptual framework for predicting the consequences of global change. *Annals of Botany*, 91, 455–463.

Chapin III, F.S., Matson, P.A., Vitousek, P. (2011). *Principles of Terrestrial Ecosystem Ecology*. Springer Science and Business Media, Berlin/Heidelberg.

Chase, J.M. and Leibold, M.A. (2003). *Ecological Niches: Linking Classical and Contemporary Approaches*. University of Chicago Press, Chicago.

Chase, J.M., Abrams, P.A., Grover, J.P. *et al.* (2002). The interaction between predation and competition: A review and synthesis. *Ecology Letters*, 5, 302–315.

Collins, S.L. (2000). Disturbance frequency and community stability in native tallgrass prairie. *The American Naturalist*, 155, 311–325.

Daily, G.C., Alexander, S., Ehrlich, P.R. *et al.* (1997). Ecosystem services: Benefits supplied to human societies by natural ecosystems. *Issues in Ecology*, 2, 1–16.

Denno, R.F. and Fagan, W.F. (2003). Might nitrogen limitation promote omnivory among carnivorous arthropods? *Ecology*, 84, 2522–2531.

Downing, A.L., Jackson, C., Plunkett, C. *et al.* (2020). Temporal stability vs. community matrix measures of stability and the role of weak interactions. *Ecology Letters*, 23, 1468–1478.

Ehrenfeld, J.G., Ravit, B., Elgersma, K. (2005). Feedback in the plant–soil system. *Annual Review of Environment and Resources*, 30, 75–115.

Elser, J.J., Bracken, M.E.S., Cleland, E.E. *et al.* (2007). Global analysis of nitrogen and phosphorus limitation of primary producers in freshwater, marine and terrestrial ecosystems. *Ecology Letters*, 10(12), 1135–42.

Elton, C.S. (1958). *The Ecology of Invasions by Animals and Plants*. Methuen, London.

Evans-White, M.A. and Halvorson, H.M. (2017). Comparing the ecological stoichiometry in green and brown food webs – A review and meta-analysis of freshwater food webs. *Frontiers in Microbiology*, 8, 1184.

Field, C.B. (1998). Primary production of the biosphere: Integrating terrestrial and oceanic components. *Science*, 281, 237–240.

Goldberg, D.E. (1990). Components of resource competition in plant communities. In *Perspectives on Plant Competition*, Grace, J.B. and Tilman, D. (eds). Academic Press, New York.

Graham, M.H. (2004). Effects of local deforestation on the diversity and structure of southern California giant kelp forest food webs. *Ecosystems*, 7, 341–357.

Gravel, D., Guichard, F., Loreau, M., Mouquet, N. (2010). Source and sink dynamics in meta-ecosystems. *Ecology*, 91, 2172–2184.

Hall, S.R., Shurin, J.B., Diehl, S., Nisbet, R.M. (2007). Food quality, nutrient limitation of secondary production, and the strength of trophic cascades. *Oikos*, 116, 1128–1143.

Hay, M.E., Parker, J.D., Burkepile, D.E. *et al.* (2004). Mutualisms and aquatic community structure: The enemy of my enemy is my friend. *Annual Review of Ecology, Evolution, and Systematics*, 35, 175–197.

Hays, G.C. (2003). A review of the adaptive significance and ecosystem consequences of zooplankton diel vertical migrations. *Hydrobiologia*, 503, 163–170.

van der Heijden, M.G.A., Martin, F.M., Selosse, M.-A., Sanders, I.R. (2015). Mycorrhizal ecology and evolution: The past, the present, and the future. *New Phytologist*, 205, 1406–1423.

Hering, D., Carvalho, L., Argillier, C. *et al.* (2015). Managing aquatic ecosystems and water resources under multiple stress – An introduction to the MARS project. *Science of the Total Environment*, 503–504, 10–21.

HilleRisLambers, J., Adler, P.B., Harpole, W.S., Levine, J.M., Mayfield, M.M. (2012). Rethinking community assembly through the lens of coexistence theory. *Annual Review of Ecology, Evolution, and Systematics*, 43, 227–248.

Hooper, D.U., Chapin III, F.S., Ewel, J.J. *et al.* (2005). Effects of biodiversity on ecosystem functioning: A consensus of current knowledge. *Ecological Monographs*, 75(1), 3–35.

Hutchinson, G.E. (1961). The paradox of the plankton. *The American Naturalist*, 95, 137–145.

Jenny, H. (1941). *Factors of Soil Formation*. McGraw-Hill, New York.

Joseph, M.B., Mihaljevic, J.R., Arellano, A.L. *et al.* (2013). Taming wildlife disease: Bridging the gap between science and management. *Journal of Applied Ecology*, 50, 702–712.

Kardol, P., Deyn, G.B.D., Laliberté, E., Mariotte, P., Hawkes, C.V. (2013). Biotic plant–soil feedbacks across temporal scales. *Journal of Ecology*, 101, 309–315.

Kayranli, B., Scholz, M., Mustafa, A., Hedmark, Å. (2010). Carbon storage and fluxes within freshwater wetlands: A critical review. *Wetlands*, 30, 111–124.

Kéfi, S., Domínguez-García, V., Donohue, I., Fontaine, C., Thébault, E., Dakos, V. (2019). Advancing our understanding of ecological stability. *Ecology Letters*, 22, 1349–1356.

Loreau, M., Mouquet, N., Holt, R.D. (2003). Meta-ecosystems: A theoretical framework for a spatial ecosystem ecology. *Ecology Letters*, 6, 673–679.

Lorenz, K. and Lal, R. (2010). *Carbon Sequestration in Forest Ecosystems*. Springer Netherlands, Dordrecht.

Martin, J.H., Coale, K.H., Johnson, K.S. *et al.* (1994). Testing the iron hypothesis in ecosystems of the equatorial pacific ocean. *Nature*, 371(6493), 123–129.

McCauley, D.J., Gellner, G., Martinez, N.D. *et al.* (2018). On the prevalence and dynamics of inverted trophic pyramids and otherwise top-heavy communities. *Ecology Letters*, 21, 439–454.

Mitchell, R.J., Irwin, R.E., Flanagan, R.J., Karron, J.D. (2009). Ecology and evolution of plant–pollinator interactions. *Annals of Botany*, 103, 1355–1363.

Moore, J.C., Berlow, E.L., Coleman, D.C. *et al.* (2004). Detritus, trophic dynamics and biodiversity. *Ecology Letters*, 7, 584–600.

Moore, C.M., Mills, M.M., Arrigo, K.R. *et al.* (2013). Processes and patterns of oceanic nutrient limitation. *Nature Geoscience*, 6, 701–710.

Paine, R.T. (1974). Intertidal community structure: Experimental studies on the relationship between a dominant competitor and its principal predator. *Oecologia*, 15, 93–120.

Paine, R.T. (1979). Disaster, catastrophe, and local persistence of the sea palm *Postelsia palmaeformis*. *Science*, 205, 685–687.

Paul, E.A. and Clark, F.E. (1996). *Soil Microbiology and Biochemistry*. Academic Press, San Diego.

Pearce, D.W. (2001). The economic value of forest ecosystems. *Ecosystem Health*, 7(4), 284–96.

Pickett, S.T.A. and White, P.S. (1985). *The Ecology of Natural Disturbance and Patch Dynamics*. Academic Press, San Diego.

Rabalais, N.N. (2002). Nitrogen in aquatic ecosystems. *AMBIO: A Journal of the Human Environment*, 31, 102.

Raffaelli, D. and Hawkins, S.J. (2012). *Intertidal Ecology*. Springer Science and Business Media, Berlin/Heidelberg.

Remoundou, K., Koundouri, P., Kontogianni, A., Nunes, P.A.L.D., Skourtos, M. (2009). Valuation of natural marine ecosystems: An economic perspective. *Environmental Science and Policy*, 12(7), 1040–51.

Rip, J.M.K. and McCann, K.S. (2011). Cross-ecosystem differences in stability and the principle of energy flux. *Ecology Letters*, 14, 733–740.

Ritchie, H. and Roser, M. (2013). Land Use [Online]. Available at: https://ourworldindata.org/land-use [Accessed 30 January 2021].

Schlesinger, W.H. and Bernhardt, E.S. (2020). *Biogeochemistry*, 4th edition. Academic Press, San Diego.

Schmitz, O.J., Hawlena, D., Trussell, G.C. (2010). Predator control of ecosystem nutrient dynamics. *Ecology Letters*, 13, 1199–1209.

Schoener, T.W. and Spiller, D.A. (1996). Devastation of prey diversity by experimentally introduced predators in the field. *Nature*, 381, 691–694.

Schröder, A., Persson, L., De Roos, A.M. (2005). Direct experimental evidence for alternative stable states: A review. *Oikos*, 110, 3–19.

Shurin, J.B. and Seabloom. E.W. (2005). The strength of trophic cascades across ecosystems: Predictions from allometry and energetics. *Journal of Animal Ecology*, 74, 1029–1038.

Sollenberger, L.E., Kohmann, M.M., Dubeux, J.C.B., Silveira, M. L. (2019). Grassland management affects delivery of regulating and supporting ecosystem services. *Crop Science*, 59(2), 441–59.

Sousa, W.P. (1979). Disturbance in marine intertidal boulder fields: The nonequilibrium maintenance of species diversity. *Ecology*, 60(6), 1225.

Staver, A.C., Archibald, S., Levin, S. (2011). Tree cover in sub-Saharan Africa: Rainfall and fire constrain forest and savanna as alternative stable states. *Ecology*, 92, 1063–1072.

Vitousek, P.M. and Farrington, H. (1997). Nutrient limitation and soil development: Experimental test of a biogeochemical theory. *Biogeochemistry*, 37(1), 63–75.

Vitousek, P.M., Walker, L.R., Whiteaker, L.D., Mueller-Dombois, D., Matson, P.A. (1987). Biological invasion by *Myrica faya* alters ecosystem development in Hawaii. *Science*, 238(4828), 802–804

Volterra, V. (1926). Fluctuations in the abundance of a species considered mathematically. *Nature*, 118, 558–560.

Wetzel, R.G. (2001). *Limnology: Lake and River Ecosystems*, 3rd edition. Academic Press, San Diego.

White, A.F. and Blum, A.E. (1995). Effects of climate on chemical: Weathering in watersheds. *Geochimica and Cosmochimica Acta*, 59(9), 1729–47.

Wolkovich, E.M., Allesina, S., Cottingham, K.L., Moore, J. C., Sandin, S.A., de Mazancourt, C. (2014). Linking the green and brown worlds: The prevalence and effect of multichannel feeding in food webs. *Ecology*, 95, 3376–3386.

Wright, A., Schnitzer, S.A., Reich, P.B. (2014). Living close to your neighbors: The importance of both competition and facilitation in plant communities. *Ecology*, 95, 2213–2223.

PART 2

How Biodiversity Affects Ecosystem Functioning

4

Biodiversity and Ecosystem Functioning: Theoretical Foundations

Shaopeng WANG
Institute of Ecology, Peking University, Beijing, China

4.1. Introduction

The accelerated rate of species extinction calls for an advanced understanding of not only the causes, but also the consequences, of biodiversity loss. Attempts to understand the maintenance and loss of biodiversity have brought about a large set of hypotheses which emphasize different biotic and abiotic factors (Willig *et al.* 2003; Hillebrand 2004). Among them, the productivity hypothesis predicts that the amount of available energy or nutrients limits the upper level of species diversity that an ecosystem can support (Currie *et al.* 2004). Such a hypothesis underlies the classic paradigm for the relationship between biodiversity and ecosystem productivity, where productivity is the cause and biodiversity the effect (Figure 4.1). However, the amount of available energy or nutrients represents only the potential productivity; realized productivity also depends on the biotic processes in ecosystems. One example is the relationship between food-chain length and productivity. Following Lindeman (1942), the traditional scientific wisdom predicts that food chain length is constrained by ecosystem potential productivity, due to the low transfer efficiency between trophic levels. Such a bottom-up perspective was later accompanied by a top-down perspective, which emphasizes the cascading effects of higher trophic levels on lower ones (Harrison *et al.* 1960) and their

The Ecological and Societal Consequences of Biodiversity Loss,
coordinated by Michel LOREAU, Andy HECTOR, and Forest ISBELL.

implications for realized primary productivity (Loreau 2010; Wang and Brose 2018).

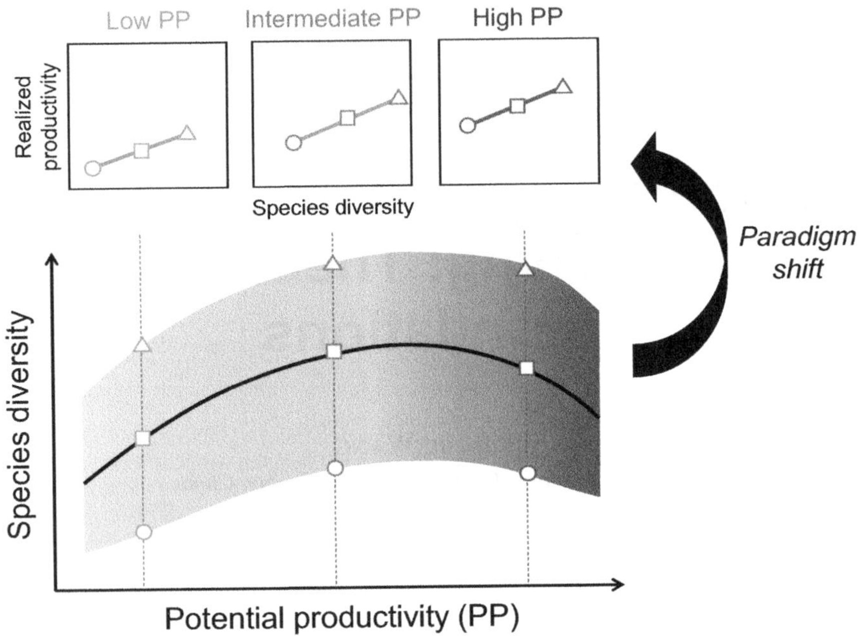

Figure 4.1. *Paradigm shift in the cause and effect between biodiversity and ecosystem functioning. The classic paradigm predicts that species diversity first increases and then saturates or even decreases along the gradient of ecosystem potential productivity (PP). Despite such an average trend, different ecosystems exhibit large variation in species diversity, given any level of PP. The new paradigm predicts that given the PP, the realized ecosystem productivity increases with an increasing species diversity. The light-, intermediate-, and dark-green colors indicate low, intermediate, and high levels of PP, respectively, under which the relationships between species diversity and realized productivity are presented. For a color version of this figure, see www.iste.co.uk/loreau/biodiversity.zip*

The recognition that biodiversity can regulate ecosystems generated a paradigm shift in the 1990s that motivated the development of a new research field, now referred to as biodiversity and ecosystem functioning (BEF; Figure 4.1) (Loreau 2001; Tilman 2014; van der Plas 2019). Note, however, that the conceptualization of BEF has a much longer history as interest in BEF goes back to Darwin and has guided many plantation practices in agriculture and forestry (de Wit 1960; Hector and Hooper 2002).

The past three decades of studies have made BEF a mature research field, which has greatly benefited from close communications between theorists and empirical scientists. While theory clarifies different classes of potential mechanisms that awaits quantification by experimental data, experimental and observational studies raise new patterns and stimulate the development of new theory to clarify the underlying mechanisms. Moreover, new statistical approaches have been developed to bridge theory and experimental data (see section 4.3 and Box 4.1). Such developments of BEF have contributed to reconciling community and ecosystem ecology, which historically diverged to have different concepts and methodologies (Loreau 2010).

In this chapter, I illustrate the theoretical advances achieved in the BEF field, which laid the foundation for understanding the mechanisms that underlie biodiversity effects on ecosystems and their links to mechanisms that underlie species coexistence. These theoretical advances are largely driven by the development of new mathematical models, although I explain only details of some of the most basic models. Specifically, I first use simple-competition models to illustrate the mechanisms underlying species coexistence and their implications for ecosystem functioning. While early studies mainly focused on local single-trophic systems, recent efforts have extended BEF theory to multitrophic communities across multiple spatiotemporal scales in order to match the complexity of natural ecosystems and the scale of management. I also summarize theoretical advances from such efforts.

4.2. Biodiversity: from causes to consequences

To understand the maintenance of biodiversity in ecological communities, early studies developed different types of competition models. Such models provide key insights for understanding the coexistence of species within a trophic level. Moreover, they also have direct implications for understanding the functional consequences of biodiversity and how they link to the mechanisms underlying species coexistence (Loreau 2010). In this section, we will use two types of competition models to explore whether mechanisms underlying the coexistence of species (*cause*) imply a higher functioning (*consequence*).

The first type of models formulate competition in an implicit way, where competition is captured by negative interspecific interactions, without specifying the processes underlying competition (e.g. resource consumption). The classic Lotka–Volterra model of competitors belongs to this type. For illustration, we consider the two-species Lotka–Volterra model:

$$\frac{dN_1}{dt} = r_1 N_1 \left(1 - \frac{N_1 + \alpha_{12} N_2}{K_1}\right) \quad [4.1a]$$

$$\frac{dN_2}{dt} = r_2 N_2 \left(1 - \frac{N_2 + \alpha_{21} N_1}{K_2}\right) \quad [4.1b]$$

where N_1 and N_2 represent the population size or biomass of species 1 and 2, respectively. r_i and K_i are the intrinsic per capita growth rate and carrying capacity of species i, respectively. α_{ij} is the competition coefficient, which measures the per capita impact of species j on the per capita growth rate of species i relative to that of species i on itself. As formulated, competition is not specified as any specific process but can arise from resource competition, interference competition, or both (Case 2000). The Lotka–Volterra model, however, has been shown to provide an approximation for more complicated, biologically realistic models; for example, MacArthur's niche model (MacArthur and Levins 1967; Loreau 2010) and resource competition model (Tilman 1982; Latten *et al.* 2017; see below).

The two-species Lotka–Volterra model can be analyzed using a graphical approach (Figure 4.2a). Specifically, based on equation [4.1], we can easily obtain the conditions under which the net growth rates of both species are zero (i.e. $\frac{dN_1}{dt} = \frac{dN_2}{dt} = 0$). Such conditions correspond to lines in the state space of the two species (N_1, N_2), which are referred to as zero net growth isoclines (isoclines for short). Each species has one non-trivial isocline, which intersects with both axes of N_1 and N_2. The two species can coexist if the non-zero isocline of each species intersects with their own axis at a smaller value compared with that of its competitor, for example the intersection of the isocline of species 1 with the x-axis is smaller than that of species 2 (Figure 4.2a). This requires: $\alpha_{12} < \frac{K_1}{K_2} < \frac{1}{\alpha_{21}}$. Therefore, coexistence between the two competitors is promoted by weaker intraspecific competition and/or smaller differences in species' carrying capacities.

The above condition for species coexistence also has implications for understanding the functional impact of species diversity. The common practice to assess the effect of biodiversity on ecosystem functioning is to compare the observed community functioning (e.g. total biomass of our two-species community) with some expectation based on monoculture functions. One such metric is the relative yield total (RYT) or the sum of species relative yields. Here, a species' relative yield is defined as the ratio of its biomass in mixture to its biomass in monoculture. A mixture is considered to exhibit *overyielding* if RYT > 1. Note that the RYT metric allows species to shift their relative yields (e.g. the two extremes where species 1 or 2 goes extinct both lead to RYT = 1), but adds a constraint on the summed relative yield. In the two-species Lotka–Volterra model, species coexistence ensures overyielding, which can be easily seen from a graphical analysis (Loreau 2004). In the state space, the null expectation that $RYT = 1$ corresponds to the line connecting (K_1, 0) and (0, K_2), that is, any point along this line satisfies $\frac{N_1}{K_1} + \frac{N_2}{K_2} = 1$. The coexistence condition implies that the intersection of the two isoclines must lie above this null expectation, that is, $RYT = \frac{N_1^*}{K_1} + \frac{N_2^*}{K_2} > 1$. Note, however, that in other models where species' isoclines are nonlinear, the intersection of isoclines may lie below the null expectation and hence coexistence of two species does not lead to overyielding (Loreau 2010).

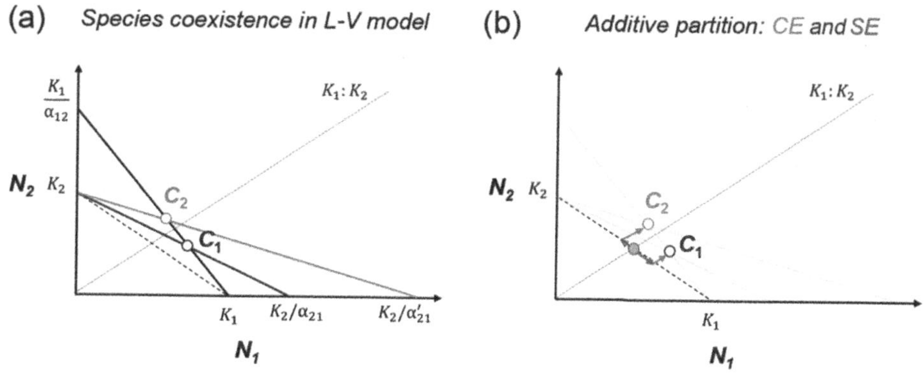

Figure 4.2. *Species coexistence and ecosystem functioning in Lotka–Volterra models. (a) Coexistence of two competitors illustrated by a graphical analysis of Lotka–Volterra models. Black line represents the null isocline of species 1 (i.e. the conditions where species 1 has zero growth), and blue and orange lines represent the isoclines of species 2, respectively, under high and low competition effects of species 1 on 2. The intersections between the isoclines of species 1 and 2 (C_1 and C_2) determine the equilibrium species biomass. When the two species coexist, this intersection is always above the dashed line that connects (K_1, 0) and (0, K_2), which implies an overyielding effect. (b) Additive partition of net biodiversity effects (NBEs) into selection (SE; red arrows) and complementarity (CE; green arrows) effects. In both panels, the black and gray dashed lines and the two intersections (C_1 and C_2) are the same, where the gray dashed line indicates the direction of K_1:K_2. (a) is adapted from Loreau (2004). For a color version of this figure, see www.iste.co.uk/loreau/biodiversity.zip*

Moreover, overyielding does not imply that the two-species mixture has a higher biomass than the maximum monoculture biomass, a phenomenon called *transgressive overyielding*. Overyielding does not even imply that the two-species mixture has a higher biomass than the mean monoculture biomass, the difference between the two being called the *net biodiversity effect* (*NBE*; see next section). Overyielding, however, is closely related to the NBE. As we will explain in the next section, overyielding is one mechanism that generates a positive NBE. Although these different ways to quantify biodiversity effects do not coincide in general, in the special case where the two species have the same monoculture biomass ($K_1 = K_2$) in the Lotka–Volterra model, their coexistence always implies overyielding, transgressive overyielding, and a positive NBE.

The second type of model represents interspecific competition mediated by resource consumption. Such models have been studied extensively in the literature and here we adopt Tilman's (1982) graphical treatment. Consider two species that compete for a single resource, where the per capita growth rate of each species increases with resource concentration but they have resource-independent mortality rates (Figure 4.3a). In monoculture, each species can reduce the resource concentration to a level where population growth is balanced by mortality, called its $R*$ (Tilman 1982). This value has implications for understanding both species coexistence and ecosystem functioning. On the one hand, the species with the lowest R* can tolerate a lower level of resource concentration and thus it excludes the other species with a higher R* when planted together. On the other hand, the species with the lowest R* requires a lower level of resource concentration and has a higher efficiency of resource uptake. In other words, for two species competing for a single resource, the consequence of competition is a higher community functioning, as defined by the efficiency of resource uptake (Tilman 1997). Although only one species can persist locally in this model, spatial heterogeneity in environmental conditions allows a higher diversity where different locations are dominated by species with the locally lowest R* (Figure 4.3b). Local dominance by more efficient species then increases the overall efficiency at larger scales (see Part 5 of this book).

While two species cannot coexist when they compete for a single resource, coexistence is achieved if there are two limiting resources and the two species exhibit trade-offs in their capacity to tolerate low concentrations of the two resources and in the way they consume them (Figure 4.3c). More specifically, coexistence requires that each species tolerates a lower level of one resource (e.g. resource 1 for species 1, resource 2 for species 2 in Figure 4.3c) and that its growth decreases its limiting resource more than the other resource (e.g. species 1 is more limited by resource 2 and the growth of species 1 decreases resource 2 more than resource 1; see Figure 4.3c). This coexistence condition also implies higher efficiencies in resource uptake when species are together, compared with monocultures of either species. Specifically, given the resource supply point that allows coexistence (e.g. the green-red point in Figure 4.3c), the two-species mixture can maximize resource uptake and reduce the concentrations of both resources to low levels (red circle), but a single species 1 or 2 can reduce only one of the resources to a low level (black or blue dashed circle) (Figure 4.3c). Similarly, although at most two species can coexist locally when competing on two resources, spatial heterogeneity in resource supply rate can support many more species as long as they exhibit similar trade-offs as described above (Figure 4.3d). The coexistence of such species also allows higher resource uptake from the heterogeneous landscape (Hodapp *et al.* 2016).

While different competition models tend to emphasize different mechanisms of species coexistence, the so-called "modern coexistence theory" attempts to develop a common language to summarize mechanisms underlying species coexistence (Chesson 2000). This theory states that species coexistence is promoted by a higher niche difference (ND) and/or a lower relative fitness difference (RFD). Intuitively, a lower RFD requires species to have a similar capacity to cope with the local environment (e.g. small variation in K_i in the Lotka–Volterra model), thereby reducing their competitive advantage over others; a higher ND requires species to develop complementary strategies when competing with others (e.g. smaller α_{ij} in the Lotka–Volterra model). Such an ND–RFD framework provides a language to intuitively understand the various mechanisms involved in different models, for example Lotka–Volterra and resource competition models (Letten *et al.* 2017). Recent efforts have been made to understand the functional consequences of biodiversity in the ND–RFD framework, which we will revisit in the next section.

4.3. Why does biodiversity promote ecosystem functioning?

The above models show clearly that species interactions can drive the functioning of ecosystems. While biodiversity increases ecosystem functioning in various ways in different models, theory has clarified two broad sets of processes that underlie biodiversity effects on ecosystem functioning, namely selection and complementarity effects (Loreau 2001). Below we explain these two effects using the resource competition models introduced above. Note, however, that both selection and complementarity effects represent general mechanisms arising from various biological processes and are broader than our resource competition examples.

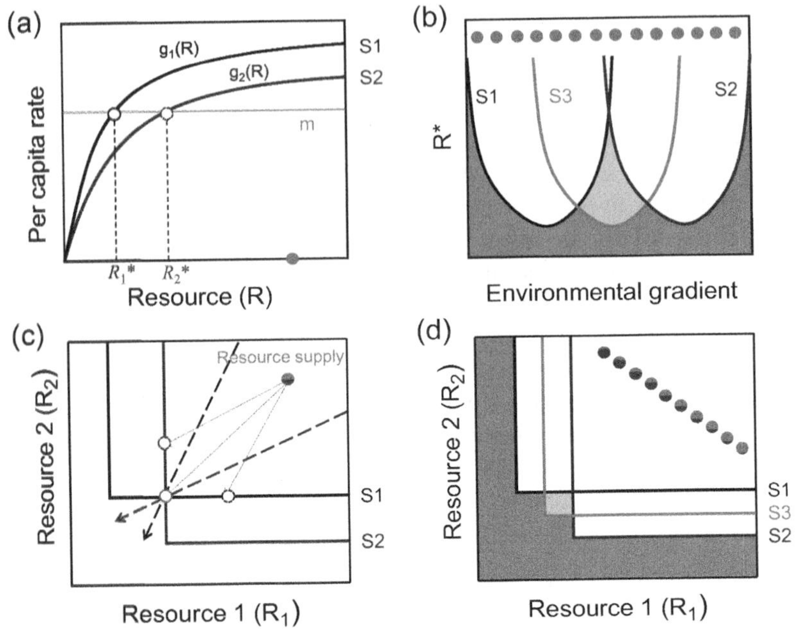

Figure 4.3. *Species coexistence and ecosystem functioning in resource competition models: (a,b) competition on a single resource and (c,d) competition on two resources. (a) Given one limiting resource (R), the equilibrium for each species is achieved when its resource-dependent growth rate (g(R)) is balanced by the death rate (m). The corresponding resource concentration is called the R* for the species. As species with a lower R* (e.g. S1) is more efficient in resource uptake and can exclude its competitor with higher R* (e.g. S2). (b) Along an environmental gradient, species may vary in their R*s on a given resource. At each location, the equilibrium resource concentration is determined by the species with the lowest R*, which excludes all other species locally. The upper boundary of the gray area denotes the equilibrium resource concentration along the environmental gradient, in the presence (the dark-gray area) or absence (the combined dark- and light-gray area) of species 3. (c) Two species can coexist when competing on two limiting resources (R_1 and R_2) if they exhibit trade-offs in their null isoclines (black and blue solid lines for S1 and S2, respectively) and consumption vector (black and blue dashed arrows). Given the resource supply, the two-species community can reduce both resources to low levels (red circle), whereas a single species can reduce only one resource to a low level (black and blue dashed circles). (d) Spatial heterogeneities in the resource supply ratio can support a larger number of species. In panels (a) and (b), the green points indicate the resource supply; in panels (c) and (d), the hybrid green-red points indicate the supply of the two resources (green-R_1 and red-R_2). Parts (a) and (c) are adapted from Tilman (1982), (b) from Tilman (2004), and (d) from Tilman et al. (1997). For a color version of this figure, see www.iste.co.uk/loreau/biodiversity.zip*

Selection effects (SEs) emerge when a diverse community includes and becomes dominated by species with a high productivity. A useful example is the above model of two species competing for a single resource. Whatever the initial conditions, this model predicts the extinction of all species except that with the lowest R*, which completely dominates the community and maintains a high productivity or resource uptake efficiency. As a community with a higher initial number of species has a higher probability of including and being dominated by species with a low R*, we can expect that a higher initial biodiversity would cause a higher productivity (Tilman 1997). Note, however, that in general cases the most efficient and productive species does not necessarily dominate the community, and the dominant species can be those with either higher- or lower-than-average productivity depending on the trait that confers competitive ability (Loreau 1998). In the cases in which the less productive species dominate the community, theory predicts a negative SE will emerge (Loreau 1998). Therefore, SEs should be distinguished from the "sampling effect", which denotes the fact that a more diverse community has a higher probability of including more productive species. Early studies suggested such a sampling effect as a purely statistical explanation of positive biodiversity effects (Huston 1997). However, simply including the more productive species in the community does not ensure a higher productivity, unless such more productive species become dominating in the community. Taken together, SE captures the combined outcome of both sampling and dominance effects, which indicates that biodiversity effects are not artificial but require species interactions for dominance to emerge.

Complementarity effects (CEs) emerge when species perform better on average (e.g. they have higher per capita rates of biomass production) when growing with other species than in monoculture. Such enhanced performance can arise from interspecific niche differences, which causes an overall reduced competition. The model of two competitors with two resources provides a good example for understanding CEs (Figure 4.3d). In this model, the two species differ in their resource uptake efficiencies, and thus mixtures benefit from the complementary capacities of the two species and reach a larger overall resource uptake efficiency (Tilman 1997). In Figure 4.3d, the two species together reduce the resources to the boundary of the combined dark- and light-gray area. The inclusion of an additional species 3 that exhibits trade-offs with both species 1 and 2 will further benefit the community in its resource uptake efficiency, such that the resources in the light-gray area are consumed and the overall resource concentrations are further reduced (Figure 4.3d). Facilitation can also promote complementarity if the resource uptake efficiency of one species increases in the presence of another species. This can be understood graphically as a case in which the presence of species 1 (or 2) lowers the isocline of species 2 (or 1), that is, the black or blue line is shifted towards the origin in Figure 4.3d. This also reduces resource concentration, thereby leading to a higher efficiency of the community.

The two mechanisms underlying biodiversity effects are not exclusive, and we expect both CEs and SEs to operate in reality. One important question is which mechanism plays a more important role in driving biodiversity effects in real ecosystems. This issue is not only of theoretical concern but is also relevant to biodiversity conservation. Specifically, if CEs are the dominant mechanism, then species diversity *per se* should be preserved for the maintenance of ecosystem functioning; but if SEs are the dominant mechanism, conservation should be targeted at the most efficient species (e.g. species with the lowest R* in the resource competition model). This question has motivated the development of statistical approaches to quantifying CEs and SEs in empirical data.

Loreau and Hector (2001) developed an additive partition approach to quantify CEs and SEs in experimental data. This approach requires information on the functioning of both the mixture and monoculture of each species, which is often available in biodiversity experiments. The key to this approach is to partition the net biodiversity effect into two additive components, which correspond conceptually to CEs and SEs (see Box 4.1). Briefly, the CE is quantified by the species-averaged difference between observed and expected relative yields, which is positive whenever there is overyielding (i.e. RYT>1); the SE is quantified by the covariance between species monoculture yield and the difference between observed and expected relative yields, which is positive (negative) if more productive species in monocultures also exhibit a higher (lower) relative yield. In a two-species case, the additive partition can be intuitively understood using a graphical approach (see Box 4.1; Pillai and Touhier 2018). This additive approach has been widely applied in biodiversity experiments and has been extended to capture more detailed processes (e.g. Fox 2005).

Theory identifies two sets of processes underlying biodiversity effects on ecosystem functioning, namely selection effects and complementarity effects. Loreau and Hector (2001) developed an additive partitioning approach to quantify these two effects in experiments or theoretical models. Based on information of the functioning of both mixtures and monocultures of each species, this approach partitions the net biodiversity effect into two additive components, which correspond conceptually to the SE and CE. Below we explain it in details.

Consider n different species $i = 1,2, \ldots, n$. The monoculture yield of species i is M_i, and the observed mixture yield is Y_i. The expected mixture yield of i is $M_i e_i$, where e_i denotes the expected relative yield in the mixture (e.g. proportion of sown seeds), which satisfies $\sum e_i = 1$. The net biodiversity effect is defined as the difference between observed and expected mixture yields:

$$NBE = \sum_i (Y_i - M_i e_i) = \sum_i M_i (z_i - e_i) = \sum_i M_i \Delta z_i$$

where $z_i = Y_i/M_i$ is the observed relative yield of species i, and $\Delta z_i = z_i - e_i$. Based on these definitions, Loreau and Hector (2001) provided an additive partition that expresses the NBE as the sum of two components corresponding to the CE and SE:

$$NBE = \sum_i M_i \Delta z_i = \underbrace{n \cdot \overline{M} \cdot \overline{\Delta z}}_{Complementarity\ (CE)} + \underbrace{n \cdot COV(M, \Delta z)}_{Selection\ (SE)}$$

where n is the number of species, $\overline{M}$ and $\overline{\Delta z}$ represent the simple averages, and the covariance term is defined as: $COV(M, \Delta z) = \frac{1}{n} \sum_i (M_i - \overline{M})(\Delta z_i - \overline{\Delta z})$. Note that this covariance term is normalized by n, rather than n–1.

In a two-species case, Pillai and Gouhier (2018) developed a graphical approach to visualize the additive partition (see Figure 4.4). Specifically, the NBE can be represented by the vector *EZ* that connects the expected (*E*) and observed (*Z*) species yield. Such a vector can be decomposed into the sum of two vectors, that is, *EC* and *CZ*, where the point *C* is on the line connecting (M_1, 0) and (0, M_2), with its location determined by the constraint that the vector *CZ* is parallel to the line M_1:M_2. By definition, *EC* and *CZ* correspond to the SE and CE, respectively. The CE is positive (negative) if the vector *CZ* points away from (towards) the origin, and the SE is positive (negative) if the vector *EC* points towards the axis representing the more (less) productive species.

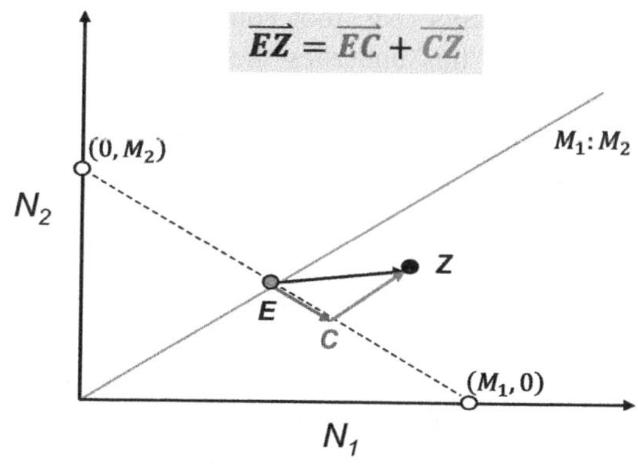

Figure 4.4. *A graphical illustration of the additive partition of net biodiversity effects into selection effects and complementarity effects, indicated by the vectors EC and CZ, respectively*

Box 4.1. *Additive partition of net biodiversity effects*

Here we apply the additive partition to our two-species Lotka–Volterra model. As explained above, species coexistence always implies overyielding in such models, and thus a positive CE (i.e. green arrows pointing away from the origin; Figure 4.2b). However, the SE can be either positive or negative, depending on which species has a relatively higher biomass. When the less productive species 2 receives strong competition from species 1 (α_{21}), it reaches a relatively lower biomass in the mixture (C_1) and thus the community has an overall positive SE (i.e. the red arrow pointing toward the more productive species 1; Figure 4.2b). In contrast, if species 2 is less affected by species 1 (α'_{21}), it reaches a relatively high biomass (C_2) and thus the community has an overall negative SE (i.e. the red arrow pointing toward the less productive species 2; Figure 4.2b). In the latter case, the NBE can be either positive or negative depending on whether the magnitude of the CE is larger or smaller than that of the SE. Overall, the coexistence of two competitors in the Lotka–Volterra model implies a positive CE (or overyielding), but either a positive or negative SE and NBE.

We add two notes on the interpretation of the CE and SE. First, both the CE and SE, as derived from the additive partition developed by Loreau and Hector (2001), capture emergent outcomes of species interactions, which can result from the combined effects of various biological processes. This may be taken as a limitation of the additive partition as it does not allow us to identify the specific biological processes underlying CEs or SEs (e.g. root complementarity; Barry *et al.* 2019). However, on the other hand, this underlies the generality of CEs and SEs as a general framework to summarize biodiversity effects and to make comparisons across systems and studies. As we will see in the next section, these concepts provide useful tools to understand the effects of horizontal and vertical diversity in food webs, although they are mediated by more complicated processes. Second, the quantification of CEs and SEs provides opportunities for better understanding the link between species coexistence and ecosystem functioning. In particular, theoretical efforts have attempted to explore whether the CE and SE correspond, respectively, to the ND and RFD under the framework of modern coexistence theory. Theory has shown that although the CE is most closely related to the ND and the SE to the RFD, the CE is also related to the RFD and the SE to the ND (Carroll *et al.* 2011; Loreau *et al.* 2012). Such theoretical predictions have been supported by a recent experiment (Godoy *et al.* 2020). The lack of one-to-one correspondence between mechanisms underlying species coexistence and ecosystem functioning suggests that the two approaches of post hoc statistics (i.e. ND–RFD and CE–SE) summarize species interactions in related, but different ways.

4.4. Trophic diversity and ecosystem functioning

While theoretical and experimental studies on BEF have predominantly focused on single-trophic systems, particularly plant communities (Tilman 2014), significant efforts have also been devoted to extending BEF theory to multitrophic systems. This enhances the relevance of BEF theory because species at higher trophic levels are more susceptible to extinction compared with basal species (Purvis *et al.* 2000; Estes *et al.* 2011; Wang and Brose 2018). In multitrophic communities, biodiversity can be characterized along two dimensions, namely horizontal diversity (e.g. species richness within a trophic level) and vertical diversity (e.g. number of trophic levels) (Duffy *et al.* 2007; Loreau 2010). In such a context, theoretical studies have offered new insights into two questions: how does the effect of horizontal diversity vary across trophic levels, and how does vertical diversity regulate ecosystem functioning?

Along the horizontal axis, horizontal diversity within any particular trophic levels is expected to have similar positive effects on the resource uptake efficiency of this trophic level, because the processes of complementarity and selection still operate, though in more complicated ways due to trophic interactions. However, the effect size of horizontal diversity at a focal trophic level can be modulated by the presence of its predators, due to feedback between predators and prey that can affect the outcome of competition (Thebault and Loreau 2003; Duffy *et al.* 2007). In particular, the strength of horizontal diversity effects may weaken from low to high trophic levels. This is because, in contrast to the effects of plant diversity on the uptake efficiency of their abiotic resources (which are donor controlled), an increased resource uptake efficiency of a predator trophic level will impose strong top-down control and induce population decline and instability in their prey levels (Rosenzweig 1971), which can feedback to the predator level and dampen their efficiency (Duffy et al. 2007). Such a prediction, however, was not supported by meta-analyses of 111 biodiversity experiments across trophic levels, which showed consistent and equally strong positive effects of horizontal diversity across trophic levels (e.g. plants, herbivores, predators, decomposers) (Cardinale *et al.* 2006). This might have been because in some experiments the dynamic feedback between predators and prey was almost intentionally broken by experimental design, as these experiments mainly consisted of invertebrate predators that have relatively weaker top-down control (Borer *et al.* 2005), or because the presence of intraguild predation within the predator level dampens its top-down effects (Wang *et al.* 2019).

In multitrophic communities, the horizontal diversity of a trophic level does not only increase its resource uptake efficiency but may also increase its resistance to predation. One explanation is that a higher horizontal diversity increases the probability that a resistant species exists in the community, which can become advantageous in the face of the predation pressure and weaken the effects of higher trophic levels on total prey abundance or biomass (Hillebrand and Cardinale 2004). Such a process is analogous to the SE, but it is mediated by top-down effects of predators rather than resource competition. Alternatively, a higher horizontal diversity at the host level can decrease the transmission rate of parasitism through a dilution effect (Civitello *et al.* 2015); that is, the different host species cooperate to control the population and spread of parasites, which is analogous to the CE.

Along the vertical axis, vertical diversity can also regulate the functioning of the whole ecosystem through direct and indirect top-down control of top predators. In food chains, trophic cascade theory predicts that the top predator suppresses species at an odd number of trophic levels below the top predator and releases those at an even number of levels below it (Oksanen *et al.* 1981; Loreau 2010). As a consequence, the biomass of the basal species and primary production both exhibit oscillating patterns as food chain length increases (Figure 4.5; Loreau 2010). However, this conclusion does not extend to complex food webs, where the prevalence of omnivory can significantly dampen the strength of trophic cascades (Polis and Strong 1996; Wang and Brose 2018). Complex food-web models predict that primary productivity increases exponentially with vertical diversity, as defined by the maximum trophic level of the food web (Figure 4.5; referred to as the vertical diversity hypothesis; Wang and Brose 2018).

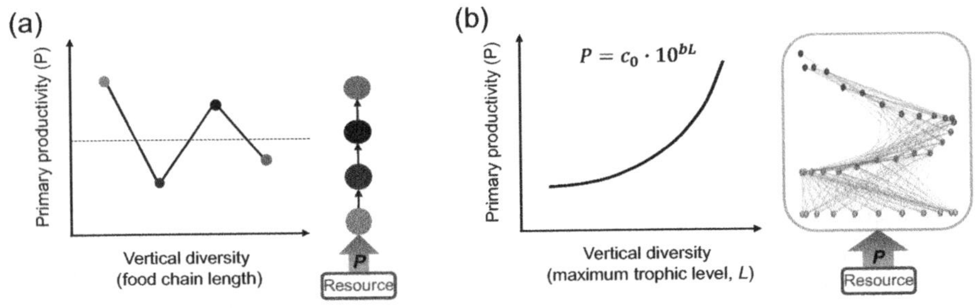

Figure 4.5. *Relationship between vertical diversity and primary productivity in food chains (a) and complex food webs (b). The vertical diversity is measured by the food chain length in (a) and the maximum trophic level of the food web in (b). Visualized relationships are based on results from Loreau (2010) and Wang and Brose (2018). For a color version of this figure, see www.iste.co.uk/loreau/biodiversity.zip*

How can the effects of vertical diversity, which characterizes the maximum trophic level across all animal species, on primary productivity, which characterizes the efficiency of plant communities in absorbing resources and converting them to biomass, be explained? This question can still be understood from processes of complementarity and selection. In multitrophic communities, CE can arise from interspecific differences between plants, not only in exploiting resources (bottom-up processes) but also in the way they are exploited by predators (top-down processes). The latter process is called trophic complementarity (Poisot *et al.* 2013). In complex food webs, a higher vertical diversity creates a higher heterogeneity in the trophic niches of plants and thus enhances trophic complementarity (Wang and Brose 2018). On the other hand, predation by animal communities can alter plant species composition by selecting species with particular traits. Specifically, a higher vertical diversity exerts a selection force that benefits larger-bodied plant species, which have lower per-mass metabolic rates and thus more efficient energy use (Schneider *et al.* 2016; Wang and Brose 2018).

Thus, in multitrophic communities, horizontal diversity can enhance the performance of the corresponding trophic level by increasing its resource uptake efficiency and/or its resistance to predators, while vertical diversity can regulate ecosystem functioning, for example primary productivity, through top-down regulation. The effects of horizontal and vertical diversity can be understood from the CE–SE framework, but both mechanisms are mediated by complex trophic interactions.

4.5. BEF over time and space

While BEF theory has developed into a mature research field, its implications have remained restricted because most theoretical and empirical studies have addressed the local scale and the short term. So there is an urgent need to extend existing knowledge to larger spatiotemporal scales to match the scale of management and conservation (Gonzalez *et al.* 2020; Qiu and Cardinale 2020). Recent theoretical efforts have provided new insights into how biodiversity effects, including the CE and SE, change through time and across space. In this section, I focus on theoretical progress in understanding the scale dependence of biodiversity effects and briefly mention related empirical work, but we refer readers to Chapters 5 and 6 for more details.

The strength of biodiversity effects can change through time. Based on a simple two-species Lotka–Volterra model, Pacala and Tilman (2002) showed that over succession, the mechanism underlying biodiversity effects can shift from selection to complementarity. This can be understood from a Lotka–Volterra competition model consisting of two species with very different population growth rates (e.g. $r_1 >> r_2$ in equation [4.1]). Starting from a low density of both species, the community will initially experience a dramatic increase in the abundance of the fast species (i.e. species 1), and this dominance by species 1 results in a positive SE; but as the community gets close to its equilibrium (e.g. the point C_1 in Figure 4.2a), the biomass of species 1 decreases and there is a more balanced biomass of the two species, causing a positive CE but a varying SE. In other words, the CE increases over time but is preceded by a transient SE (Pacala and Tilman 2002). This transient SE is positive if the species with the higher carrying capacity is also faster (e.g. species 1 has both a larger r and a larger K), and it is negative if there is a trade-off between carrying capacity and growth rate (e.g. species 1 has a lower r but a higher K). In contrast, if the two competitors cannot coexist in the long term but can co-occur in a transient stage due to small population growth (e.g. small r_1, r_2), an increasing SE preceded by a transient CE should be expected (Turnbull *et al.* 2013). While both switches between CEs and SEs are possible theoretically, experimental data have generally revealed an increasing CE in plant communities (van Ruijven and Berendse 2005; Reich *et al.* 2012; Huang *et al.* 2018). In addition to transient dynamics, an alternative explanation is that species interactions drive character displacement between competitors and thus an enhanced complementarity through time (Reich *et al.* 2012). Such eco-evolutionary dynamics seem to explain, at least partly, the results of a recent experiment (Zuppinger-Dingley *et al.* 2014) and are receiving attention by theoreticians (Aubree *et al.* 2020).

The two biodiversity effects (i.e. CEs and SEs) could also shift in relative importance as the study area increases from local to regional scales. Consider an environmental gradient where each local community is dominated by species that have the highest fitness (e.g. the lowest R* in resource competition; Figure 4.3b). Such species-sorting processes can cause selection effects in local communities. However, seen at larger scales, the regional community (or metacommunity) consists of different sets of resources, each being consumed by species that are most efficient in extracting resources in each local environmental condition. Consequently, selection at local scales promotes complementarity at larger scales (Isbell *et al.* 2018). Similarly, complementarity at local scales can give rise to selection at larger scales (Isbell *et al.* 2018). To characterize the different sources of complementarity and selection effects at larger scales, Isbell *et al.* (2018) extended Loreau and Hector's (2001) original partitioning framework to quantify biodiversity effects on ecosystem functioning at multiple spatial and temporal scales. In particular, selection effects were partitioned into four components arising from different spatiotemporal processes, namely average species selection, spatial selection, temporal selection, and spatiotemporal selection. Note, however, that in Isbell *et al.* (2018), the latter three terms were referred to as "insurance effects", though the "insurance" concept generally applies to the effect of biodiversity on the stability of ecosystem functioning (see Chapter 7). Here we use the term "selection" because these processes operate like the original SE

(Loreau and Hector 2001). Just as the original SE is positive when species with a higher-than-average productivity dominate a mixture, temporal or spatial selection effects are positive when a species dominates mixtures at the times or places where it has the highest monoculture productivity. This framework provides new opportunities to understand how environmental heterogeneity interacts with biodiversity to regulate ecosystem functioning.

Overall, just as spatiotemporal heterogeneities create more opportunities for species coexistence, larger spatial and temporal scales provide more room for different sources of biodiversity effects to operate (Figure 4.3). Recent theoretical and experimental studies suggest that biodiversity effects can increase at larger spatial or temporal scales (Thompson *et al.* 2018, 2021; Qiu and Cardinale 2020), although the opposite can also occur (Thompson *et al.* 2018). Future studies are needed to clarify the mechanisms underlying the scale dependency of biodiversity effects and the scenarios under which biodiversity effects increase at a particular scale. Such knowledge will be valuable for designing efficient strategies for ecosystem management. In particular, we have mainly considered the level of ecosystem functioning achieved in the relatively short term under constant environmental conditions, but whether such a higher level of functioning also implies a higher stability of ecosystems in the face of external perturbations (see Chapter 7) is unclear. Recent theory predicts that whereas complementarity promotes stability, selection impairs it; consequently, ecosystem functioning and stability can exhibit either a synergy or a trade-off depending on the drivers of species interactions (Wang *et al.* 2021). In the case of a trade-off between functioning and stability, ecosystems achieving a higher biomass in constant environments might be more fragile in the face of large perturbations that are likely to occur in the long term.

4.6. Conclusion

Three decades of studies have now established biodiversity and ecosystem functioning (BEF) as a vivid research field, which has implications both in theory, by integrating community and ecosystem ecology, and in practice, by offering insights into the functional consequences of biodiversity loss. The theoretical foundations of BEF are built upon classic competition models, as well as, more recently, upon food web models and metacommunity theory. These recent developments provide important steps to better match BEF theory to the scenarios of biodiversity loss in natural ecosystems and the scale at which conservation and management are implemented. Future efforts in this direction should take advantage of the well-established tradition in the field of BEF whereby theoretical and empirical scientists work together closely and thus contribute to making BEF a practical theory.

4.7. Acknowledgements

I thank the editors for inviting me to write this chapter and Michel Loreau and Mary I. O'Connor for useful comments. This work was supported by the National Natural Science Foundation of China (31988102, 31870505).

4.8. References

Borer, E.T., Seabloom, E.W., Shurin, J.B. *et al.* (2005). What determines the strength of a trophic cascade? *Ecology*, 86, 528–537.

Cardinale, B.J., Srivastava, D.S., Duffy, J.E. *et al.* (2006). Effects of biodiversity on the functioning of trophic groups and ecosystems. *Nature*, 443, 989–992.

Carroll, I.T., Cardinale, B.J., Nisbet, R.M. (2011). Niche and fitness differences relate the maintenance of diversity to ecosystem function. *Ecology*, 92, 1157–1165.

Case, T.J. (2000). *An Illustrated Guide to Theoretical Ecology*. Oxford University Press, New York.

Chesson, P. (2000). Mechanisms of maintenance of species diversity. *Annual Review of Ecology, Evolution and Systematics*, 31, 343–366.

Civitello, D.J., Cohen, JM., Fatima, H. *et al.* (2015). Biodiversity inhibits parasites: Broad evidence for the dilution effect. *Proceedings of the National Academy of Sciences*, 112, 8667–8671.

Currie, D.J., Mittelbach, G.G., Cornell, H.V. *et al.* (2004). A critical review of species-energy theory. *Ecology Letters*, 7, 1121–1134.

De Wit, C.T. (1960). *On Competition*. Verslag Landbouwkundig Onderzoek, Wageningen.

Duffy, J.E., Cardinale, B.J., France, K.E., McIntyre, P.B., Thébault, E., Loreau, M. (2007). The functional role of biodiversity in ecosystems: Incorporating trophic complexity. *Ecology Letters*, 10, 522–538.

Estes, J.A., Terborgh, J., Brashares, J.S. *et al.* (2011). Trophic downgrading of planet Earth. *Science*, 333, 301–306.

Godoy, O., Gómez-Aparicio, L., Matías, L., Pérez-Ramos, I.M., Allan, E. (2020). An excess of niche differences maximizes ecosystem functioning. *Nature Communications*, 11, 4180.

Gonzalez, A., Germain, R.M., Srivastava, D.S. *et al.* (2020). Scaling up biodiversity–ecosystem functioning research. *Ecology Letters*, 23, 757–776.

Hairston, N.G., Smith, F.E., Slobodkin, L.B. (1960). Community structure, population control, and competition. *The American Naturalist*, 94, 421–425.

Hector, A. and Hooper, R. (2002). Darwin and the first ecological experiment. *Science*, 295, 639–640.

Hector, A., Schmid, B., Beierkuhnlein, C. *et al.* (1999). Plant diversity and productivity experiments in European grasslands. *Science*, 286, 1123–1127.

Hillebrand, H. (2004). On the generality of the latitudinal diversity gradient. *The American Naturalist*, 163, 192–211.

Hillebrand, H. and Cardinale, B.J. (2004). Consumer effects decline with prey diversity. *Ecology Letters*, 7, 192–201.

Hodapp, D., Hillebrand, H., Blasius, B., Ryabov, A.B. (2016). Environmental and trait variability constrain community structure and the biodiversity–productivity relationship. *Ecology*, 97, 1463–1474.

Huang, Y., Chen, Y., Castro-Izaguirre, N. *et al.* (2018). Impacts of species richness on productivity in a large-scale subtropical forest experiment. *Science*, 362, 80–83.

Huston, M.A. (1997). Hidden treatments in ecological experiments: Re-evaluating the ecosystem function of biodiversity. *Oecologia*, 110, 449–460.

Isbell, F., Cowles, J., Dee, L.E. *et al.* (2018). Quantifying effects of biodiversity on ecosystem functioning across times and places. *Ecology Letters*, 21(6), 763–778.

Letten, A.D., Ke, P.-J., Fukami, T. (2017). Linking modern coexistence theory and contemporary niche theory. *Ecological Monograph*, 87, 161–177.

Lindeman, R.L. (1942). The trophic-dynamic aspect of ecology. *Ecology*, 23(4), 399–417.

Loreau, M. (2004). Does functional redundancy exist? *Oikos*, 104, 606–611.

Loreau, M. (2010). *From Populations to Ecosystems: Theoretical Foundations for a New Ecological Synthesis*. Princeton University Press, New Jersey.

Loreau, M. and Hector, A. (2001). Partitioning selection and complementarity in biodiversity experiments. *Nature*, 412(6842), 72–76.

Loreau, M., Naeem, S., Inchausti, P. *et al.* (2001). Biodiversity and ecosystem functioning: Current knowledge and future challenges. *Science*, 294, 804–808.

Loreau, M., Sapijanskas, J., Isbell, F., Hector, A. (2012). Niche and fitness differences relate the maintenance of diversity to ecosystem function: Comment. *Ecology*, 93(6), 1482–1487.

MacArthur, R.H. and Levins, R. (1967). The limiting similarity, convergence and divergence of coexistence species. *American Naturalist*, 101, 377–385.

Manning, P., Loos, J., Barnes, A.D. *et al.* (2019). Transferring biodiversity-ecosystem function research to the management of "real-world" ecosystems. *Advances in Ecological Research*, 61, 323–356

Oksanen, L., Fretwell, S.D., Arruda, J., Niemela, P. (1981). Exploitation ecosystems in gradients of primary productivity. *American Naturalist*, 118, 240–261.

Pacala, S.W. and Tilman, D. (2002). The transition from sampling to complementarity. In *The Functional Consequences of Biodiversity*, Kinzig, A.P., Pacala, S.W., Tilman, D. (eds). Princeton University Press, Princeton.

Pillai, P. and Gouhier, T.C. (2018). On the use and abuse of Price equation concepts in ecology. arXiv:1812.10930 [q-bio].

van der Plas, F. (2019). Biodiversity and ecosystem functioning in naturally assembled communities. *Biological Reviews*, 94, 1220–1245.

Poisot, T., Mouquet, N., Gravel, D. (2013). Trophic complementarity drives the biodiversity–ecosystem functioning relationship in food webs. *Ecology Letters*, 16, 853–861.

Polis, G.A. and Strong, D.R. (1996). Food web complexity and community dynamics. *American Naturalist*, 147, 813–846.

Purvis, A., Gittleman, J.L., Cowlishaw, G., Mace, G.M. (2000). Predicting extinction risk in declining species. *Proceedings of the Royal Society B: Biological Sciences*, 267, 1947–1952.

Qiu, J. and Cardinale, B.J. (2020). Scaling up biodiversity–ecosystem function relationships across space and over time. *Ecology*, 101, e03166.

Reich, P.B., Tilman, D., Isbell, F. *et al.* (2012). Impacts of biodiversity loss escalate through time as redundancy fades. *Science*, 336, 589–592.

Rosenzweig, M.L. (1971). Paradox of enrichment: Destabilization of exploitation ecosystems in ecological time. *Science*, 171, 385–387.

van Ruijven, J. and Berendse, F. (2005). Diversity–productivity relationships: Initial effects, long-term patterns, and underlying mechanisms. *Proceedings of the National Academy of Sciences*, 102, 695–700.

Schneider, F.D., Brose, U., Rall, B.C., Guill, C. (2016). Animal diversity and ecosystem functioning in dynamic food webs. *Nature Communications*, 7, 12718.

Smith, D.M. (1986). *The Practice of Silviculture*. John Wiley and Sons, New York.

Thébault, E. and Loreau, M. (2003). Food-web constraints on biodiversity–ecosystem functioning relationships. *Proceedings of the National Academy of Sciences*, 100, 14949–14954.

Thompson, P.L., Isbell, F., Loreau, M., O'Connor, M.I., Gonzalez, A. (2018). The strength of the biodiversity–ecosystem function relationship depends on spatial scale. *Proceedings of the Royal Society B: Biological Sciences*, 285, 20180038.

Thompson, P.L., Kéfi, S., Zelnik, Y.R. *et al.* (2021). Scaling up biodiversity–ecosystem functioning relationships: The role of environmental heterogeneity in space and time. *Proceedings of the Royal Society B: Biological Sciences*, 288: 20202779.

Tilman, D. (1982). *Resource Competition and Community Structure.* Princeton University Press, New Jersey.

Tilman, D., Lehman, C.L., Thomson, K.T. (1997). Plant diversity and ecosystem productivity: Theoretical considerations. *Proceedings of the National Academy of Sciences*, 94(5), 1857–1861.

Tilman, D., Isbell, F., Cowles, J.M. (2014). Biodiversity and ecosystem functioning. *Annual Review of Ecology, Evolution, and Systematics*, 45, 471–493.

Turnbull, L.A., Levine, J.M., Loreau, M., Hector, A., (2013). Coexistence, niches and biodiversity effects on ecosystem functioning. *Ecology Letters*, 16, 116–127.

Wang, S. and Brose, U. (2018) Biodiversity and ecosystem functioning in food webs: The vertical diversity hypothesis. *Ecology Letters*, 21, 9–20.

Wang, S., Brose, U., Gravel, D. (2019). Intraguild predation enhances biodiversity and functioning in complex food webs. *Ecology*, 100, e02616.

Wang, S., Isbell, F., Deng, W. *et al.* (2021). How complementarity and selection affect the relationship between ecosystem functioning and stability. *Ecology*, 102, e03347.

Willig, M.R., Kaufmann, D.M., Stevens, R.D. (2003). Latitudinal gradients of biodiversity: Pattern, process, scale and synthesis. *Annual Review of Ecology, Evolution, and Systematics*, 34, 273–309.

Zuppinger-Dingley, D., Schmid, B., Petermann, J.S., Yadav, V., De Deyn, G.B., Flynn, D.F.B. (2014). Selection for niche differentiation in plant communities increases biodiversity effects. *Nature*, 515, 108–111.

5

Experimental Evidence for How Biodiversity Affects Ecosystem Functioning

Mary I. O'CONNOR[1], Joey R. BERNHARDT[2], Keila STARK[1], Jacob USINOWICZ[1], and Matthew A. WHALEN[1,3]

[1] Biodiversity Research Centre,
University of British Columbia, Vancouver, Canada
[2] Yale University, New Haven, USA
[3] Hakai Institute, Heriot Bay, Canada

As with most major scientific advances, our understanding of the causal relationship between biodiversity and ecosystem functioning has developed largely through a joint process of theoretical development and testing of critical components of that theory using experiments. Naeem *et al.*'s classic 1994 experiment catalyzed modern interest in this question; building on earlier experiments in agricultural systems, it laid the groundwork for demonstrating how diverse plant assemblages can be more productive and stable than their less diverse counterparts. By 2020, thousands of experiments have tested effects of manipulating diversity on ecosystem functioning and stability, many testing hypotheses explicitly derived from theory (Chapter 4) and many others manipulating diversity to simulate its consequences at larger scales. Still others have tested how changes in diversity concurrent with other environmental changes may affect ecosystem functioning and ecosystem services. Here, we briefly review the major themes in this work, with specific reference to the

The Ecological and Societal Consequences of Biodiversity Loss,
coordinated by Michel LOREAU, Andy HECTOR, and Forest ISBELL.

core hypothesis tests that have propelled the field forward. We discuss major findings and conclude with remaining challenges and opportunities.

5.1. The role of experiments

Experiments, together with theory and observations (including historical observations) (Chapter 6), play a critical role in advancing our collective scientific knowledge. The power of experiments lies in their relationship to scientific theory and their capacity to provide evidence for or against specific and well-stated causal hypotheses. In this chapter, we review the role experiments have played in advancing BEF knowledge. We highlight a few specific experiments that made major contributions to our understanding of BEF, and we also introduce experiments that are representative of major themes in BEF research and understanding.

5.1.1. *The experiment that launched a thousand experiments*

Occasionally, but not often, the course of science is altered by unexpected and surprising new evidence. This happened in 1994, when the results of an experiment were published on the subject of how declining biodiversity alters the performance of ecosystems. Shahid Naeem and his colleagues Lindsey Thompson, Sharon Lawler, John Lawton, and Richard Woodfin used the Ecotron facility at Silwood Park in the UK to experimentally test the hypothesis that biodiversity loss, as observed in many communities in response to human activities, reduces ecosystem functioning (Naeem *et al.* 1994) (Figure 5.1). They experimentally manipulated diversity in a plant–soil foodweb and found that lower-diversity foodwebs had less biomass, lower ecosystem respiration rates, and reduced nutrient uptake rates relative to more diverse foodwebs (Figure 5.1). This experiment marked the beginning of a major transformation in how ecologists understood biodiversity, historically viewed as a consequence of the abiotic environment and now considered as a driver of ecosystem processes (Naeem 2002). This shift in perspective was, like many major shifts, highly controversial yet extremely important to the role ecological science plays in society; and it had roots in scientific thinking that predated the shift by decades, even centuries (see the Introduction). However, one of the events that catalyzed thousands of experimental studies, millions of invested research dollars, and major policy developments was a carefully designed experimental test of an old hypothesis that produced new empirical evidence about the causal role of biodiversity in ecosystems.

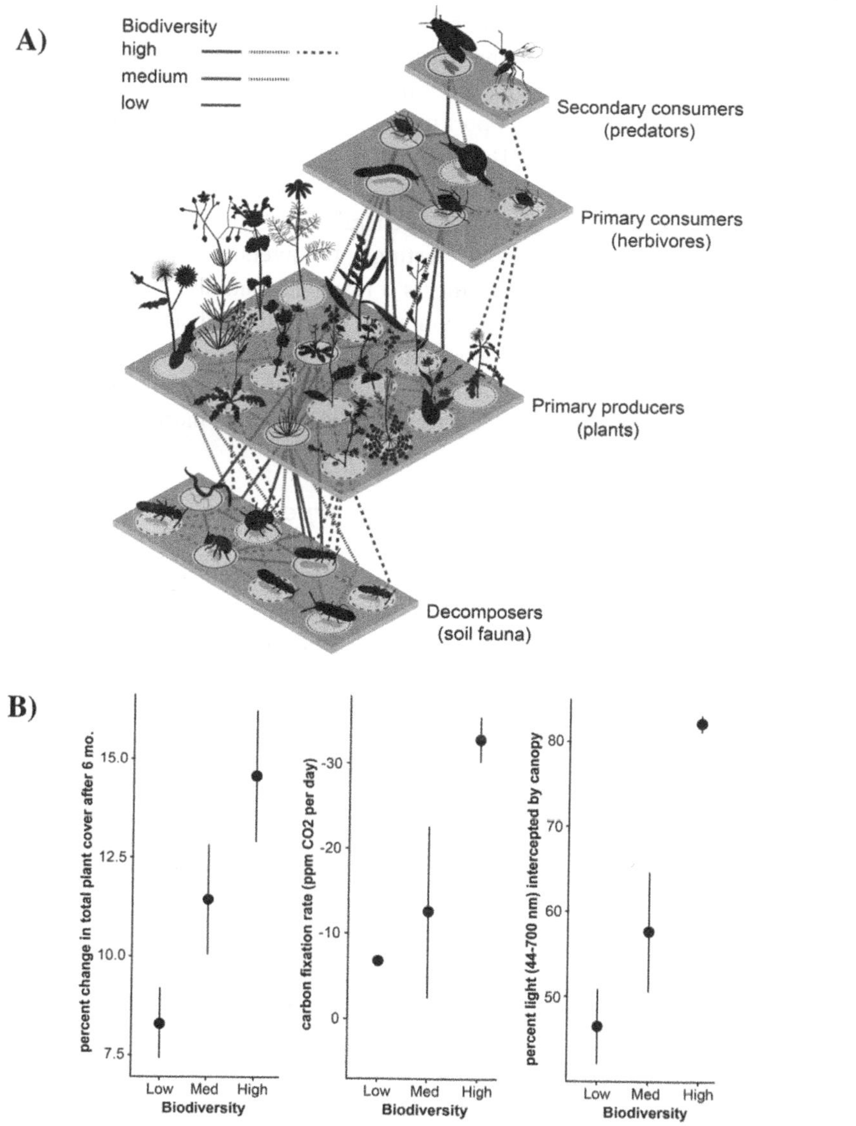

Figure 5.1. *The Ecotron experiment tested the hypothesis that diversity loss reduces ecosystem functioning (Naeem et al. 1994). A) The experiment manipulated diversity by creating a high diversity (all species) and two lower diversity treatments that included only some of the species (image: Sylvia Heredia). b) Effects of biodiversity on three ecosystem functions: plant abundance, carbon fixation rate, and amount of light intercepted by plants (Naeem et al. 1994)*

Following Naeem *et al.*'s work, researchers tested the same hypotheses in a wide variety of communities, including grassland communities, freshwater, and marine systems, and involving organisms from microbes to fish. In just a few years, enough evidence had accumulated to support meta-analyses of these experiments (Cardinale *et al.* 2006; Griffin *et al.* 2013; Balvanera *et al.* 2006). Meta-analyses are analyses of results across many experiments that share the same experimental design and can help to determine whether a result is general across many experimental systems. Meta-analyses of experiments testing biodiversity–ecosystem functioning hypotheses (hereafter "BEF experiments") revealed that, although there were many interesting and important differences among the studies, there is statistical support for the general biological conclusion that reducing biodiversity within a community reduces several ecosystem functions in a continuous and curvilinear way in which the impact on functioning is initially weak but increases as more diversity is lost.

5.1.2. *How do we gain knowledge from experiments?*

Experiments are a method of gaining information that allow researchers to control conditions to isolate a single or small set of factor(s) thought to cause changes in a system. In BEF experiments, the defining feature is that some measure of diversity – species richness, functional diversity, etc. – is manipulated as the independent fixed variable and response variables related to ecosystem functioning are measured as dependent random variables (Figure 5.2). Experimenters have many decisions to make about how to manipulate diversity and what other potentially confounding factors – such as species density or composition – they will control. These are all important considerations for the architecture of the experiment relative to its epistemological (knowledge-producing) purpose, or the experimental design. Design elements include which treatments are used, the levels of those treatments, the number of replicates, which potentially confounding factors are controlled or manipulated, and which assumptions of the theory are demonstrably met in the experimental system. One important consideration particular to BEF experiments is whether or how effects of species composition and species "identity" will be distinguished from effects of species diversity (agnostic to species composition). Early BEF experiments such as the Ecotron experiment did not distinguish between composition and richness (Naeem *et al.* 1995; Niklaus *et al.* 2001), though the next generation of major experiments did. Randomization of species composition among replicated diversity levels has since become an important feature of BEF experiments (e.g. BIODEPTH).

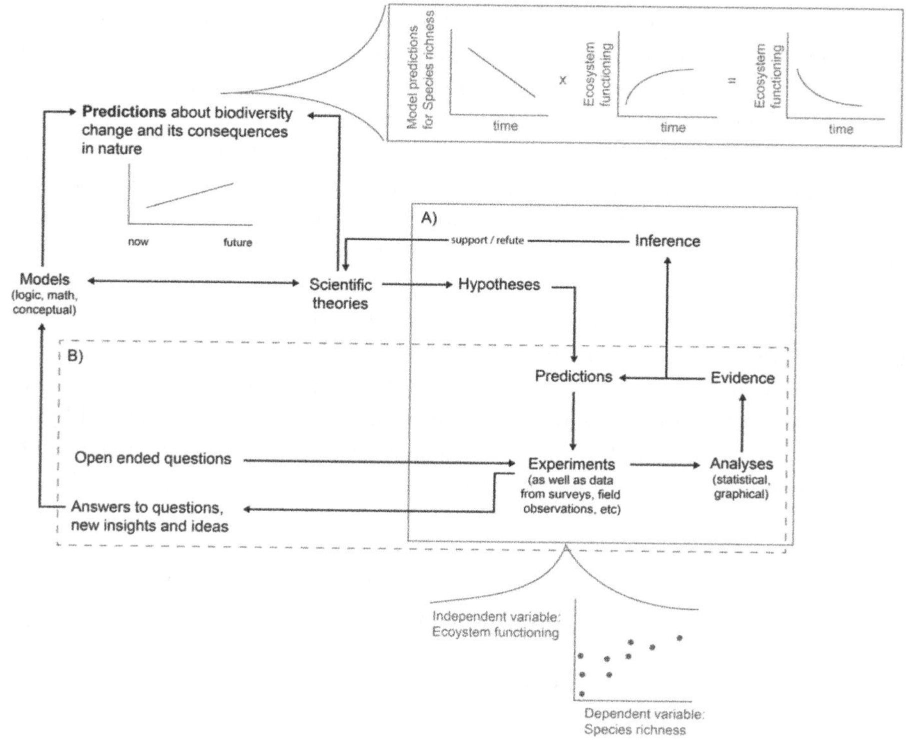

Figure 5.2. *Experiments contribute knowledge through the production and evaluation of evidence in the context of theories, analyses, and models. There are two main pathways of thinking that are often used: a) a hypothesis-testing approach explicitly related to the theoretical frameworks scientists use to make predictions at scales beyond what is directly observed and b) an exploratory approach in which experiments are used to stimulate or create conditions that allow researchers to explore possible outcomes*

How diversity is manipulated depends on the biology of the species and the physics of the ecosystem in question – aquatic researchers tend to use mesocosms or aquaria with mobile organisms, while plant researchers have more often used defined "plots" in which soils and plants can be manipulated. Experiments have traditionally been set in highly controlled indoor lab systems (e.g. chemostats, growth chambers) and in a variety of outdoor settings exposing systems to local abiotic and biotic conditions, including in fields, greenhouses, and ponds. Even though there are many details to consider when designing an experiment, BEF experiments share their main features of manipulating biodiversity.

There are a variety of epistemological approaches to how we use experiments to deepen our understanding about the relationship between diversity and function. Approaches differ in how we use information gained from the results of experiments. One approach is to design experiments to test a clearly defined hypothesis derived from a larger ecological theory for how nature works. This type of experiment involves manipulations and controls of a factor or factors that are hypothesized to have a particular and measurable consequence on a response variable, also defined by the theory. Results of hypothesis-testing experiments are then used to support inferences that bolster, refute, or refine the theoretical framework, and it is this framework that is used to make statements or projections about the world and to guide application of the experimental results to other systems or scales (Figure 5.2). Any statements or projections based on that theory are relevant only to the conditions (assumptions, scales, resolutions) proscribed by the theory (Figure 5.2). Examples of this type of experiment in BEF include the Cedar Creek Experiments, the BIODEPTH Experiment, the Jena experiments, and many others.

A complementary approach is to design experiments that explore or answer an open-ended question. These experiments are considered simulations of nature in controlled contexts. Questions that can be asked using this kind of experimental design might be less restricted in their form because they do not require a clear and resolved theoretical framework to guide them. An example of such creative experiments in BEF is the one by Kristin France and Emmett Duffy (2006), who tested the role of dispersal (movement among habitat patches) by grazers and whether grazer movement enhanced or reduced local effects of grazer diversity. They found that movement reduced diversity effects and increased variability in diversity effects over time, providing a then-perplexing example of what is now better understood in terms of metacommunity dynamics (e.g. Thompson *et al.* 2020). Such experiments can stimulate scientific understanding and development with exciting new perspectives. O'Connor and Bruno (2009) built on this study to show that high immigration rates by prey can erase effects of predator diversity. However, as a consequence of this freedom for design, the results of exploratory experiments are more difficult to generalize to other systems or scales besides those directly tested without a theory to guide such generalization.

Our understanding is deepened by considering the insights gained from a variety of experimental approaches because each has strengths and limitations. For example, one consequence of the strength of experiments for testing theory is that these studies can involve conditions that are highly abstracted from reality – for example, well mixed soils or biodiversity treatments that do not clearly reflect likely extinctions. One drawback of these necessarily abstracted conditions is that they can appear "unrealistic" when considered outside the context of the theory they are designed to test. This perceived lack of realism has been raised as a concern regarding the relevance of BEF experiments to understanding consequences of biodiversity change in nature (Srivastava and Vellend 2005). However, to be fair, most experiments designed to test hypotheses that higher diversity is associated with higher ecosystem functioning were not at all designed to test hypotheses about, or to simulate, biodiversity *change* in response to drivers of biodiversity change, so additional theory is required to relate the experiments to the problem of human driven biodiversity change. Furthermore, experiments are designed to be interpreted in the context of the theory they were meant to test; to attempt to interpret them outside this context is to misuse their results. Other experiments have been designed with the "simulation" approach to target more realistic representations of biodiversity change as it is perceived to be happening in nature (Bracken *et al.* 2008). These experiments may inform understandings about biodiversity change in response to one or two specific ecological mechanisms or human drivers, but even these experiments face challenges in terms of lacking clarity about how to scale up or extend to other systems when we lack the theoretical framework to transcend their scales and scope (Gonzalez *et al.* 2020), as noted above.

5.2. BEF experiments as tests of theory

5.2.1. *Diversity as a driver of change in ecosystem function*

Though the Ecotron experiment (Naeem *et al.* 1994) answered the motivating question that yes, it is possible for less diverse systems to function at lower levels (Figure 5.1), it actually raised many more questions: how? Why? The experimental design of the Ecotron experiment was ideal for exploring the answer to the initial question, but alone it was not suited to explaining why the result was observed in terms of a specific ecological theory. Another experimental system was being developed in the mid-1990s by a team of researchers at Cedar Creek at the University of Minnesota in the US, and these experiments were designed explicitly to find the ecological theory for species coexistence that best explained why more diverse communities might function at higher levels (Chapter 4). The Cedar Creek grassland experiments drew upon long-term studies of how diversity and productivity had changed concurrently over time in these systems (Tilman and Downing 1994) and also on emerging theory to test the diversity–productivity hypothesis (Hooper and Vitousek 1997) that diverse mixtures of species should be more productive than depauperate mixtures, including monocultures (Chapter 4). Tilman and colleagues (1996) manipulated species diversity in levels from 1 to 24 species, replicating treatments at each species richness level but varying species composition among replicates, aiming to distinguish the effects of species richness and composition (a total of 147 experimental plots) and thus going beyond the much smaller Ecotron experiment. They used a substitutive experimental design (Figure 5.3), allowing them to infer that without changing the total density of individuals at planting, the number of species in the assemblage would be the driver of any changes in their ecosystem function variables (total biomass, resource update rates). Their earliest results (1996; Figure 5.3) demonstrated clearly that experimental communities were less productive when they included fewer species regardless of species composition, and they also used less of the available nitrogen in the soils, which was the limiting resource to plant growth in this system.

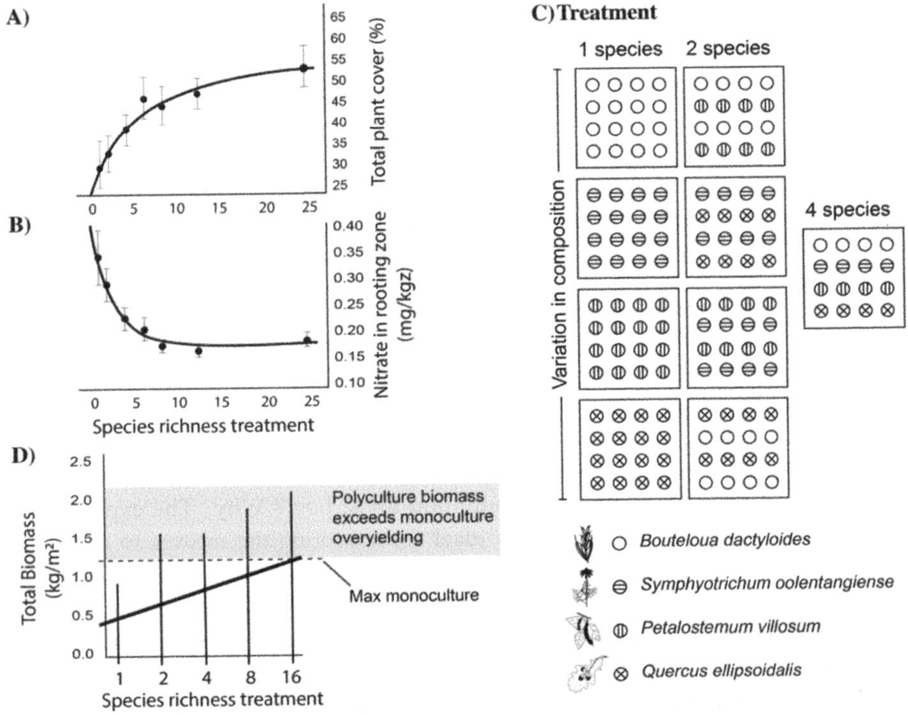

Figure 5.3. *Tilman and colleagues (1997) manipulated plants in a tallgrass prairie community in an experiment that randomized species composition while manipulating species richness. They found that at higher species richness, plant cover (a measure of abundance) was higher and nitrogen had been more completely used from the soils. They used a substitutive experimental design (c) in which the total density of individuals at sowing was held constant across treatments while the number of species varied. They also detected overyielding, which is when diverse mixtures outperform the best monoculture (Tilman et al. 2001)*

In this same period, a team of researchers in Europe developed the BIODEPTH experiment in European grasslands to broaden the scope of BEF experiments further and include an additional level of diversity variation. BIODEPTH shared many features of the Cedar Creek design, but included replicated species composition treatments to further disentangle the importance of diversity (independent of species composition) from species composition effects on ecosystem functioning (Hector *et al.* 1999; Hector *et al.* 2002). Further, this experiment was replicated in seven countries using a standard experimental design. They found that lower species richness was associated with lower ecosystem functioning at most but not all sites, but across the whole experiment, the effects of species richness and species composition were important drivers of total ecosystem functioning (Hector *et al.* 2011).

Research in aquatic systems has supported the role of species richness as a driver of variation in function, in addition to species composition. Using freshwater mesocosms, Amy Downing and Matthew Leibold (2002) tested the hypothesis that species richness, not composition, explains BEF effects on ecosystem functioning in communities with multiple trophic levels (algae, grazers, predators, and decomposers). Building on the experimental approach of Naeem *et al.* (1994), Downing and Liebold manipulated species richness within functional groups, allowing functional diversity to remain constant while species diversity and species composition varied across experimental treatments. Many of their findings were consistent with what had been reported in plant-only systems, such as the finding that productivity was higher at higher species richness. They reported novel observations too specifically that foodweb structure (relative amounts of different functional groups) shifted with species richness. This study demonstrated the importance of indirect effects of species, such that in diverse mixtures the interactions of species and their effects on other species' abundances and functions play an important role in overall BEF effects. Several experiments indicated that species richness plays an important role in explaining variation in ecosystem functioning even when there is a particularly influential species or functional group that also has a strong effect on functioning (Bruno and O'Connor 2005; Roscher *et al.* 2005; Reich *et al.* 2012). One of the most innovative was the treehole bacteria experiment designed by Bell and colleagues (2005), which distinguished composition and richness in a comprehensive experimental design (Figure 5.4). Still, very few experimental designs have robustly separated species richness and composition, so in fact evidence to fully distinguish species richness and composition effects remains limited.

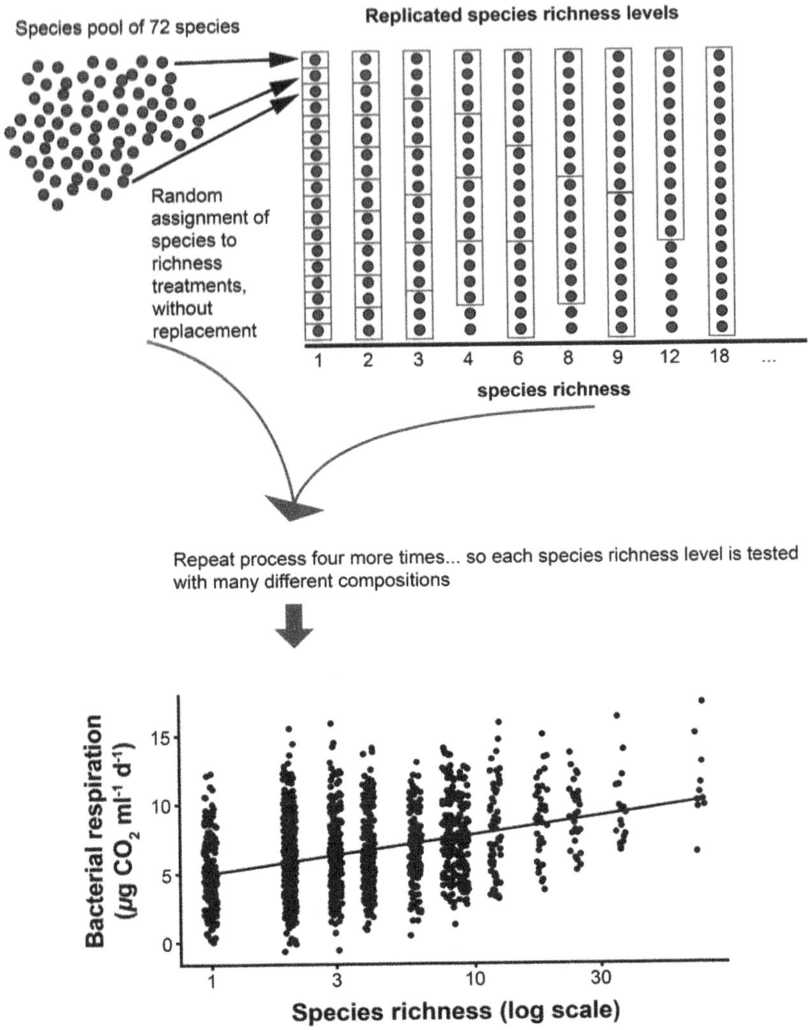

Figure 5.4. *An experimental design that decoupled species richness and composition in a novel way, using regression statistics (Bell et al. 2005). Within levels of species richness (1–72 species; the first 18 shown for clarity), species were assigned to treatments randomly and without replacement, so that each species appeared only once in each level of richness. This procedure was then repeated to generate "reshuffled" replicates with different compositions at the same levels of richness. This design is complex but also elegant and powerful. Indeed, it is feasible compared to manipulating richness with every combination of species (>4.7 × 10^{21} total combinations)*

Other experiments around that time demonstrated effects on the productivity of different dimensions of diversity – functional diversity (diversity of functional traits), trait diversity (diversity of many traits), and species composition (particular sets of species). In plant communities, the compositional and functional effects on ecosystem rates reflected different ways of accessing resources, including nitrogen (from the soil via grasses and forbs vs. from the atmosphere via N-fixing legumes), water (rooting depth), or carbon (C3 and C4 photosynthesis). These experiments showed that the effects of treatments that manipulated functional groups based on how resources were accessed were stronger and more repeatable than manipulations of species richness alone (Tilman *et al.* 1997). Some studies even reported that effects of particular species and functional traits outweighed effects of functional richness (Hooper and Vitousek 1997), though a multi-level analysis of the multi-site BIODEPTH experiment found that species richness and compositional effects were roughly equal (Hector *et al.* 2011). These early studies in grasslands, and many others to follow in aquatic systems (Cardinale and Palmer 2002; Duffy *et al.* 2005; Bruno *et al.* 2005; Stachowicz *et al.* 2008), clearly established the role of functional differences among species as being an important element of how diversity affects ecosystem functioning. This was not the end of the problem, however, because even though diversity in functional traits may explain species richness effects statistically in analyses of experiments, the maintenance of biodiversity is based on theories of population dynamics and how they vary with interspecific interactions. Population dynamics occur within species (by definition), while functional or trait groups often include more than one species, leaving a gap between the pattern of BEF attributed to functional groups and the underlying processes as described by theory (Chapter 4). Understanding biodiversity and ecosystem functioning across scales of space and time in a changing world arguably cannot be based on functional traits alone without links to dynamics.

5.2.2. *Evidence for selection and complementarity*

The BEF experiments in the 1990s clearly demonstrated a fruitful and policy-relevant avenue of scientific inquiry, but the causes of diversity effects were still strongly debated. This debate about which ecological processes best explain diversity effects was important for how experimental results are "scaled up", or applied, to broader contexts. As explained in section 5.1.2, theoretical frameworks are the tools we use for scaling up the results observed in experiments, and sometimes there is more than one possible theoretical framework, or more than one way to scale up using even a single framework. For BEF, the candidate theoretical frameworks assumed specific ecological explanations, or causes, underpinning diversity effects.

Two groups of biodiversity effects emerged from this debate: those caused by the *selection* of influential species, and those caused by *complementarity* in species' traits, including traits that determine species interactions (Chapter 4 for definitions) (Hector 1998; Loreau 1998). The challenge was that selection and complementarity effects produced a similar empirical pattern of a positive, saturating relationship between functioning and diversity (Figure 5.3). This similarity meant that even once the first wave of experimental results were in and the curvilinear pattern was apparent, the underlying processes could not be decisively proven as selection or complementarity. It became clear that to test for and distinguish the roles of selection and complementarity effects required specific statistical techniques (additive partitioning (Chapter 4) and later regression models (Bell *et al.* 2005)), and specific experimental designs (Figure 5.4). BIODEPTH was the first study to report an explicit test of selection and complementarity effects. As expected, productivity in grasslands varied with climate and region, but within regions diversity enhanced productivity beyond what was observed in typical monocultures (called "overyielding" Figure 5.3d) due to complementarity among species (Hector *et al.* 2002).

Within a few years, evidence mounted from experiments conducted in many ecosystems with many different kinds of species to support the common occurrence of both complementarity and selection effects. Testing hypotheses in very different ecosystems is important because it establishes that BEF mechanisms are general, even when different types of ecological interactions mediate community dynamics in different environments. For example, in grassland studies, the interactions driving complementarity and selection effects in early experiments were related to competition for nitrogen and water (Tilman *et al.* 1997; Hooper and Vitousek 1997). In aquatic (freshwater and marine) systems, predation pressure is strong and ubiquitous for invertebrates, and algae compete for water column nutrients (Stachowicz *et al.* 2008; Bruno *et al.* 2005; Duffy 2003). Yet across these systems, similar BEF effects are reported from similarly designed experiments (Cardinale *et al.* 2006, O'Connor *et al.* 2017).

5.2.3. *Experimental evidence for key assumptions of BEF theory*

Central to the core idea of BEF is an underlying assumption that species in a community can coexist in a way that is stable in the long-term. This assumption originates with the idea of niche complementarity – the idea that coexisting species limit niche overlap by partitioning existing pools of resources. If species were not limited in their niche overlap, then competitively dominant species would exclude others and there would be little diversity at the scale of species interactions and no reason to expect diversity to enhance ecosystem functioning. In other words, one species could be as effective as many species with respect to a particular ecosystem function. The foundations of this idea can be found in ecology's most basic models, such as the Lotka–Volterra competition model, and can even be attributed to Darwin's Principle of Divergence (Chapter 4). In this framework, the conditions for over-yielding that define BEF are identical to the conditions for coexistence. In particular, the conditions for over-yielding and for coexistence require that intraspecific competition be greater than interspecific competition for all pairs of coexisting species (Loreau 2004). Greater intra- relative to interspecific competition is generally seen as a formal test for niche partitioning. Tests of coexistence in many different diverse plant communities have found strong evidence for stable coexistence via stronger intraspecific relative to interspecific competition, but without relating the community-wide potential for coexistence to ecosystem function (Adler *et al.* 2010). However, it is unclear whether this relationship between over-yielding and coexistence is also true in even slightly more complex models that include nonlinear growth–competition relationships and environmental variability.

Direct empirical tests of the conditions for coexistence were never made in original BEF experiments. Instead, evidence for niche partitioning was assumed, based on trait differences between co-occurring species. A frequently invoked example is the idea that a greater variety of root depths and types occurring in more biodiverse plots (Berendse 1981; van Ruijven and Berendse 2005; Dimitrakopoulos and Schmid 2004) is evidence of a finer partitioning of niche space. Meanwhile, experiments in multi-trophic systems have found that a finer partitioning of resource usage can occur in diverse systems, which is consistent with the underlying assumptions of BEF (Ives *et al.* 2005), but this relationship may only occur when the diversity of specialist – and not generalist species (in this case, consumers) – increases (Finke and Snyder 2008).

5.2.4. *Testing for diversity effects under broader abiotic and biotic conditions*

Even while evidence mounted that diversity per se could have clear and measurable effects on ecosystem functions via selection and complementarity, these results emerged under highly controlled experimental conditions that matched the assumptions of theoretical models they were aiming to test (section 5.1.2). The ecological theories that explain ecological dynamics underlying BEF (Chapter 4), like all theories, are by design limited in their ability to incorporate the complexity of natural systems. Experiments, however, can introduce some complexity by testing for diversity effects under a range of contexts – with and without consumers, after disturbance, or with resource limitation, for example. These experiments can provide tests of hypotheses in empirical systems, consider multiple conditions, or even deviate somewhat from the strict assumptions of theories.

Many experiments have shown that foodweb structure can modify how diversity affects ecosystem functions (O'Connor *et al.* 2017). While Naeem *et al.* (1994) manipulated species diversity at multiple trophic levels (from decomposers to parasitoids), later experiments in marine and freshwater systems were instrumental in expanding our understanding of how species interactions – both negative and positive – within multitrophic communities modify BEF relationships (Griffin *et al.* 2013; Stachowicz *et al.* 2008; Cardinale *et al.* 2002). In marine systems, interactions involving diversity within and among trophic levels have been explored in carefully designed mesocosm and field experiments. Emmett Duffy and colleagues showed that the presence of a predator made diversity effects among grazers more likely, exacerbating trade-offs in grazer traits between predation resistance and competitive ability for food (Duffy *et al.* 2005). Another example is the work led by Deborah Finke in salt marsh systems, which used an experimental manipulation of spider diversity to demonstrate that BEF concepts apply to predator guilds as well (Finke and Denno 2004). Bruno and O'Connor (2005) showed that omnivory can weaken cascading predator diversity effects, and Byrnes and colleagues showed that predator diversity effects can depend on prey behavioral traits (Byrnes *et al.* 2005).

Janneke Hille Ris Lambers and colleagues tested whether species interactions leading to coexistence or competitive exclusion contributed to BEF patterns in the Cedar Creek experimental sites (Hille Ris Lambers *et al.* 2004). They manipulated species richness and composition among experimental plots and assessed species' yield in all treatments, as well as the ability to compete for nitrogen in monoculture. They found that overyielding was positively associated with an ability to compete for nitrogen and a negative effect of strong nitrogen competitors on weak nitrogen competitors: evidence for competition. However, this did not explain why under-yielding species were not completely competitively excluded, nor could they explain why over-yielding species did not all have the highest performance in monoculture.

Another important contribution of experimental tests of the BEF relationship has been to investigate how it varies with environmental context. Some experiments manipulate environmental variables in addition to diversity in order to determine the effect of the abiotic environment on the BEF relationship. In Bodega Harbor, California, Matt Whalen and Jay Stachowicz compared the relationship between suspension feeder species richness (tunicates and bryozoans) and seawater filtration rate (phytoplankton consumption) in marine fouling communities that varied in the presence and absence of water flow. They found that community-level water filtration, an ecosystem function, increased with species richness under both flow conditions, and the strength of this relationship did not differ among flow conditions. Their result shows that the diversity effect – complementarity in this case – persisted even while one aspect of the abiotic environment changed. In contrast, sometimes environmental context can change diversity effects. In freshwater stream mesocosms, Bradley Cardinale and Margaret Palmer (2002) showed how interactions among three suspension feeding caddisfly larvae were expressed differently when a disturbance (simulated larval mortality) was applied, reducing selection effects and changing the relationship between caddisfly species richness and fluxes of particulate organic matter and algal productivity.

5.2.5. *Diversity effects in space and time*

Some ecological effects of diversity emerge over time. In 1997, Jill McGrady-Steed and colleagues demonstrated in an experimental microbial system that more diverse species assemblages show more predictable ecosystem function (respiration) over time (McGrady-Steed *et al.* 1997). This result supported ideas about diversity that dated back to Elton (Chapter 7), but predated the modern insurance hypothesis (Yachi and Loreau 1999). Stachowicz *et al.* (2008) demonstrated in an intertidal field manipulation that more diverse algal assemblages outperformed monocultures, but this result took over a year to emerge. In another example, the Cedar Creek experiment has clearly shown how differences among species manifest in different ways in different years (e.g. drought years vs. wet years), elevating the performance of diverse mixtures over time in variable environments (Reich *et al.* 2012). Experiments that are able to test for and observe responses to biodiversity treatments over time periods long enough to include environmental variation (either year to year, or seasonal) are relatively rare. They can yield findings showing that effects of diversity align well with hypotheses during one period, but not others (O'Connor and Crowe 2005; Boyer *et al.* 2009). Importantly, these experiments suggest that selection effects at one period may translate to niche complementarity over time, as different environmental conditions allow for different species to contribute to maximizing ecosystem functioning.

Another component of the abiotic environment that can influence the BEF relationship is spatial heterogeneity. This is because heterogeneity, whether temporal or spatial, generates conditions for species differences to be expressed. Cardinale (2011) experimentally tested how water flow heterogeneity in diverse stream biofilm communities influences biomass accumulation and nitrogen uptake – a key ecosystem function for maintaining water quality. He manipulated algal species richness and identity (diatoms and chlorophytes) in the presence or absence of spatial variation in water flow and disturbance. He found that different morphological forms of algae were established in different flow microhabitats, leading to overyielding of high diversity treatments. In the absence of flow heterogeneity, a single diatom species became dominant, leading to a strong selection effect and underyielding at all levels of diversity. While this study supports the theory that habitat heterogeneity enhances biodiversity–ecosystem functioning relationships by allowing niche partitioning among species, it is worth noting that total rates in high diversity treatments were similar between homogeneous and heterogenous flow environments. However, no treatment outperformed one species of filamentous green algae grown in homogeneous conditions, something others have observed (Wacker *et al.* 2008; Weis *et al.* 2008).

5.3. Experiments that extend classic theory

5.3.1. *Does extinction order matter?*

Much of the controversy surrounding early BEF experiments focused on criticisms of experimental design and the extent to which communities assembled in experimental plots reflected natural processes (Huston and Aarssen *et al.* 2000), including the ideas that drivers of biodiversity change lead to extinction at higher trophic levels and of rarer species (Duffy 2003; Vermeij and Grosberg 2018). In a ground-breaking study, Erika Zavaleta and Kristin Hulvey (2004) manipulated plant diversity after first determining the nested order of diversity change across space in a California grassland. Using outdoor mesocosms, they found that resistance to invasive star thistles decreased dramatically with loss of native plant diversity, owing in large part to loss of functional groups that compete with the invader. This study also nicely demonstrates that rare species can substantially influence ecosystem functioning.

5.3.2. *Experiments that bridge BEF and modern coexistence theory (MCT)*

Modern coexistence theory (MCT) is perhaps the most utilized theoretical framework for formal tests of coexistence in diverse communities of competing species. Additionally, much of the development and application of MCT has been in the context of plant communities, as we have seen with tests of BEF. MCT is a framework for quantifying coexistence as a function of two generalized components of population dynamics: species relative fitness differences and their stabilizing niche differences (see Chapter 4 for further details). As discussed earlier, niche differences are measured by the rate of intraspecific competition relative to interspecific competition. Fitness differences, on the other hand, compare species' intrinsic performance in a given environment: the species with a better intrinsic fitness will outcompete the other in the case that niche differences cease to exist. These concepts should, to some degree, match the theoretical basis of BEF, and it has been suggested that complementarity effects are predicted by niche differences, while sampling effects map onto fitness differences.

Godoy *et al.* (2020) set out to examine the links between BEF and MCT by testing whether increased coexistence promotes higher ecosystem functioning. They employed a classic BEF experimental design by manipulating the richness of an annual plant community to a maximum of 10 species, under two different moisture levels (ambient and drought). They also used empirical measurements of demography taken from the same experiment to parameterize mathematical models of population growth. From the BEF experiment, they parsed complementarity and selection effects on biomass production, leaf litter decomposition, and soil nitrogen. Using MCT, they parsed fitness and niche differences from the parameterized population models. They found that complementarity positively correlated with niche differences and selection effect differences correlated with fitness differences. However, this correspondence was not perfect and they also found that niche differences contributed to selection effects, and fitness differences to complementarity effects. Overall, experimental communities with greater niche differences than what was necessary for coexistence always produced more biomass and had faster decomposition rates under drought. This provides empirical evidence that the mechanisms determining coexistence correlate with those maximizing certain ecosystem functions, but it still leaves many questions unresolved.

5.3.3. *Experimental evidence for effects of biodiversity on ecosystem services*

Our understanding of the connections between biodiversity and ecosystem services is complicated by the fact that ecosystem services are often regulated by multiple underlying functions that may each respond to changes in biodiversity differently, and mechanistic links between biodiversity and ecosystem services are generally poorly understood (Cardinale *et al.* 2012). While most of the work linking biodiversity to ecosystem services is correlative, experiments have been used to estimate the effects of biodiversity on ecosystem service provisioning such as water purification, crop production, and wood production (Balvanera *et al.* 2006). For example, Cardinale's (2011) experiment discussed above showed that, using a stream mesocosm experiment, more diverse stream algae communities produced greater biomass and captured more of the available nitrogen in the water, thereby purifying the water and reducing nutrient pollution. Using tree plantations as "natural experiments" has demonstrated that more diverse tree plantations have higher wood production rates than monocultures (Piotto 2008). Experiments have shown that plant diversity in agricultural systems reduces herbivory and suppresses crop damage (Letourneau *et al.* 2011). These examples demonstrate that experiments can be used to advance our understanding of the relationships between biodiversity and ecosystem services, though much work remains to be done to achieve a mechanistic understanding of the links between ecosystem functions and services.

5.4. Conclusion

Since the mid 1990s, experiments have been at the forefront of our understanding of the causal relationships between biodiversity and ecosystem functioning. Experiments have provided rigorous support for the theoretical frameworks we use to broadly consider the effects of biodiversity change and explored possible ecological patterns that theory has yet to formalize. Importantly, the work is not finished! Exciting prospects include experiments that bridge BEF theory and other ecological frameworks to extend our understanding of the functional role of biodiversity. Another important area for development in experiments is the relationship between biodiversity, ecosystem functions, and ecosystem services. Furthermore, as noted in our chapter, there is always a need for experimental tests of the critical assumptions embedded in our theories. The future decade of BEF experiments promises to be as productive and influential to our understanding of nature and global change as the past few decades have been.

5.5. References

Adler, P.B., Ellner, S.P., Levine, J.M. (2010). Coexistence of perennial plants: An embarrassment of niches. *Ecology Letters*, 13, 1019–1029.

Balvanera, P., Pfisterer, A.B., Buchmann, N. *et al.* (2006). Quantifying the evidence for biodiversity effects on ecosystem functioning and services. *Ecology Letters*, 9, 1–11.

Bell, T., Newman, J.A., Silverman, B.W., Turner, S.L., Lilley, A.K. (2005). The contribution of species richness and composition to bacterial services. *Nature*, 436(7054), 1157–1160.

Berendse, F. (1981). Competition between plant populations with different rooting depths II. Pot experiments. *Oecologia*, 48, 334–341.

Boyer, K.E., Kertesz, J.S., Bruno, J.F. (2009). Biodiversity effects on productivity and stability of marine macroalgal communities: The role of environmental context. *Oikos*, 118(7), 1062–1072.

Bracken, M.E.S., Friberg, S.E., Gonzalez-Dorantes, C.A., Williams, S.L. (2008). Functional consequences of realistic biodiversity changes in a marine ecosystem. *Proceedings of the National Academy of Sciences of the United States of America*, 105(3), 924–028.

Bruno, J.F. and O'Connor, M.I. (2005). Cascading effects of predator diversity and omnivory in a marine food web. *Ecology Letters*, 8(10), 1048–1056.

Bruno, J.F., Boyer, K.E., Duffy, J.E., Lee, S.C., Kertesz, J.S. (2005). Effects of macroalgal species identity and richness on primary production in benthic marine communities. *Ecology Letters*, 8(11), 1165–1174.

Byrnes, J., Stachowicz, J.J., Hultgren, K.M., Randall Hughes, A., Olyarnik, S.V., Thornber, C.S. (2006). Predator diversity strengthens trophic cascades in kelp forests by modifying herbivore behaviour. *Ecology Letters*, 9(1), 61–71, 051109031307002.

Cardinale, B.J. (2011). Biodiversity improves water quality through niche partitioning. *Nature*, 472(7341), 86–U113.

Cardinale, B.J. and Palmer, M.A. (2002). Disturbance moderates biodiversity–ecosystem function relationships: Experimental evidence from caddisflies in stream mesocosms. *Ecology*, 83(7), 1915–1927.

Cardinale, B.J., Palmer, M.A., Collins, S.L. (2002). Species diversity enhances ecosystem functioning through interspecific facilitation. *Nature*, 415, 426–429.

Cardinale, B.J., Srivastava, D.S., Emmett Duffy, J. *et al.* (2006). Effects of biodiversity on the functioning of trophic groups and ecosystems. *Nature*, 443(7114), 989–992.

Cardinale, B.J., Duffy, J.E., Gonzalez, A. *et al.* (2012). Biodiversity loss and its impact on humanity. *Nature*, 486(7401), 59–67.

Dimitrakopoulos, P.G. and Schmid, B. (2004). Biodiversity effects increase linearly with biotope space. *Ecology Letters*, 7(7), 574–583.

Downing, A.L. and Leibold, M.A. (2002). Ecosystem consequences of species richness and composition in pond food webs. *Nature*, 416, 837–840.

Duffy, J.E. (2003). Biodiversity loss, trophic skew and ecosystem functioning. *Ecology Letters*, 6(8), 680–687.

Duffy, J.E., Richardson, J.P., France, K.E. (2005). Ecosystem consequences of diversity depend on food chain length in estuarine vegetation. *Ecology Letters*, 8, 301–309.

Finke, D.L. and Snyder, W.E. (2008). Niche partitioning increases resource exploitation by diverse communities. *Science*, 321(5895), 1488–1490.

France, K.E. and Duffy, J.E. (2006). Diversity and dispersal interactively affect predictability of ecosystem function. *Nature*, 441, 1139–1143.

Godoy, O., Gómez-Aparicio, L., Matías, L., Pérez-Ramos, I.M., Allan, E. (2020). An excess of niche differences maximizes ecosystem functioning. *Nature Communications*, 11(1), 4180.

Gonzalez, A., Germain, R.M., Srivastava, D.S. *et al.* (2020). Scaling-up biodiversity–ecosystem functioning research. *Ecology Letters*, 23(4), 757–776.

Griffin, J.N., Byrnes, J.E.K., Cardinale, B.J. (2013). Effects of predator richness on prey suppression: A meta-analysis. *Ecology*, 94(10), 2180–2187.

Hector, A. (1998). The effect of diversity on productivity: Detecting the role of species complementarity. *Oikos*, 83(3), 597–599.

Hector, A., Schmid, B., Beierkuhnlein, C. *et al.* (1999). Plant diversity and productivity experiments in European grasslands. *Science*, 286, 1123–1127.

Hector, A., Bazeley-White, E., Loreau, M., Otway, S., Schmid, B. (2002). Overyielding in grassland communities: Testing the sampling effect hypothesis with replicated biodiversity experiments. *Ecology Letters*, 5(4), 502–511.

Hector, A., Bell, T., Hautier, Y. *et al.* (2011). BUGS in the analysis of biodiversity experiments: Species richness and composition are of similar importance for grassland productivity. *PLoS ONE*, 6(3), e17434.

Hooper, D.U. and Vitousek, P.M. (1997). The effects of plant composition and diversity on ecosystem processes. *Science*, 277, 1302–1305.

Ives, A.R., Cardinale, B.J., Snyder, W.E. (2005). A synthesis of subdisciplines: Predator–prey interactions, and biodiversity and ecosystem functioning. *Ecology Letters*, 8, 102–116.

Letourneau, D., Armbrecht, I., Rivera, B.S. *et al.* (2011). Does plant diversity benefit agroecosystems? A synthetic review. *Ecological Applications*, 21(1), 9–21.

Loreau, M. (1998). Separating sampling and other effects in biodiversity experiments. *Oikos*, 83(3), 600–602.

McGrady-Steed, J., Harris, P.M., Morin, P.J. (1997). Biodiversity regulates ecosystem predictability. *Nature*, 390, 162–165.

Naeem, S. (2002). Ecosystem consequences of biodiversity loss: The evolution of a paradigm. *Ecology*, 83(6), 1537–1552.

Naeem, S., Thompson, L.J., Lawler, S.P., Lawton, J.H., Woodfin, R.M. (1994). Declining biodiversity can alter the performance of ecosystems. *Nature*, 368, 734–737.

Naeem, S., Thompson, L.J., Lawler, S.P., Lawton, J.H., Woodfin, R.M. (1995). Empirical evidence that declining species diversity may alter the performance of terrestrial ecosystems. *Philosophical Transactions of the Royal Society of London Series B : Biological Sciences*, 347(1321), 249–262.

Niklaus, P.A., Kandeler, E., Leadley, P.W., Schmid, B., Tscherko, D., Korner, C. (2001). A link between plant diversity, elevated CO2 and soil nitrate. *Oecologia*, 127(4), 540–548.

O'Connor, M.I. and Bruno, J.F. (2009). Predator richness has no effect in a diverse marine food web. *Journal of Animal Ecology*, 78(4), 732–740.

O'Connor, N.E. and Crowe, T.P. (2005). Biodiversity loss and ecosystem functioning: Distinguishing between number and identity of species. *Ecology*, 86(7), 1783–1796.

O'Connor, M.I., Gonzalez, A., Byrnes, J.E.K. *et al.* (2017). Toward predictive scaling coefficients for biodiversity and ecosystem functioning relationships. *Oikos*, 126, 18–31.

Piotto, D. (2008). A meta-analysis comparing tree growth in monocultures and mixed plantations. *Forest Ecology and Management*, 255(3–4), 781–786.

Reich, P.B., Tilman, D., Isbell, F. *et al.* (2012). Impacts of biodiversity loss escalate through time as redundancy fades. *Science*, 336(6081), 589–592.

Roscher, C., Temperton, V.M., Scherer-Lorenzen, M. *et al.* (2005). Overyielding in experimental grassland communities – Irrespective of species pool or spatial scale. *Ecology Letters*, 8(4), 419–429.

van Ruijven, J. and Berendse, F. (2005). Diversity-productivity relationships: Initial effects, long-term patterns, and underlying mechanisms. *Proceedings of the National Academy of Sciences of the United States of America*, 102(3), 695–700.

Srivastava, D.S. and Vellend, M. (2005). Biodiversity-ecosystem function research: Is it relevant to conservation? *Annual Review of Ecology, Evolution and Systematics*, 36, 267–294.

Stachowicz, J.J., Best, R.J., Bracken, M.E.S., Graham, M.H. (2008). Complementarity in marine biodiversity manipulations: Reconciling divergent evidence from field and mesocosm experiments. *Proceedings of the National Academy of Sciences of the United States of America*, 105(48), 18842–18847.

Thompson, P.L., Guzman, L.M., De Meester, L. *et al.* (2020). A process-based metacommunity framework linking local and regional scale community ecology. *Ecology Letters*, 23(9), 1314–1329.

Tilman, D. and Downing, J.A. (1994). Diversity and stability in grasslands. *Nature*, 367, 363–366.

Tilman, D., Wedin, D., Knops, J. (1996). Productivity and sustainability influenced by biodiversity in grassland ecosystems. *Nature*, 379, 718–720.

Tilman, D., Knops, J., Wedin, D., Reich, P.B., Ritchie, M., Siemann, E. (1997). The influence of functional diversity and composition on ecosystem processes. *Science*, 277, 1300–1302.

Vermeij, G.J. and Grosberg, R.K. (2018). Rarity and persistence. *Ecology Letters*, 21, 3–8.

Wacker, L., Baudois, O., Eichenberger-Glinz, S., Schmid, B. (2008). Environmental heterogeneity increases complementarity in experimental grassland communities. *Basic and Applied Ecology*, 9(5), 467–474.

Weis, J.J., Madrigal, D.S., Cardinale, B.J. (2008). Effects of algal diversity on the production of biomass in homogeneous and heterogeneous nutrient environments: A microcosm experiment. *PLoS ONE*, 3(7), e2825.

Yachi, S. and Loreau, M. (1999). Biodiversity and ecosystem productivity in a fluctuating environment: The insurance hypothesis. *Proceedings of the National Academy of Sciences*, 96, 1463–1468.

Zavaleta, E.S. and Hulvey, K.B. (2004). Realistic species losses disproportionately reduce grassland resistance to biological invaders. *Science*, 306(5699), 1175–1177.

6

Biodiversity and Ecosystem Functioning in Observational Analyses

Laura E. DEE[1], Kaitlin KIMMEL[2], and Meghan HAYDEN[1]

[1] *University of Colorado, Boulder, USA*

[2] *Johns Hopkins University, Baltimore, USA*

6.1. Introduction

Understanding the consequences of accelerating biodiversity change for ecosystem functioning and services is now a central aim for both ecological science and practice (IPBES 2019). Historically, biodiversity research in natural ecosystems, using observational studies, focused on understanding how so many species coexist (e.g. Hutchinson 1959). Through this lens, productivity was viewed primarily as a driver of biodiversity (Connell and Orias 1964; Waide *et al.* 1999). However, recognition of the importance of the reverse relationship, in which biodiversity affects ecosystem functioning, has grown (Naeem *et al.* 1994; Tilman *et al.* 1997; Loreau *et al.* 2001; Hooper *et al.* 2005; Balvanera *et al.* 2006), particularly given concerns over the consequences of global biodiversity change for ecosystems and people (Díaz *et al.* 2006, 2015; Isbell *et al.* 2017). The predominant approach for studying relationships between biodiversity and ecosystem function (BEF) has been experiments manipulating biodiversity (Tilman *et al.* 2014) (reviewed in Chapter 5). More recently, interest in returning research to real-world ecosystems has surged, in light of ongoing, global change in biodiversity, in order to understand how changes

The Ecological and Societal Consequences of Biodiversity Loss,
coordinated by Michel LOREAU, Andy HECTOR, and Forest ISBELL.

in biodiversity impact ecosystem functioning in nature (Duffy *et al.* 2017; Isbell *et al.* 2017; Oehri *et al.* 2017; van der Plas 2019) and at landscape scales (Gonzalez *et al.* 2020). Observational studies, or studies that assess relationships in natural systems without manipulation of treatment groups (i.e. instead leveraging observations of natural variation in biodiversity), provide an opportunity to understand relationships between biodiversity and ecosystem function (BEF) in real-world ecosystems.

Without experimental manipulations, disentangling the relationships among multiple dimensions of biodiversity, environmental conditions, management context, and ecosystem function poses challenges. Quantifying whether and how much a change in biodiversity causes a change in an ecosystem function or service requires causal inference – or isolating and estimating the average effect of a change in X (e.g. biodiversity) on a change in an outcome, Y (e.g. functioning). The goal of causal inference differs from the aim of describing which factors drive variation in productivity, or of purely predicting where productivity is the highest on the landscape – without causal attribution (Ferraro *et al.* 2019; Hernán *et al.* 2018). While each is an important research objective, here we focus on the aim of causal inference for BEF from non-experimental observations (see Chapter 5 for experiments) – of interest for tests of BEF theory (Chapter 4) and conservation practice (Chapter 15).

In this chapter, we review progress, evidence, and challenges for understanding the consequences of biodiversity change for ecosystem functioning in observational data. Our review asks: what have we learned about BEF relationships in nature from observational studies? How do we learn things from observational studies and what are the unique opportunities? This chapter reviews both the benefits and challenges of observational data, as different analysis designs and considerations are needed for causal inference when controlled manipulative experiments are not possible. Then, we summarize the evidence to date for relationships between multiple dimensions of biodiversity and ecosystem functioning across ecosystem types and highlight the importance of study design in inferring these causal relationships from observational data. We conclude by highlighting research gaps and methodological opportunities to advance our understanding of biodiversity-functioning relationships in natural systems, across multiple scales.

6.2. A historical perspective: returning to observational data

Some of the earliest BEF studies were conducted by observing wild and unmanipulated ecosystems (McNaughton 1977, 1985; Tilman and Downing 1994) and found positive correlations between biodiversity and ecosystem productivity and stability. However, the results of these studies prompted debates over whether the correlations reflected causal relationships in which changes in diversity caused changes in ecosystem function, changes in ecosystem function caused changes in diversity, or in which changes in some other factor caused changes in function (reviewed in Huston *et al.* 2000; Naeem 2000; Wardle *et al.* 2000). These debates sparked research to further test causal links between biodiversity and ecosystem productivity and stability (see Chapters 4 and 7). To address critiques that these early studies did not isolate the effects of biodiversity from those of other factors long-known to affect ecosystem functioning (e.g. nutrient availability or species composition (Aarssen 1997; Huston 1997; Wardle 1999)), experiments manipulating biodiversity (Hooper and Vitousek 1997; Tilman 1997; Hector *et al.* 1999; O'Connor *et al.* 2017) and theory (Tilman *et al.* 1997; Loreau 1998, 2010) were developed to directly test the hypothesis that biodiversity change causes changes in ecosystem functioning or stability (see Chapters 5 and 8). The following 25 years of BEF research primarily focused on experimental tests of theory and, more recently, on experiments designed to simulate realistic species losses (see Chapter 5).

Now, an increasing number of studies are using observations of natural and human-dominated ecosystems to study BEF relationships (Duffy *et al.* 2017; Oehri *et al.* 2017; van der Plas 2019) because of the inherent logistical limits of factorial experiments (e.g. in scale and complexity) (Oehri *et al.* 2017) and a desire to understand BEF relationships at scales relevant to management and conservation beyond what has traditionally been included in experiments (Isbell *et al.* 2017). We review the benefit, challenges, and knowledge gained from observational studies of BEF relationships in the following sections.

6.3. Benefits of observational data

Observational studies are an important complement to theory and experiments (Figure 6.1). Observational studies can help us understand the consequences of biodiversity change for ecosystem functioning at larger scales where experimental manipulations are not feasible and under natural processes and ecosystem dynamics (e.g. as reviewed in Brose and Hillebrand (2016); Oehri *et al.* (2017)). Aiding in theory development, observational analyses may reveal important processes that are not incorporated or studied in current theory or patterns that merit further theoretical or experimental investigation of mechanism (e.g. Dee *et al.* 2016). They are also important complements to experiments, given the trade-offs in the benefits of experimental versus of observational approaches, which we briefly review.

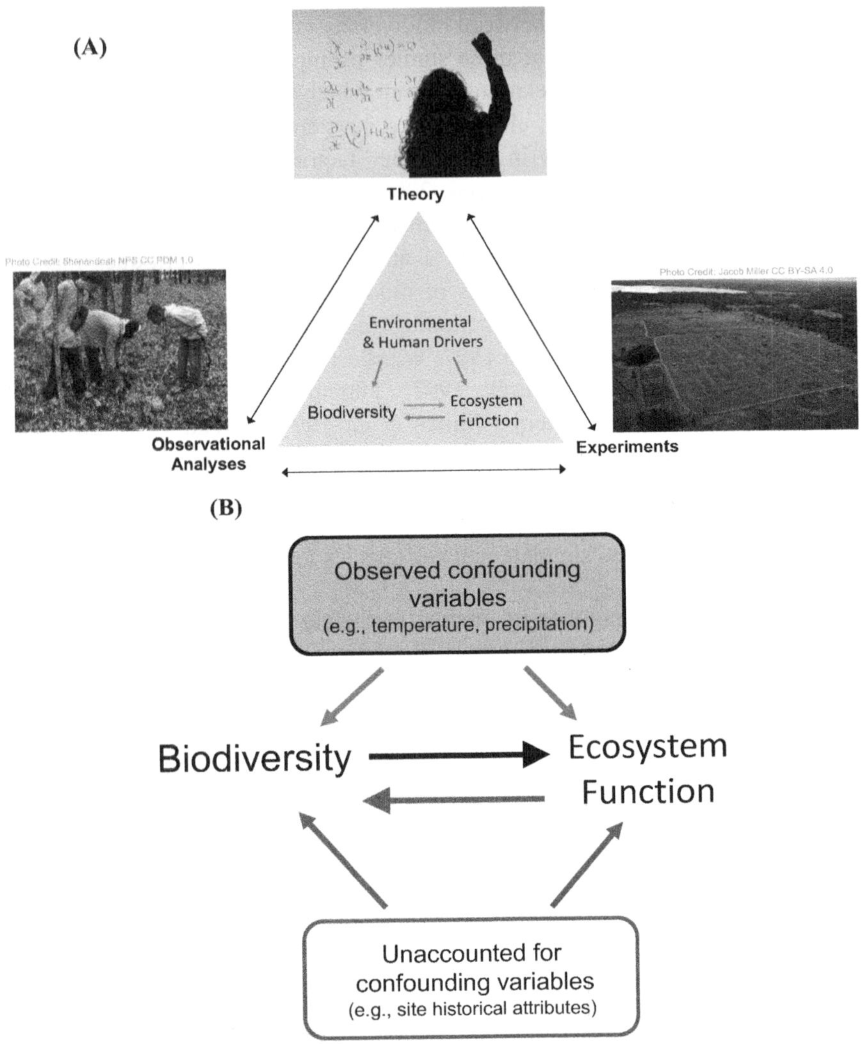

Figure 6.1. *The role and challenge of observational studies in biodiversity-function research. (A) The interplay between experiments, observations, and theory: why we need all three for biodiversity-function research. (B) The relationships between biodiversity and ecosystem function are challenging to unravel in complex systems*

NOTES ON FIGURE 6.1.– *(A) Observational analyses will need to be part of this puzzle of understanding how changes in biodiversity alter ecosystem function in natural and human-dominated ecosystems. The interplay between observations, theory, and experiments is needed to deepen our understanding of the processes and mechanisms that occur across a wide range of environmental and human conditions. Each approach has its strengths and weaknesses. To date, for biodiversity-function research, experimental tests of theory have been the predominant approach. Observations of nature can inform theory and experimental treatments (e.g. experiments that mimic biodiversity declines found in nature or other processes) and account for a broader set of conditions and changes in biodiversity than can be feasibly manipulated. Thus, observations can reveal processes not currently captured in theory and experiments, which necessarily make simplifying assumptions by which observational analyses are not constrained. For example, observations can prompt new theory or experiments that can better isolate processes and mechanisms identified in observational analyses. Observations are needed to help push the field forward to unravel causal relationships between biodiversity, function, and the biotic and abiotic environment across scales. (B) shows a simplified causal diagram, or a visualization of qualitative causal assumptions used for making causal claims from observable data. The gray arrows indicate confounding pathways that are often controlled for in observational studies. In our review, most studies aiming to quantify the effect of biodiversity on ecosystem functioning accounted for some confounding variables – namely abiotic variables like temperature, precipitation, nutrient supply, plot size, elevation, and stand age. In contrast, few studies attempted to account for the reverse relationship or other types of confounding variables, such as disease dynamics or communities of soil microbes, pollinators, and herbivores (as shown as red arrows). Several confounding variables are hard, expensive, or extremely difficult to measure or characterize (e.g. historical attributes of the site or evolutionary history). Thus, to interpret the effect of biodiversity on functioning as a causal relationship, such an analysis makes an assumption that none of the red lines matter and only the confounding variables in gray affect both biodiversity and productivity, among other assumptions.*

Observational studies can help generalize inferences across a broader set of conditions, scales, and levels of complexity than it is feasible to replicate in controlled experiments or to represent analytically in a model. Factorial experiments are costly, labor-intensive, and can pose logistical challenges (e.g. sourcing seeds of rare species to plant, manipulating combinations of long-lived tree species or mobile organisms). Thus, experiments cannot feasibly manipulate all combinations of species and all dimensions of diversity (the "full suite of complexity") across the wide range of conditions and ecosystems. Observational data are not subject to the same logistical constraints. Observations can be taken over larger spatial and/or temporal extents

(Oehri *et al.* 2017) and capture the combinations of species that co-occur in nature under current and real conditions. For example, Oehri *et al.* (2017) found that primary productivity and stability consistently increased with biodiversity, even over heterogeneous landscapes encompassing large altitudinal and environmental gradients. These findings demonstrated the potential of observational studies to reveal landscape-level relationships between biodiversity and ecosystem function at scales and levels of heterogeneity realistic to nature. Thus, observational analyses offer researchers a way to study biodiversity effects on ecosystem function across a wide set of conditions, scales, and complexity characteristics of natural ecosystems. When observations are made over a broad set of conditions and points in space and time, observational studies can help generalize inferences – known as high "*external validity*".

Observational studies have the potential to offer several new insights for BEF by capturing processes occurring in nature. First, in real-world ecosystems, BEF relationships could be driven by different mechanisms than the mechanisms commonly examined in most experiments (Srivastava and Vellend 2005; Wardle *et al.* 2011; Brose and Hillebrand 2016; Brose *et al.* 2016; Isbell *et al.* 2018; Gonzalez *et al.* 2020; Jochum *et al.* 2020). For example, experimental systems commonly link biodiversity to function through niche complementarity, but observational studies have linked biodiversity to function through soil fertility (Grace *et al.* 2016; Chen *et al.* 2021) or through increased growing season length over the past few decades, where highly diverse sites were more able to fill a growing environmental niche space (Oehri *et al.* 2017). Further, observational studies provide an opportunity to examine the different processes that may contribute to BEF relationships at larger scales (Gonzalez *et al.* 2020), when repeated observations are taken over larger spatial extents. For example, at smaller scales (like a "plot scale"), species interactions may increase functioning (van Ruijven and Berendse 2005; Kirwan *et al.* 2009; Wright *et al.* 2014) but other drivers like environmental heterogeneity may become increasingly important at larger scales (Isbell *et al.* 2018; Gonzalez *et al.* 2020; Thompson *et al.* 2021). Further, observational studies are also important for understanding BEF relationships in nature, because BEF relationships have been shown to differ from the positive relationships commonly found in highly controlled experiments when experiments begin to mimic processes more like real ecosystems. For example, experiments that allowed for re-colonization of species into experimental plots have found negative or no effects of species richness on productivity (e.g. Roscher *et al.* 2009; Veen *et al.* 2018). Likewise, experiments that reduced natural dispersal-limitation via seed addition found that while species richness significantly increased, productivity did not (Ladouceur *et al.* 2020).

Observational data capture the patterns in biodiversity loss occurring in the wild and in response to a variety of co-occurring global change scenarios. Long-term and spatially replicated data can reveal which species are more likely to be lost or gained from ecosystems (Tilman and Haddi 1992; Sax and Gaines 2003; Bracken *et al.* 2008; Biederman *et al.* 2017; Newbold *et al.* 2020). They also provide an opportunity to estimate the consequences of these natural colonization and extinction dynamics for functioning. As a result, observational studies can help overcome some of the critiques of BEF experiments reviewed in Chapter 5, namely the lack of realism (but see Zavaleta 2004; Schläpfer *et al.* 2005; Bracken *et al.* 2008). In particular, observations can help overcome the fact that most biodiversity experiments can only manipulate 1) a subset of species or elements of diversity (e.g. functional, phylogenetic, and genetic diversity (Winter *et al.* 2009; Mouillot *et al.* 2013; González-Orozco *et al.* 2016); and 2) a subset of global change drivers simultaneously due to logistical constraints. These data therefore capture the various ways these dimensions of diversity are changing under natural processes and in response to multiple global change drivers operating simultaneously.

6.4. The challenge of causal inference in observational studies

Despite the benefits of observational studies, inferring causal relationships from observations raises other well-known challenges – namely that correlations do not imply causation without strong assumptions (Rubin 1974, 2005a; Blundell and Dias 2000; Morgan and Winship 2015). For an accessible popular science book on this topic, see *The Book of Why: The New Science of Cause and Effect* by Judea Pearl and Dana Mackenzie. Experiments have the advantage of directly manipulating the cause (biodiversity here) to make inferring a causal relationship easier – at the cost of losing the advantages of observational studies described above. In contrast, causal inference in observational studies is difficult because of the potential for many confounding variables (reviewed in Blundell and Dias 2000; Ferraro and Hanauer 2014; Larsen *et al.* 2019). Confounding variables are those that influence *both* biodiversity and the ecosystem function and can include factors like soil fertility, disturbances, climate, herbivory, and land-use history (Figure 6.1B). When not accounted for, confounding variables can obscure a true causal relationship. Failing to account for confounding variables can lead to statistical bias[1] because the effect of the confounding variable is misattributed to the effect of biodiversity on function. Further, estimates of causal relationships in observational studies can have statistical bias when failing to account for the reverse causal relationship – the effect of a given function (e.g. productivity) on biodiversity. Disentangling the direction of this

1 An issue with a statistical model ("estimator") or result wherein the expected value of the result differs from the true underlying parameter being estimated (Wooldridge 2002). The direction of bias is nearly always unknown. Statistical bias differs from the concepts of precision and sampling variability and from type I and type II errors.

causal relationship from data in observational studies is challenging when both causal relationships can occur simultaneously. In contrast, cause (biodiversity change) can precede effect (function) in experiments.

Together, these challenges, along with the common potential for context dependence in ecological systems, likely contribute to debates about biodiversity–productivity relationships in natural systems (Duffy *et al.* 2017) and provide one potential explanation for mixed results (Grace *et al.* 2007, 2016; Adler *et al.* 2011; Lewandowska *et al.* 2016; van der Plas 2019). Thus, study design and the extent to which a design addresses these challenges are extremely important for interpreting results from observational studies (Figure 6.1B).

6.5. Observational studies: results and evidence to date

In the following section, we review evidence for biodiversity–ecosystem functioning relationships based on the search criteria and papers systematically reviewed in the study by van der Plas (2019), which investigated 258 observational studies (covering 1,156 biodiversity–ecosystem functioning relationships) published prior to October 2018. These studies reveal that the strength of evidence and direction of the relationship between biodiversity and ecosystem function vary across ecosystems and across functions measured (Figure 6.2).

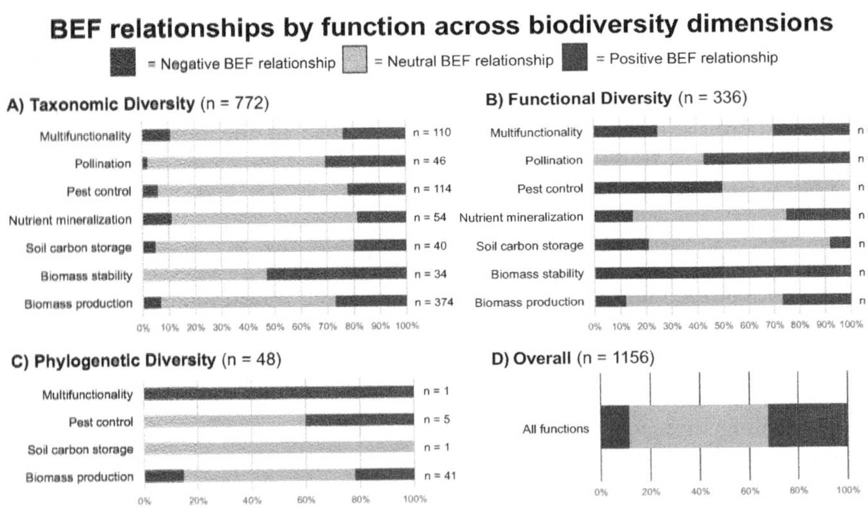

Figure 6.2. *Summarizing what is known from observational studies*

NOTES ON FIGURE 6.2.– *We compiled evidence for biodiversity–ecosystem functioning (BEF) relationships based on the search criteria and papers systematically reviewed in the study by van der Plas (2019), which investigated 258 observational studies (covering 1,156 BEF relationships) published prior to October 2018. Of the 1,156 documented BEF relationships, 67% (n=772) consider taxonomic diversity (A), 29% (n = 336) consider functional diversity (B), and 4% (n=48) consider phylogenetic diversity (C). For A, B, and C, the number of studies assessing each ecosystem function (i.e., multifunctionality, pollination, biomass production, nutrient mineralization, soil carbon storage, biomass stability, and biomass production) is recorded and the proportion (%) of these studies finding negative (red), neutral (purple), and positive (blue) relationships between the biodiversity dimension (taxonomic, functional, and phylogenetic diversity, respectively) and the given function is shown. Across all functions and biodiversity dimensions (D), most studies demonstrated neutral relationships, followed by positive and negative relationships.*

6.5.1. *Across dimensions of biodiversity*

Most observational studies to date that assessed BEF relationships have considered taxonomic diversity as the sole metric of diversity (67% of studies prior to 2018). Studies analyzing functional diversity accounted for 29% of studies, whereas those assessing phylogenetic diversity accounted for only 4% of studies (Figure 6.2D). Studies that included functional diversity found functional composition (particularly trait metrics) to be the strongest "drivers" of ecosystem functioning (van der Plas 2019). However, many studies still do not include trait metrics in their analyses: only 35% of studies prior to 2018 included community weighted mean (CWM) trait values.

Overall, most studies found no relationship between biodiversity and ecosystem function (56%), followed by positive relationships (32%) and negative relationships (11%) (Figure 6.2D). Here, we refer to positive relationships as those conferring higher values or rates of ecosystem function (i.e. increases in productivity or pollination) with increases in biodiversity. It is worth noting that we classify relationships where pest/pathogen productivity decreases as positive relationships with "pest control" rather than negative relationships with pest productivity. Evidence of negative, neutral, and positive relationships was relatively evenly dispersed across biodiversity metrics. Indeed, studies assessing taxonomic diversity found no relationship 67% of the time, while those assessing functional diversity found no relationship 60% of the time and those assessing phylogenetic diversity found no relationship 63% of the time. Likewise, positive relationships were found 24% of the time in taxonomic diversity studies, 26% of the time in functional diversity studies, and 21% of the time in phylogenetic studies (Figure 6.2).

6.5.2. *Across ecosystem functions*

The documented relationships between different dimensions of biodiversity and function depend on which function is measured (Figure 6.2). Most studies focused on a single function (i.e. biomass production or pest control). About 60% of studies assessed metrics related solely to either biomass production or stability. The remainder of the studies reviewed by van der Plas (2019) assessed pest control (11%), soil carbon storage (7%), nutrient mineralization (6%), or pollination (5%).

Studies focusing on biomass production and stability, pollination, and nutrient mineralization demonstrated more positive relationships with biodiversity than negative ones. In contrast, studies focusing on soil carbon storage and pest control found an approximately equal proportion of positive and negative relationships.

Only 11% of studies assessed the impacts of diversity on multiple functions simultaneously (i.e. multifunctionality) (Hector and Bagchi 2007; Byrnes *et al.* 2014; Manning *et al.* 2018). Evidence for relationships between biodiversity and multifunctionality demonstrate similar trends as single-function studies where neutral relationships dominate. Studies assessing functional diversity found more negative relationships (25%) between diversity and multifunctionality than those assessing taxonomic diversity (11%). However, compared to the 110 studies from van der Plas (2019) that examine the relationship between taxonomic diversity and multifunctionality, only 20 studies looked at the relationship between functional diversity and multifunctionality. Similarly, only one study had been conducted on the relationship between multifunctionality and phylogenetic diversity. Thus, it is difficult to discern any true differences in multifunctionality–biodiversity relationships across other dimensions of biodiversity beyond taxonomic diversity.

6.5.3. *Across ecosystem types*

Temperate ecosystems were the most studied in the context of BEF relationships, namely temperate (and boreal) forests and woodlands (24%) and temperate grasslands (21%). Tropical forests made up another 17% of studies, with drylands and deserts comprising 11% of studies prior to 2018. Among the least studied ecosystems were open oceans (0.25%) and deep seas (0.35%), and studies that compared BEF relationships across multiple ecosystem types are also scarce (2%).

Urban and peri-urban ecosystems have also received relatively little attention in the realm of BEF research (1% of studies). BEF studies are also biased by continent; 76% of studies took place in Europe, Asia (particularly in China), or North America, with 4% comparing across continents.

Due to the paucity of studies considering marine and urban ecosystems, it is difficult to make conclusions regarding the relationships between biodiversity and ecosystem functioning found there. Across more studied ecosystem types, such as temperate and tropical forests and grasslands, BEF relationships were neutral in 70% of temperate grassland studies and 63% of temperate forest studies. More positive relationships were demonstrated in forests (27%) than grasslands (17%). In tropical forest systems, neutral relationships dominated (67%) and positive and negative relationships were found in equal proportions (16% each). A possible explanation for the dominance of neutral relationships in tropical forest systems is their high levels of baseline diversity; theory predicts that productivity, for instance, levels off at higher levels of species richness (Waide *et al.* 1999). When broken down further by ecosystem type and ecosystem function, the proportions of negative, neutral, and positive relationships varied greatly.

6.5.4. *Summary of current evidence gaps*

Despite the increase in observational studies looking at biodiversity–ecosystem functioning relationships, evidence gaps remain and highlight opportunities for further research. For example, while observational studies have the potential to capture multiple dimensions of biodiversity simultaneously, few do so, and the majority remain focused on species richness (i.e. taxonomic diversity). Similarly, most studies remain focused on biomass production, potentially missing opportunities to learn about other key functions, such as nutrient mineralization, pollination, and soil carbon storage. Most of work on biodiversity and ecosystem functioning reviewed above took place in temperate and tropical forests and grasslands (van der Plas 2019). In contrast, aquatic, urban, savanna, and tundra ecosystems remain understudied, following similar patterns in ecosystem biases as experiments (van der Plas 2019). The absence of a sufficient number of observational studies on certain ecosystem functions, ecosystem types, and biodiversity dimensions – and the diversity of study designs employed (reviewed next) – makes it difficult to generalize BEF relationships in natural systems. Why observational studies find less conclusive evidence than experiments remains to be determined, but could relate to differences in study design and control over confounding factors, as explored in the next section.

6.6. Reviewing study design to date: how are studies analyzing observational data?

Moving out of an experimental setting, where treatments can be randomized, study designs need to account for the myriad of confounding variables and for the potential for productivity to alter diversity. For the reasons outlined in section 6.3, rigorous study designs are critical for realizing the potential benefits of observational analyses, given the challenge of inferring a causal relationship from observations without a direct manipulation. For this reason, we next review study designs used in past observational analyses.

We first assess the extent to which observational studies to date have controlled for 1) abiotic confounding variables, 2) confounding human variables (e.g. management, land-use, grazing, pollution, species' harvest), and 3) the reverse direction of causality where function also affects biodiversity. We also assess how many studies used stratified sampling designs (Gotelli and Ellison 2004), because they can help control for some confounding variables (e.g. abiotic or management conditions) by studying the BEF relationship within a stratum that shares abiotic, biotic, and human variables (e.g. Buchmann *et al.* 2018). Second, observations can be made over a broader set of times, places, and scales, enabling generalizable inferences. Thus, we also review the extent to which these studies to date are generalizing (e.g. *how many studies use multiple years of data, multiple sites, or both*?). Finally, we review the types of biodiversity studied – since multiple dimensions of biodiversity are linked and can be critical for function independently and jointly. Thus, we assessed how many papers studied multiple aspects of biodiversity (e.g. taxonomic and functional diversity and composition).

We review each of these considerations and the extent to which current observational studies address them in their study designs. To do so, we randomly sample 97 (37.5%) of the studies from a comprehensive, systematic review by van der Plas (2019) to look at relationships between biodiversity and some measure of productivity or stability. We assess whether each paper's goal is to quantify a biodiversity–ecosystem function relationship; whether the study was conducted over multiple years, multiple locations, or both; whether abiotic factors, other elements of composition, or human variables were taken into consideration[2]; whether the study used a stratified design; and whether the relationship between productivity and biodiversity was also quantified.

2 Note that there is some bias based on the sample of papers, as studies that did not take any of these factors into consideration were excluded from the original van der Plas review (2019).

From our sample of papers, about 80% aimed to quantify a relationship between some dimension of biodiversity and ecosystem functioning as their objective. Of these papers, about 34% of them used multiple years of data, 52% were across multiple sites, and 20% used both multiple years and were across multiple sites. Most studies accounted for at least one abiotic variable (~81%) or one aspect of species composition (~61%). Abiotic variables included temperature, precipitation, a range of edaphic characteristics, nutrient supply, plot size, elevation, and stand age, for example. Composition was considered by including community weighted means of traits, species identity, dominance of species, or the proportion of certain species groups among others in analyses. Almost half of the studies (~48%) included at least one abiotic variable, together with at least one other aspect of species composition. Indeed, in the full sample of papers, only 20% included both abiotic factors and variables related to functional composition in their analyses (van der Plas 2019). Further, only 26% of the studies controlled for human variables, such as land management (e.g. burning, or logging). Likewise, only 21% of studies used a stratified design that could account for confounders that may have occurred across different sites or types of ecosystems. Finally, almost none (~2.5%) of the studies considered the reverse relationship between biodiversity and the ecosystem function of choice (but see Paquette and Messier (2011); Grace *et al.* (2016); Chen *et al.* (2018)).

Observational studies face many obstacles when attempting to infer causal relationships from correlations. While some of these challenges are currently recognized, we also identify other opportunities to improve study designs to obtain more robust and accurate estimates from observational analyses. For example, in our analyses of previously published papers, it was found that many accounted for various confounding variables, such as soil pH (Chen *et al.* 2018) or logging (van der Sande *et al.* 2017), that impact productivity and biodiversity. However, there are likely other confounding variables that are not included in analyses either because they are not measured, are challenging to measure, or are not even known to the researcher. For example, some variables, like precipitation, temperature, and certain soil characteristics, are commonly measured and included in analyses as control variables. However, other variables are hard to characterize and/or would require intensive field sampling, like herbivore pressure, disease dynamics, or historical aspects of a site. Therefore, these and other confounding variables are not often controlled for (Figure 6.1B). While not yet the norm, some analytical techniques can be used to address missing confounding variables in observational data, without requiring that they are measured and directly controlled for (Dee *et al.* 2016; for more details about these methods, see Wooldridge (2002); Angrist and Pischke (2009); Greenstone and Gayer (2009); Imbens and Wooldridge (2009); Kendall (2015); Athey and Imbens (2017); Butsic *et al.* (2017); and Larsen *et al.* (2019)).

Moreover, few studies explicitly account for the effect of productivity on biodiversity in their analyses, though this reverse relationship is well-documented (Grime 1973; Chase *et al.* 2002; Chase 2010; Grace *et al.* 2014). Not accounting for this reverse causality can lead to biased estimates of the effects of biodiversity on ecosystem function (e.g. confounded by the reverse relationship). In these cases, to interpret the estimated effects causally, one must assume there are no remaining known or unknown confounding variables and no reverse causality, among other assumptions. Moving forward, we urge researchers to be explicit about the assumptions embedded in study designs using observational data, as these are critical for readers to assess the strength and validity of causal claims.

Ecologists can make more credible and generalized inferences about the causal effects of biodiversity on ecosystem functioning from observational data (reviewed in Dee *et al.*, 2016) by using more *longitudinal data* (e.g. repeated observations of the same unit or place over time) that have greater temporal and spatial coverage. Thus, longitudinal data provide a step towards understanding BEF relationships over the spatial and temporal scales most relevant to management and conservation applications. Longitudinal data also provide an opportunity to better address and uncover heterogeneity in the effect of biodiversity on function: namely how and why this relationship differs across sites or years, depending on moderating factors like environmental or human context. Indeed, moderating factors are expected and documented in some cases; for example, Wang *et al.* (2019) show that the effect of biodiversity on productivity depends on whether the site has high or low levels of productivity. Further, theory suggests that the effect of biodiversity should be stronger in more heterogeneous environments (Isbell *et al.* 2018; Gonzalez *et al.* 2020). A test of this theory in a global observational study of marine fisheries found consistent evidence; Dee *et al.* (2016) found that the functional diversity of global fishery catches offsets ~7% of the annual losses in fishery yields resulting from increased temperature variability.

Finally, species and functional composition are inextricably linked to richness: species richness cannot change without a change in composition (the loss or gain of a distinct species) (Smith *et al.* 2009). The effect of one dimension of biodiversity may be dependent on another. For example, once these other components of community composition (e.g. which species are present, their abundance and functional make up as community-weighted means) are included in better analyses, the effect of richness may be weakened or trend towards zero (van der Plas 2019). While many studies account for some other dimensions of composition, they may not be capturing all of the changes that can happen simultaneously within a community (e.g. Smith *et al.* 2009; Kimmel *et al.* 2019). A loss of species richness and functional richness may have a different effect on function than a decline in species richness without a loss in functional richness. To understand mechanisms underlying the effect of biodiversity on function, researchers will need to consider other dimensions of composition when examining the causal effects of biodiversity.

6.6.1. *Moving forward: improving study designs for observational data and analyses*

Important advances in causal inference from observational data, largely from public health, biomedicine, and economics, can help with challenges posed by observational data – but, to date, they are not applied widely in ecology (but see e.g. Dee *et al.* (2016); MacDonald *et al.* (2019)). As controlled experiments are not always feasible or ethical, these other fields have a long history of observational statistics, with well-established frameworks for causal inference from observations (Rubin 1974, 2005a; Angrist and Pischke 2009; Pearl 2009, 2011; Morgan and Winship 2015; Athey and Imbens 2017). This revolution in causal inference was, in part, born out of experimental design thinking (Splawa-Neyman 1990; Fisher 1935) and generalized to observational data (Rubin 1974). These advances in causal thinking, often called the "Neyman–Rubin Causal Model", are encompassed in the Potential Outcomes framework (or quantitative counterfactual theory[3]), which we briefly review next.

The core idea of the Potential Outcomes (PO) framework is to quantify the difference between what is observed versus what would have happened under a counterfactual scenario, holding all else equal. This difference is the causal effect. However, the fundamental challenge in quantifying that difference, and the causal effect, is that we cannot observe a unit (whether a plot, individual, species, or site) in both actual and counterfactual states (Holland 1986). This is known as the fundamental challenge in causal inference (Holland 1986). Said another way, in a common example, we cannot observe both what would happen if a person received a medical treatment *and* the health outcome for that person if they did not. In our context, we cannot observe the amount of a function of the *same* plot if we both lost and retained a given species. Thus, each plot has several *potential outcomes*, or what is observed for a specified level of biodiversity. However, only one of the potential outcomes is observed, and the unobserved potential outcomes are the *counterfactuals*. For example, imagine that we observe that a plot currently has three species and 10 g/m^2 of biomass. One of the counterfactuals would be the amount of biomass if that same plot had four species present (which we cannot observe). More generally, counterfactuals can be a comparison in space or time; for example, the

3 More subtle distinctions between the Potential Outcomes framework and quantitative counterfactual theory are outside of the scope of this introduction to core concepts.

counterfactual may function in a plot compared to itself in different years when at different levels of diversity (e.g. as in Dee *et al.* 2016), which we can observe. In experiments, assuming that treatments are randomized, plots with different levels of planted richness serve as counterfactuals to the other plots. Randomization ensures that the driver of diversity (the experimenter) is not systematically related to the outcome (the function being measured). However, observational studies do not have this luxury, because researchers do not know what drives differences in biodiversity between sampling units and whether that driver is also correlated with productivity (thus a confounding variable).

The PO framework was extended to observational data, given the many settings in which experimental randomization is not feasible or ethical, by Donald Rubin in 1974. In observational analyses, the PO framework is used to rule out other potential causes besides change in biodiversity. This is done through study and analysis designs that eliminate the other causes and leave only sampling variability and the causal effect of biodiversity as competing explanations for the estimated relationship. In doing so, this approach also clarifies assumptions required for interpreting a correlation as a causal relationship. For more information, see overviews by Ferraro and Hanauer (2014) and Larsen *et al.* (2019).

Another advantage of this framework for thinking about causal effects is that it forces the researcher to be very clear about the effect being estimated, including its spatial and temporal scale (Hernán 2016). For example, do we aim to study the year-to-year changes in functioning at a site when biodiversity changes from X to Y in a manner that mimics natural processes? Or do we aim to study the difference in ecosystem functioning across regions over hundreds of years when species richness changes from X to Z due to naturally occurring processes (e.g. speciation)? We suggest that researchers clarify the intended causal effect being estimated.

This framework offers underutilized solutions to clarify and quantify key causal questions in ecology, including the effect of biodiversity on ecosystem functioning in natural systems. This topic is written about extensively and thus we refer readers to this existing literature (e.g. Rubin 2005b; Angrist and Pischke 2009; Pearl 2011; Ferraro and Hanauer 2014; Morgan and Winship 2015; Hernán *et al.* 2018), as details on implementing this framework in different data contexts are outside of the scope of this chapter. For more detail, Larsen *et al.* (2019) provide an overview of the Potential Outcomes framework for ecologists as applied to study designs for observational data.

6.7. Future directions

While biodiversity–function research has more recently focused on the interplay between theory and experiments, observational studies offer a complementary approach to complete the triangle shown in Figure 6.1. We highlight several research priorities, informed by gaps in current research:

– multiple dimensions of biodiversity can work together to drive changes in ecosystem function (e.g. trait diversity and species identity). Thus, we suggest that observational studies and sampling account for multiple measures of diversity. For example, functional and phylogenetic measures of diversity are likely underrepresented in observational studies because of the logistical constraints of collecting these data. Counting the number of unique species is less time- and resource-consuming than measuring the many traits necessary to characterize functional diversity. However, advances being made in remote sensing (e.g. Pau and Dee 2016; Schweiger *et al.* 2018; Wang *et al.* 2020) and trait databases (e.g. Kattge *et al.* 2020) are making obtaining these data more feasible;

– observational studies are most effectively connected to other kinds of studies when the assumptions being made in observational analyses are transparent (e.g. Figure 6.1B). Clear statements about assumptions required for interpreting estimates as causal can also help resolve or understand discrepancies in estimates – are differences due to reasons related to the biology, sampling (e.g. site or year), context (i.e. moderating variables), or statistical issues (e.g. statistical bias)? This clarity will help build a stronger body of evidence and better enable studies to build on one another to advance knowledge. Causal diagrams[4] (i.e. directed acyclic graphs or DAGs) and the PO framework are ways to make assumptions transparent and explicit. For more details on the PO framework, see Larsen *et al.* (2019). For more details on DAGs, see Pearl (2011), Fieberg and Ditmer (2012), Grace and Irvine (2020), and Schoolmaster *et al.* (2020);

– current theory is largely consistent with experimental studies (Tilman *et al.* 2014) but could be updated to consider processes underlying mixed results from observational studies (van der Plas 2019) and those occurring in natural systems and at larger scales (as reviewed in Gonzalez *et al.* 2020). For example, theory should consider cases (e.g. ecosystem, function, type of species) in which biodiversity may have no impact or a negative impact on function and consider when other dimensions of biodiversity (functional or phylogenetic diversity) may have different relationships to functioning than taxonomic diversity;

4 DAGs should represent and include not only measured variables, but also unmeasured variables that are potentially confounding.

– observational analyses and inferences are inherently limited by what has occurred, whereas experiments can manipulate future scenarios. To make the most of exchanges between experiments, theory, and observations, theory and experiments can focus on anticipated or forecasted future scenarios (e.g. experimental removals of the species that will be lost in the future due to climate change (Urban 2015));

– statistical designs and methods appropriate for observational studies can be applied to global change experiments to understand how changes in function are driven by changes in biodiversity (a post-treatment effect[5] of the global change experiment, as in Tilman and Downing (1994));

– to capitalize on synergies between observations and experiments, long-term observations can inform realistic biodiversity manipulations that mimic the patterns of species losses observed in nature. Such experiments are becoming more prevalent and can be expanded (see Chapter 5), such as by removing species with higher likelihoods of local extinction (e.g. Lyons and Schwartz 2001; Zavaleta 2004; Lyons *et al.* 2005; Bracken *et al.* 2008; Bracken and Low 2012), based on their relative abundances, trophic levels, or traits. Further, to strengthen exchanges between experimental and observational studies, there is a need to design experiments that manipulate the *mechanisms* underlying effects of biodiversity on ecosystem functioning found in observations (e.g. competition for light (Hautier *et al.* 2009)).

6.8. Conclusion

In many places, biodiversity is changing (Dirzo *et al.* 2014; Ceballos *et al.* 2015; Elahi *et al.* 2015; Newbold *et al.* 2016; Young *et al.* 2016) or predicted to change with warming climates (Urban 2015), human overexploitation, or expanding agriculture and development (Maxwell *et al.* 2016; Venter *et al.* 2016). To understand the consequences of these changes for ecosystem functioning and the services they support at policy relevant-scales, observational data will be critical (Figure 6.1). Fortunately, there is a growing library of observations of ecosystems to leverage.

However, as interest in and opportunities for analyzing observational data grow in ecology, there are also new challenges that will require new tools and considerations, as we have reviewed here. Fortunately, other fields with a deep history of

5 A consequence of a treatment or manipulation, but not the treatment that is randomized as part of the experiment.

observational analyses, like public health and economics, offer frameworks and tools to address some of these challenges posed by observational data analyses. Established frameworks for causal inference, as discussed here, present fruitful avenues for future investigation of biodiversity–ecosystem function relationships in natural systems and are an important complement to theory and experiments.

6.9. References

Aarssen, L.W. (1997). High productivity in grassland ecosystems: Effected by species diversity or productive species? *Oikos*, 80, 183.

Adler, P.B., Seabloom, E.W., Borer, E.T. *et al.* (2011). Productivity is a poor predictor of plant species richness. *Science*, 333, 1750–1753.

Angrist, J.D. and Pischke, J. (2009). *Mostly Harmless Econometrics: An Empiricist's Companion*. Princeton University Press, Princeton.

Athey, S. and Imbens, G.W. (2017). The state of applied econometrics: Causality and policy evaluation. *J. Econ. Perspect.*, 31, 3–32.

Balvanera, P., Pfisterer, A.B., Buchmann, N. *et al.* (2006). Quantifying the evidence for biodiversity effects on ecosystem functioning and services. *Ecol. Lett.*, 9, 1146–1156.

Biederman, L., Mortensen, B., Fay, P. *et al.* (2017). Nutrient addition shifts plant community composition towards earlier flowering species in some prairie ecoregions in the U.S. Central Plains. *PLoS ONE*, 12, 1–15.

Blundell, R. and Dias, M.C. (2000). Evaluation methods for non-experimental data. *Fisc. Stud.*, 21, 427–468.

Bracken, M.E.S. and Low, N.H.N. (2012). Realistic losses of rare species disproportionately impact higher trophic levels. *Ecol. Lett.*, 15, 461–467.

Bracken, M.E.S., Friberg, S.E., Gonzalez-Dorantes, C.A., Williams, S.L. (2008). Functional consequences of realistic biodiversity changes in a marine ecosystem. *Proc. Natl. Acad. Sci.*, 105, 924–928.

Brose, U. and Hillebrand, H. (2016). Biodiversity and ecosystem functioning in dynamic landscapes. *Philosophical Transactions of the Royal Society B: Biological Sciences*, 371, 20150267.

Brose, U., Hillebrand, H., Brose, U. (2016). Biodiversity and ecosystem functioning in dynamic landscapes. *Philosophical Transactions of the Royal Society B*, 371, 20150267.

Buchmann, T., Schumacher, J., Ebeling, A. *et al.* (2018). Perspectives in plant ecology, evolution and systematics connecting experimental biodiversity research to real-world grasslands. *Perspect. Plant Ecol. Evol. Syst.*, 33, 78–88.

Butsic, V., Lewis, D.J., Radeloff, V.C., Baumann, M., Kuemmerle, T. (2017). Quasi-experimental methods enable stronger inferences from observational data in ecology. *Basic Appl. Ecol.*, 19, 1–10.

Byrnes, J.E.K., Gamfeldt, L., Isbell, F. *et al.* (2014). Investigating the relationship between biodiversity and ecosystem multifunctionality: Challenges and solutions. *Methods Ecol. Evol*, 5, 11–124.

Ceballos, G., Ehrlich, P.R., Barnosky, A.D., García, A., Pringle, R.M., Palmer, T.M. (2015). Accelerated modern human-induced species losses: Entering the sixth mass extinction. *Sci. Adv.*, 1, e1400253.

Chase, J.M. (2010). Stochastic community assembly causes higher biodiversity in more productive environments. *Science*, 328, 1388–1391.

Chase, J.M., Abrams, P.A., Grover, J.P. *et al.* (2002). The interaction between predation and competition: A review and synthesis. *Ecol. Lett.*, 5, 302–315.

Chen, S., Wang, W., Xu, W. *et al.* (2018). Plant diversity enhances productivity and soil carbon storage. *Proc. Natl. Acad. Sci. U.S.A.*, 115, 4027–4032.

Chen, L., Jiang, L., Jing, X. *et al.* (2021). Above- and belowground biodiversity jointly drive ecosystem stability in natural alpine grasslands on the Tibetan Plateau. *Glob. Ecol. Biogeogr*, 30, 1418–1429.

Connell, J.H. and Orias, E. (1964). The ecological regulation of species diversity. *Am. Nat.*, 98, 399–414.

Dee, L.E., Miller, S.J., Peavey, L.E. *et al.* (2016). Functional diversity of catch mitigates negative effects of temperature variability on fisheries yields. *Proceedings of the Royal Society B*, 283, 20161435.

Díaz, S., Tilman, D., Fargione, J. *et al.* (2006). Biodiversity regulation of ecosystem services. In *Ecosystems and Human Well-being: Current State and Trends*, Hassan, R. and Scholes, R. (eds). Island Press, Washington DC.

Díaz, S., Demissew, S., Carabias, J. *et al.* (2015). The IPBES conceptual framework – Connecting nature and people. *Curr. Opin. Environ. Sustain.*, 14, 1–16.

Dirzo, R., Young, H.S., Galetti, M., Ceballos, G., Isaac, N.J.B., Collen, B. (2014). Defaunation in the Anthropocene. *Science*, 345, 401–406.

Duffy, J.E., Godwin, C.M., Cardinale, B.J. (2017). Biodiversity effects in the wild are common and as strong as key drivers of productivity. *Nature*, 549, 261–264.

Elahi, R., O'Connor, M.I., Byrnes, J.E.K. *et al.* (2015). Recent trends in local-scale marine biodiversity reflect community structure and human impacts. *Curr. Biol.*, 25, 1938–1943.

Ferraro, P.J. and Hanauer, M.M. (2014). Advances in measuring the environmental and social impacts of environmental programs. *Annu. Rev. Environ. Resour.*, 39, 495–517.

Ferraro, P.J., Sanchirico, J.N., Smith, M.D. (2019). Causal inference in coupled human and natural systems. *Proc. Natl. Acad. Sci.*, 116, 5311–5318.

Fieberg, J. and Ditmer, M. (2012). Understanding the causes and consequences of animal movement: A cautionary note on fitting and interpreting regression models with time-dependent covariates. *Methods Ecol. Evol.*, 3, 983–991.

Fisher, R.A. (1935). The logic of inductive inference. *J. R. Stat. Soc.*, 98, 39–82.

Gonzalez, A., Germain, R.M., Srivastava, D.S. *et al.* (2020). Scaling-up biodiversity–ecosystem functioning research. *Ecol. Lett.*, 23, 757–776.

González-Orozco, C.E., Pollock, L.J., Thornhill, A.H. *et al.* (2016). Phylogenetic approaches reveal biodiversity threats under climate change. *Nat. Clim. Chang.*, 6, 1110–1114.

Gotelli, N.J. and Ellison, A.M. (2004). *A Primer of Ecological Statistics*. Sinauer Associates, Inc., Sunderland.

Grace, J.B. and Irvine, K.M. (2020). Scientist's guide to developing explanatory statistical models using causal analysis principles. *Ecology*, 101, 1–14.

Grace, J.B., Anderson, T.M., Smith, M.D. *et al.* (2007). Does species diversity limit productivity in natural grassland communities? *Ecol. Lett.*, 10, 680–689.

Grace, J.B., Adler, P.B., Stanley Harpole, W., Borer, E.T., Seabloom, E.W. (2014). Causal networks clarify productivity-richness interrelations, bivariate plots do not. *Funct. Ecol.*, 28, 787–798.

Grace, J.B., Borer, E.T., Adler, P.B. *et al.* (2016). Integrative modeling reveals mechanisms linking productivity and plant species richness. *Nature*, 529, 390–393.

Greenstone, M. and Gayer, T. (2009). Quasi-experimental and experimental approaches to environmental economics. *J. Environ. Econ. Manag.*, 57, 21–44.

Grime, J.P. (1973). Competitive exclusion in herbaceous vegetation. *Nature*, 242, 344–347.

Hautier, Y., Niklaus, P.A., Hector, A. (2009). Competition for light causes plant biodiversity loss after eutrophication. *Science*, 324, 636–638.

Hector, A. and Bagchi, R. (2007). Biodiversity and ecosystem multifunctionality. *Nature*, 448, 188–190.

Hector, A., Schmid, B., Beierkuhnlein, C. *et al.* (1999). Plant diversity and productivity experiments in European grasslands. *Science*, 286, 1123–1127.

Hernán, M.A. (2016). Does water kill? A call for less casual causal inferences. *Ann. Epidemiol.*, 26, 674–680.

Hernán, M.A., Hsu, J., Healy, B. (2018). Data science is science's second chance to get causal inference right: A classification of data science tasks. *Chance*, 31, 42–49.

Holland, P.W. (1986). Statistics and causal inference. *J. Am. Stat. Assoc.*, 81, 945–960.

Hooper, D.U. and Vitousek, P.M. (1997). The effect of plant diversity and composition on ecosystem processes. *Ecology*, 277, 1302–1305.

Hooper, D.U., Chapin, F.S., Ewel, J.J. *et al.* (2005). Effects of biodiversity on ecosystem functioning: A consensus of current knowledge. *Ecol. Monogr.*, 75, 3–35.

Huston, M.A. (1997). Hidden treatments in ecological experiments: Re-evaluating the ecosystem function of biodiversity. *Oecologia*, 110, 449–460.

Huston, M.A., Aarssen, L.W., Austin, M.P. *et al.* (2000). No consistent effect of plant diversity on productivity. *Science*, 289, 1255.

Hutchinson, G.E. (1959). Homage to Santa Rosalia or why are there so many kinds of animals? *Am. Nat.*, 93, 145–159.

Imbens, G.W. and Wooldridge, J.M. (2009). Recent developments in the econometrics of program evaluation. *J. Econ. Lit.*, 47, 5–86.

IPBES (2019). Global Assessment Report on Biodiversity and Ecosystem Services of the Intergovernmental Science-Policy Platform on Biodiversity and Ecosystem Services. Bonn, Germany.

Isbell, F., Gonzalez, A., Loreau, M. *et al.* (2017). Linking the influence and dependence of people on biodiversity across scales. *Nature*, 546, 65–72.

Isbell, F., Cowles, J., Dee, L.E. *et al.* (2018). Quantifying effects of biodiversity on ecosystem functioning across times and places. *Ecol. Lett.*, 21, 763–778.

Jochum, M., Fischer, M., Isbell, F. *et al.* (2020). The results of biodiversity–ecosystem functioning experiments are realistic. *Nat. Ecol. Evol.*, 4, 1485–1494.

Kattge, J., Bönisch, G., Díaz, S. *et al.* (2020). TRY plant trait database – Enhanced coverage and open access. *Glob. Chang. Biol.*, 26, 119–188.

Kendall, B.E. (2015). A statistical symphony: Instrumental variables reveal causality and control measurement error. In *Ecological Statistics: Contemporary Theory and Application*, Fox, G.A., Negrete-Yankelevich, S., Sosa, V.J. (eds). Oxford University Press, Oxford.

Kimmel, K., Dee, L., Tilman, D. *et al.* (2019). Chronic fertilization and irrigation gradually and increasingly restructure grassland communities. *Ecosphere*, 10, e02625.

Kirwan, L., Connolly, J., Finn, J.A. *et al.* (2009). Diversity-interation modeling: Estimating contributions of species identities and interactions to ecosystem function. *Ecology*, 90, 2032–2038.

Ladouceur, E., Stanley Harpole, W., Blowes, S.A. *et al.* (2020). Reducing dispersal limitation via seed addition increases species richness but not above-ground biomass. *Ecol. Lett.*, 23, 1442–1450.

Larsen, A.E., Meng, K., Kendall, B.E. (2019). Causal analysis in control-impact ecological studies with observational data. *Methods Ecol. Evol.*, 2041–210X.13190.

Lewandowska, A.M., Biermann, A., Borer, E.T. *et al.* (2016). The influence of balanced and imbalanced resource supply on biodiversity–functioning relationship across ecosystems. *Philosophical Transactions of the Royal Society B: Biological Sciences*, 371, 20150283.

Loreau, M. (1998). Biodiversity and ecosystem functioning: A mechanistic model. *Proc. Natl. Acad. Sci. U.S.A.*, 95, 5632–5636.

Loreau, M. (2010). Linking biodiversity and ecosystems: Towards a unifying ecological theory. *Philosophical Transactions of the Royal Society B: Biological Sciences*, 365, 49–60.

Loreau, M., Naeem, S., Inchausti, P. *et al.* (2001). Biodiversity and ecosystem functioning: Current knowledge and future challenges. *Science*, 294, 804–808.

Lyons, K.G. and Schwartz, M.W. (2001). Rare species loss alters ecosystem function – Invasion resistance. *Ecol. Lett.*, 4, 358–365.

Lyons, K.G., Brigham, C.A., Traut, B.H., Schwartz, M.W. (2005). Rare species and ecosystem functioning. *Conserv. Biol.*, 19, 1019–1024.

MacDonald, A.J., Larsen, A.E., Plantinga, A.J. (2019). Missing the people for the trees: Identifying coupled natural–human system feedbacks driving the ecology of Lyme disease. *J. Appl. Ecol.*, 56, 354–364.

Manning, P., van der Plas, F., Soliveres, S. *et al.* (2018). Redefining ecosystem multifunctionality. *Nat. Ecol. Evol.*, 2, 427–436.

Maxwell, S.L., Fuller, R.A., Brooks, T.M., Watson, J.E.M. (2016). Biodiversity: The ravages of guns, nets and bulldozers. *Nature*, 536, 143–145.

McNaughton, S.J. (1977). Diversity and stability of ecological communities: A comment on the role of empiricism in ecology. *Am. Nat.*, 111, 515–525.

McNaughton, S.J. (1985). Ecology of a grazing ecosystem: The Serengeti. *Ecol. Monogr.*, 55, 259–294.

Morgan, S.L. and Winship, C. (2015). *Counterfactuals and Causal Inference*. Cambridge University Press, Cambridge.

Mouillot, D., Graham, N.A.J., Villéger, S., Mason, N.W.H., Bellwood, D.R. (2013). A functional approach reveals community responses to disturbances. *Trends Ecol. Evol.*, 28, 167–177.

Naeem, S. (2000). Biodiversity and ecosystem function: An issue in ecology – Reply to Wardle *et al. Bull. Ecol. Soc. Am.*, 81, 241–246.

Naeem, S., Thompson, L.J., Lawler, S.P., Lawton, J.H., Woodfin, R.M. (1994). Declining biodiversity can alter the performance of ecosystems. *Nature*, 368, 734–737.

Newbold, T., Hudson, L.N., Hill, S.L.L. *et al.* (2016). Global patterns of terrestrial assemblage turnover within and among land uses. *Ecography*, 39, 1151–1163.

Newbold, T., Bentley, L.F., Hill, S.L.L. *et al.* (2020). Global effects of land use on biodiversity differ among functional groups. *Funct. Ecol.*, 34, 684–693.

O'Connor, M.I., Gonzalez, A., Byrnes, J.E.K. *et al.* (2017). A general biodiversity–function relationship is mediated by trophic level. *Oikos*, 126, 18–31.

Oehri, J., Schmid, B., Schaepman-Strub, G., Niklaus, P.A. (2017). Biodiversity promotes primary productivity and growing season lengthening at the landscape scale. *Proc. Natl. Acad. Sci.*, 114, 10160–10165.

Paquette, A. and Messier, C. (2011). The effect of biodiversity on tree productivity: From temperate to boreal forests. *Global Ecology and Biogeography*, 20, 170–180.

Pau, S. and Dee, L.E. (2016). Remote sensing of species dominance and the value for quantifying ecosystem services. *Remote Sens. Ecol. Conserv.*, 2, 141–151.

Pearl, J. (2009). Causal inference in statistics: An overview. *Stat. Surv.*, 3, 96–146.

Pearl, J. (2011). *Causality: Models, Reasoning, and Inference*, 2nd edition. Cambridge University Press, Cambridge.

van der Plas, F. (2019). Biodiversity and ecosystem functioning in naturally assembled communities. *Biol. Rev.*, 94, 1220–1245.

Roscher, C., Temperton, V.M., Buchmann, N., Schulze, E.D. (2009). Community assembly and biomass production in regularly and never weeded experimental grasslands. *Acta Oecologica*, 35, 206–217.

Rubin, D.B. (1974). Estimating causal effects of treatments in randomized and nonrandomized studies. *J. Educ. Psychol.*, 66, 688–701.

Rubin, D.B. (2005a). Causal inference using potential outcomes: Design, modeling, decisions. *J. Am. Stat. Assoc.*, 100, 322–331.

Rubin, D.B. (2005b). Causal inference using potential outcomes. *J. Am. Stat. Assoc.*, 100, 322–331.

van Ruijven, J. and Berendse, F. (2005). Diversity–productivity relationships: Initial effects, long-term patterns, and underlying mechanisms. *Proc. Natl. Acad. Sci. U.S.A.*, 102, 695–700.

van der Sande, M.T., Peña-Claros, M., Ascarrunz, N. *et al.* (2017). Abiotic and biotic drivers of biomass change in a Neotropical forest. *J. Ecol.*, 105, 1223–1234.

Sax, D.F. and Gaines, S.D. (2003). Species diversity: From global decreases to local increases. *Trends Ecol. Evol.*, 18, 561–566.

Schläpfer, F., Pfisterer, A.B., Schmid, B. (2005). Non-random species extinction and plant production: Implications for ecosystem functioning. *J. Appl. Ecol.*, 42, 13–24.

Schoolmaster, D.R., Zirbel, C.R., Cronin, J.P. (2020). A graphical causal model for resolving species identity effects and biodiversity–ecosystem function correlations. *Ecology*, 101, e03070.

Schweiger, A.K., Cavender-Bares, J., Townsend, P.A. *et al.* (2018). Plant spectral diversity integrates functional and phylogenetic components of biodiversity and predicts ecosystem function. *Nat. Ecol. Evol.*, 2, 976–982.

Smith, M.D., Knapp, A.K., Collins, S.L. (2009). A framework for assessing ecosystem dynamics in response to chronic resource alterations induced by global change. *Ecology*, 90, 3279–3289.

Splawa-Neyman, J. (1990). On the application of probability theory to agricultural experiments. Essay on principles. *Stat. Sci.*, 5(4), 465–472.

Srivastava, D.S. and Vellend, M. (2005). Biodiversity–ecosystem function research: Is it relevant to conservation? *Annu. Rev. Ecol. Evol. Syst.*, 36, 267–294.

Thompson, P.L., Kéfi, S., Zelnik, Y.R. *et al.* (2021). Scaling up biodiversity–ecosystem functioning relationships: The role of environmental heterogeneity in space and time. *Proceedings of the Royal Society B: Biological Sciences*, 288, 20202779.

Tilman, D. (1997). The influence of functional diversity and composition on ecosystem processes. *Science*, 277, 1300–1302.

Tilman, D. and Downing, J. (1994). Biodiversity and stability in grasslands. *Nature*, 367, 363–365.

Tilman, D. and Haddi, A. (1992). Drought and biodiversity in grasslands. *Oecologia*, 89, 257–264.

Tilman, D., Tilman, D., Lehman, C.L., Lehman, C.L., Thomson, K.T., Thomson, K.T. (1997). Plant diversity and ecosystem productivity: Theoretical considerations. *Proc. Natl. Acad. Sci. U.S.A.*, 94, 1857–61.

Tilman, D., Isbell, F., Cowles, J.M. (2014). Biodiversity and ecosystem functioning. *Annu. Rev. Ecol. Evol. Syst.*, 45, 471–493.

Urban, M.C. (2015). Accelerating extinction risk from climate change. *Science*, 348, 571–573.

Veen, G.F., van der Putten, W.H., Bezemer, T.M. (2018). Biodiversity–ecosystem functioning relationships in a long-term non-weeded field experiment. *Ecology*, 99, 1836–1846.

Venter, O., Sanderson, E.W., Magrach, A. *et al.* (2016). Sixteen years of change in the global terrestrial human footprint and implications for biodiversity conservation. *Nat. Commun.*, 7, 1–11.

Waide, R.B., Willig, M.R., Steiner, C.F. *et al.* (1999). The relationship between productivity and species richness. *Annu. Rev. Ecol. Syst.*, 30, 257–300.

Wang, Y., Cadotte, M.W., Chen, Y. *et al.* (2019). Global evidence of positive biodiversity effects on spatial ecosystem stability in natural grasslands. *Nat. Commun.*, 10, 1–9.

Wang, Z., Chlus, A., Geygan, R. *et al.* (2020). Foliar functional traits from imaging spectroscopy across biomes in eastern North America. *New Phytol.*, 228, 494–511.

Wardle, D.A. (1999). Is "sampling effect" a problem for experiments investigating biodiversity–ecosystem function relationships? *Oikos*, 87, 403.

Wardle, D.A., Huston, M.A., Grime, J.P. *et al.* (2000). Biodiversity and ecosystem function: An issue in ecology. *Bull. Ecol. Soc. Am.*, 81, 235–239.

Wardle, D.A., Bardgett, R.D., Callaway, R.M., van der Putten, W.H. (2011). Terrestrial ecosystem responses to species gains and losses. *Science*, 332, 1273–1277.

Winter, M., Schweiger, O., Klotz, S. *et al.* (2009). Plant extinctions and introductions lead to phylogenetic and taxonomic homogenization of the European flora. *Proc. Natl. Acad. Sci. U.S.A.*, 106, 21721–21725.

Wooldridge, J.M. (2002). *Econometric Analysis of Cross Section and Panel Data*. MIT Press, Cambridge.

Wright, A., Schnitzer, S.A., Reich, P.B. (2014). Living close to your neighbors: The importance of both competition and facilitation in plant communities. *Ecology*, 95, 2213–2223.

Young, H.S., McCauley, D.J., Galetti, M., Dirzo, R. (2016). Patterns, causes and consequences of Anthropocene defaunation. *Annu. Rev. Ecol. Evol. Syst.*, 47, 333–358.

Zavaleta, E.S. (2004). Realistic species losses disproportionately reduce grassland resistance to biological invaders. *Science*, 306, 1175–1177.

PART 3

How Biodiversity Affects Ecosystem Stability

7

Biodiversity and Ecosystem Stability: New Theoretical Insights

Michel LOREAU
Theoretical and Experimental Ecology Station, CNRS, Moulis, France

7.1. Introduction

The relationship between the diversity and stability of ecological systems has been the subject of a long-standing debate in ecology (McCann 2000). The early view that permeated ecology until the 1960s was that the diversity or complexity of an ecosystem begets its stability. This view was articulated by such famous ecologists as Odum (1953), MacArthur (1955), and Elton (1958) in the 1950s. Odum (1953) and Elton (1958) observed that simple communities are more easily upset than rich ones; that is, they are more subject to destructive population oscillations and invasions. Using a heuristic model, MacArthur (1955) proposed that the more pathways there are for energy to reach a consumer, the less severe the failure of any one pathway is for the consumer. These conclusions were based on either intuitive arguments or loose observations but lacked a strong theoretical and experimental foundation. Probably because they represented the conventional wisdom ("don't put all your eggs in one basket") and the prevailing philosophical view of the "balance of nature", however, they became almost universally accepted.

This "conventional wisdom" was challenged in the early 1970s by theorists such as Levins (1970), Gardner and Ashby (1970), and May (1972), who borrowed the formalism of deterministic autonomous dynamical systems from Newtonian physics and showed that the more complex and diverse a model system is, the less likely it is to be stable. In this work, stability was defined by assessing whether the system returned to equilibrium after a small perturbation – a form of stability known as local stability (see section 7.2). The intuitive explanation for this destabilizing influence of complexity is that the more diversified and the more connected a system, the more numerous and the longer the pathways along which a perturbation can propagate within the system, leading to either collapse or explosion (with no distinction between these two very different outcomes). Although this theoretical work had limitations and some empirical ecologists challenged it (McNaughton 1977), the view that diversity and complexity beget instability, not stability, quickly became the new paradigm in the 1970s and 1980s because of the mathematical rigor of the theory.

The large-scale biodiversity experiments that begun in the 1990s greatly helped to articulate a more nuanced, fact-based view of the relationship between the diversity and stability of ecological systems. In these experiments, plant species richness was manipulated experimentally and the effects of confounding environmental factors were removed through a classic randomization procedure, thereby allowing detection of the direct effect of plant diversity on ecosystem functioning. These experiments showed that plant species diversity increases the stability of ecosystem-level properties, such as total plant biomass production, while at the same time decreasing the stability of population-level properties, such as biomass production of the component species, at least in grasslands (Chapter 8). These new experimental results challenged both the "conventional wisdom" and the new theoretical paradigm since they showed that the same biodiversity metric had contrasting effects on the same stability metric at different levels of organization.

Although at first sight these results seemed to resolve the old diversity–stability debate (Tilman 1996), from a theoretical angle they raised more questions than they provided answers: what then explains the contrasting effects of diversity at the ecosystem and population levels? How can the stabilizing effect of diversity on ecosystem properties be reconciled with existing theory? These questions were the starting point for the recent development of a whole body of new theory on ecosystem stability and diversity–stability relationships. In this chapter, I review these new theoretical advances and show how they provide both a resolution of the historical debate and a new perspective on ecological stability.

7.2. What is stability?

When theoretical ecologists begun to work on stability in a systematic way, they quickly came to realize that "stability" is, in fact, an ambiguous, multifaceted concept that includes a wide range of components or dimensions (Pimm 1984; Loreau *et al.* 2002; Ives and Carpenter 2007). The same system can be viewed as being more or less stable depending on the perturbation it experiences, what is being measured in the system, and what facet of its stability is being considered. For instance, a grassland subjected to a disturbance such as a fire or a drought may have greatly reduced plant biomass while keeping the same plant species composition after the disturbance; in this case, species composition will be deemed more stable than biomass. On the opposite, a grassland experiencing invasion by exotic plant species may keep roughly the same plant biomass while having a very different plant species composition after this disturbance; species composition will then be deemed less stable than biomass. Thus, it is critical to clearly identify the type of perturbation and the variable being observed to make a meaningful statement about the stability of a system.

Once the perturbation and the observed variable have been made clear, the problem is not yet resolved because the stability concept itself needs to be further defined. Ecological theory has traditionally defined stability as either local stability or asymptotic resilience because these properties are based on (relatively) simple linear algebra, and hence are mathematically tractable. *Local stability* is a qualitative property – a system is locally stable when it returns to its local equilibrium following a small disturbance, and is unstable otherwise. In contrast, *resilience* is a quantitative property, which measures the rate or speed at which the system returns to its local equilibrium following the perturbation; *asymptotic resilience* is the value of this return rate in the very long run (in principle, after an infinitely long time, hence the term asymptotic). Both local stability and asymptotic resilience make use of a single quantity, the real part of the dominant eigenvalue of the so-called "community matrix", that is, the linear matrix whose elements describe how the population growth rate of each species (or system component) changes as the abundance of the various species (or system components) changes following a small perturbation from equilibrium. Local stability is governed by the sign of this quantity (a negative sign indicates stability, a positive sign instability), while asymptotic resilience is governed by its absolute value (more negative values indicate a higher resilience).

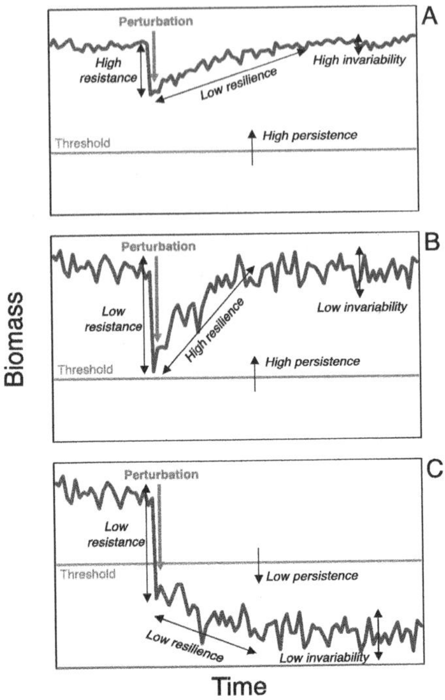

Figure 7.1. *Stability is a multifaceted concept that includes several components, the main ones of which are resistance, resilience, invariability, and persistence. The three panels show a time-series of an ecosystem property of interest – here, the biomass of some population or ecosystem component (in blue). Biomass first fluctuates around an equilibrium value, then experiences an abrupt pulse perturbation and either returns to the same equilibrium (panels A and B) or shifts to an alternative equilibrium after crossing a threshold (in orange; panel C). Panel A depicts a system with high resistance (small decrease in biomass during the perturbation), low resilience (slow return to equilibrium after the perturbation), high invariability (small fluctuations overall, in particular around the equilibrium), and high persistence (the system remains far from the threshold that defines the acceptable range). Panel B illustrates a system with low resistance (large decrease in biomass during the perturbation), high resilience (fast return to equilibrium after the perturbation), low invariability (large fluctuations overall, in particular around the equilibrium), and moderately high persistence (the system remains above the threshold). Panel C shows a system with low resistance (large decrease in biomass during the perturbation), low resilience (the system moves further away from its previous equilibrium, which can be interpreted as negative resilience), low invariability (large fluctuations overall, in particular around the equilibrium), and low persistence (the system crosses the threshold and shifts to an alternative equilibrium). For a color version of this figure, see www.iste.co.uk/loreau/biodiversity.zip*

However, this mathematically convenient solution is deceptive as it hides the true complexity of the stability concept. There are many other ways to define and measure stability (Pimm 1984; Loreau *et al.* 2002; Ives and Carpenter 2007; Donohue *et al.* 2016), including a number of definitions and metrics that are much more relevant empirically as they can readily be quantified from experimental or observational data. For instance, a continuum of non-asymptotic measures of resilience can be readily assessed experimentally by quantifying the rate at which a system returns to its equilibrium after a finite time (Arnoldi *et al.* 2016, 2018) (Figure 7.1). Shifting from asymptotic to finite-time measures of resilience has profound consequences since asymptotic resilience and finite-time resilience can have very different properties and even lead to opposite diversity–stability relationships (Arnoldi *et al.* 2018).

Resistance and persistence are other stability components that can, in principle, be readily measured and that differ strongly from resilience. *Resistance* represents the ability of a system to withstand external perturbations: if two systems are subject to the same perturbation, the more resistant system will be displaced less from its equilibrium than the less resistant one (Harrison 1979) (Figure 7.1). It is often assumed that there is a trade-off between resilience and resistance since a system that returns quickly to equilibrium should have low inertia and thus low resistance (as in Figures 7.1A and 7.1B), although the theoretical (Loreau 1994) and experimental (Chapter 8) evidence for this hypothesis remains limited. What is clear, however, is that the two stability components are very different, and hence should be largely decorrelated. A meta-analysis of the results of a large number of long-term biodiversity experiments showed that the positive effect of plant species diversity on the stability of plant biomass production in grasslands was due to a strong positive relationship between diversity and resistance, while there was no consistent relationship between diversity and resilience, suggesting that resistance, not resilience, is the stability component that mostly governs the relationship between biodiversity and ecosystem stability (Isbell *et al.* 2015).

Persistence is the ability of a system to maintain its properties within an acceptable range in spite of environmental fluctuations (Harrison 1979); *robustness* is another term that is sometimes used to capture the same idea (Donohue *et al.* 2016). The range of a variable that is deemed "acceptable" may be set purely by social norms or expectations (e.g. an acceptable level of food production), but it may also involve an ecological threshold beyond which the ecosystem shifts to an alternative stable state or dynamical regime (Scheffer *et al.* 2001) (Figure 7.1C). Some authors have used the term *ecological resilience* to denote the ability of a system to remain in the same basin of attraction and avoid shifting to an alternative dynamical regime (Holling 1973), but unfortunately this terminology has generated a great deal of confusion in the literature about the resilience concept. It seems much wiser to me to reserve the term "resilience" for its classic meaning, and use persistence or possibly other terms to capture other stability properties. Note that low persistence necessarily implies low values of resilience and resistance (which are defined as local properties in the vicinity of an equilibrium) when the system shifts to an alternative stable state (Figure 7.1C), but the converse is not true, that is, low resilience and/or low resistance do not necessarily imply low persistence (Figures 7.1A and 7.1B).

The stability component that is by far the most commonly used in empirical studies, however, is *invariability*, which describes the ability of a system to maintain a constant level of some property in spite of environmental fluctuations in either space or time. Temporal invariability is simply the inverse of temporal variability, and is generally measured either by the inverse of the temporal coefficient of variation (i.e. the ratio between the mean and the standard deviation) of a property of interest or by its square (i.e. the ratio between the squared mean and the variance) to remove effects of the mean in comparisons across systems (Haegeman *et al.* 2016; Wang *et al.* 2017; Arnoldi *et al.* 2019). Temporal invariability is so widely used in empirical studies that many authors call it simply "temporal stability". Unless it is clearly defined, however, this terminology can be confusing since the other abovementioned stability components also fall under the umbrella of the broad "temporal stability" concept.

These various components of stability capture different aspects of the dynamical response of ecosystems to perturbations (Figure 7.1); therefore, they should logically be related to each other in some way. A few recent studies have begun to tackle the relationships between stability components by subjecting ecosystems to perturbations and examining how the various stability components are correlated, either experimentally (Donohue *et al.* 2013) or theoretically (Domínguez-García *et al.* 2019). These studies showed that stability components are correlated and thus that stability has a lower dimensionality than the number of stability metrics tested. This is a promising conclusion as it supports the view that the stability concept is not as desperately diverse as terminological profusion might suggest (Grimm and Wissel 1997). Careful theory should be able to disentangle the connections between the various stability components, and recent theoretical and mathematical developments have indeed begun to do this (Arnoldi *et al.* 2016; Haegeman *et al.* 2016).

A critical limitation of the literature on ecological stability so far is the divide between empirical and theoretical studies. As noted above, much of ecological theory has focused on either local stability or asymptotic resilience, while observational or experimental studies have mostly used temporal invariability as a stability metric. This divide appears clearly in quantitative analyses of the ecological literature (Donohue *et al.* 2016). Fortunately, new theory is now providing new approaches and predictions on empirically relevant stability components such as non-asymptotic resilience and invariability (Arnoldi *et al.* 2018, 2019). A novel insight that results from this new body of theory is that stability is inherently a multidimensional concept, not only because it contains various components, but also, more fundamentally, because each of these components is governed by different species at different timescales depending on their abundance and the type of perturbation they experience. Generally speaking, abundant species tend to govern short-term rates of return to equilibrium (and thus non-asymptotic resilience), while rare species tend to govern long-term return rates (and thus asymptotic resilience) (Arnoldi *et al.* 2018).

Temporal invariability is the outcome of a complex process that involves the type, amplitude, and direction of perturbations, the response of the system to these perturbations, and the system variable that is observed (Arnoldi *et al.* 2019). Therefore, one should not expect a single universal relationship between diversity and invariability to hold. Despite this complexity, generic relationships between community invariability and species abundances do emerge from community assembly in species-rich systems (Arnoldi *et al.* 2019). Different types of perturbations, however, yield different relationships between community invariability and species abundances, which in turn predict qualitatively different diversity–stability relationships (Figure 7.2).

These new theoretical insights have broad consequences for understanding and interpreting experimental results. In particular, they might provide a simple explanation for the contrasting diversity–stability relationships observed at the population and ecosystem levels in grassland biodiversity experiments (Chapter 8). In these experiments, ecosystem stability has usually been measured by the invariability of total plant biomass. By construction, the latter gives a predominant weight to abundant species, just as environmental-type perturbations do, and this tends to generate positive diversity–invariability relationships (Figure 7.2C). In contrast, population stability has usually been measured by the average invariability of the various component species. This average gives a predominant weight to rare, highly variable species (Haegeman *et al.* 2016), just as immigration-type perturbations do, and this tends to generate negative diversity–invariability relationships (Figure 7.2A). Thus, the new multidimensional theoretical approach to ecological stability provides both a potential resolution of the old diversity–stability debate and a new perspective on stability that helps reveal the full dynamical richness of ecological systems.

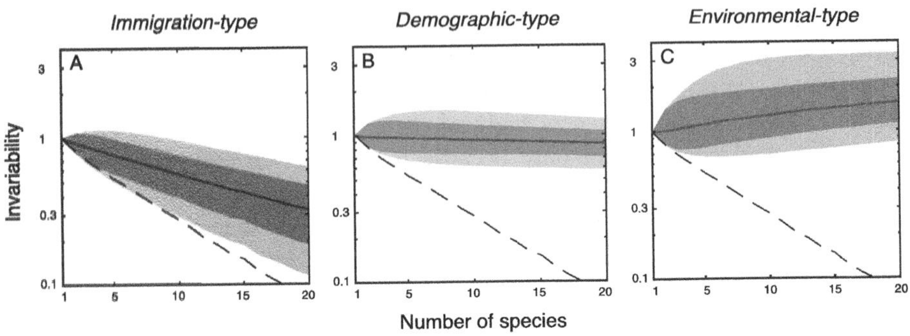

Figure 7.2. *Different types of perturbations yield contrasting diversity–stability relationships in competitive communities. Here, stability is measured by a community-wide measure of invariability that integrates all species, and biodiversity is measured by the number of species. In each panel, the solid color line shows the median invariability of randomly assembled competitive communities experiencing random perturbations (1,000 communities for each level of species richness; 1,000 perturbations for each community); the darkly shaded region shows the 5th and 95th percentiles of the distribution of invariability values; the lightly shaded region shows their minimum and maximum values; and the dashed line shows asymptotic resilience. Perturbations are of three types: immigration (i.e. they are exogenous to the system and hence independent of a species' abundance), demographic (i.e. they affect individuals independently), and environmental (i.e. they affect all the individuals of each species synchronously). After Arnoldi et al. (2019). For a color version of this figure, see www.iste.co.uk/loreau/biodiversity.zip*

7.3. Why does local biodiversity promote ecosystem stability?

While the abovementioned theory on temporal invariability and its properties provides interesting new insights into the diversity–stability relationships that can be expected under different perturbation scenarios, it does not allow a mechanistic understanding of the ecological processes that generate the positive effects of biodiversity on ecosystem stability observed in biodiversity experiments.

Two main hypotheses have been proposed to explain these effects. The first, known as the insurance hypothesis, posits that, in a variable environment, aggregate ecosystem properties (e.g. total biomass or production) will vary less in more diverse communities because declines in the performance or abundance of some species or phenotypes can be offset by increases by others due to their asynchronous responses to fluctuations in environmental conditions (Yachi and Loreau 1999). This mechanism is deeply rooted in biology since differential responses to environmental variations are ultimately based on the universal presence of trade-offs in biological systems, which constrain species to evolve towards a species-specific balance between various biological functions and thus to perform best under a species-specific set of environmental conditions. Biodiversity is viewed as providing the biological equivalent of an economic insurance because an informed decision-maker could choose to maintain a high level of biodiversity to avoid extreme lows in ecosystem functioning, even under conditions where maintaining biodiversity may entail a cost.

A second popular hypothesis, inspired by traditional competition theory, is that competition between species should generate or amplify negative covariations in their abundances because, as one species increases in abundance, it increases its negative effect on other species, thereby pushing their abundance down (Tilman 1999; Klug *et al.* 2000). Although competition often enhances the asynchrony of species abundances in models of interspecific competition, it is not necessarily the factor that generates community stability in these models. Community stability generally arises from implicit or explicit differences in the environmental preferences of the competing species (Tredennick *et al.* 2017).

Recent theory has used stochastic, discrete-time, multispecies versions of the classic Lotka–Volterra competition model (Chapter 4) to reveal and disentangle the various factors that govern local diversity–stability relationships. These models confirm that species' differential responses to environmental fluctuations, as measured by the degree of asynchrony of species' environmental responses, are the main mechanism through which biodiversity can stabilize aggregate ecosystem properties, in agreement with the insurance hypothesis (Ives *et al.* 1999; Loreau and de Mazancourt 2008, 2013; Loreau 2010). Whatever the strength of interspecific competition, when asynchrony of species' environmental responses between any two species is low, their total biomass shows the same fluctuations as does the biomass of individual species on a log scale because the fluctuations of the two species are strongly correlated positively (Figures 7.3A and 7.3C). This means that total biomass varies roughly twice as much as the biomass of each species, and since mean total biomass is also roughly twice the mean biomass of each species, the temporal invariability of total biomass, as measured by the inverse of its coefficient of variation, remains unchanged. However, when the asynchrony of species' environmental responses is high, total biomass shows strongly reduced fluctuations compared with the biomass of individual species because the fluctuations of the two species are correlated negatively and hence tend to compensate for each other (Figures 7.3B and 7.3D). In this case, the temporal invariability of total biomass is much smaller than that of the biomass of individual species because the standard deviation of total biomass decreases while its mean increases, and both these effects contribute to increasing the inverse of the coefficient of variation, which measures invariability.

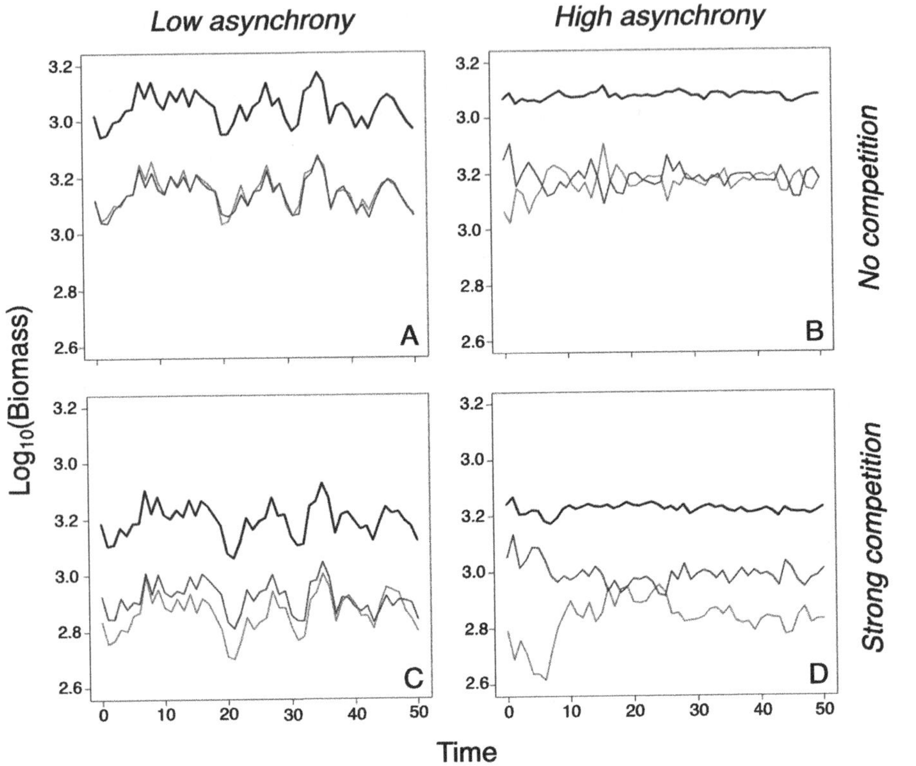

Figure 7.3. *Effects of asynchrony of species' environmental responses and interspecific competition on the temporal variability of the biomass of two species (blue and red) and their total biomass (black) in Loreau and de Mazancourt's (2013) stochastic Lotka–Volterra competition model. Left panels: minimum asynchrony between the environmental responses of the two species (maximum synchrony, $\varphi_e = 1$); right panels: maximum asynchrony between their environmental responses (minimum synchrony, $\varphi_e = 0$). Upper panels: no interspecific competition (competition coefficients $\beta_{12} = \beta_{21} = 0$); lower panels: strong interspecific competition ($\beta_{12} = \beta_{21} = 0.8$). Other parameter values: intrinsic rates of natural increase $r_{m1} = r_{m2} = 0.5$; carrying capacities $K_1 = K_2 = 1500$; demographic variances $\sigma_{d1}^2 = \sigma_{d2}^2 = 1$; environmental variances $\sigma_{e1}^2 = \sigma_{e2}^2 = 0.01$. For a color version of this figure, see www.iste.co.uk/loreau/biodiversity.zip*

By contrast, stochastic Lotka–Volterra competition models do not support the hypothesis that competition increases ecosystem stability. Interspecific competition tends to have two countervailing effects on the variability of total biomass, at least when species already differ in their environmental responses (compare Figures 7.3B and 7.3D). On the one hand, interspecific competition tends to increase the degree of asynchrony between the biomass fluctuations of individual species, which contributes to reducing the variability of total biomass. On the other hand, however, it also increases the amplitude of the biomass fluctuations of individual species, which contributes to enhancing the variability of total biomass. These two countervailing effects cancel each other in symmetric competitive systems, thereby leaving the invariability of total biomass unaffected (Ives *et al.* 1999; Loreau and de Mazancourt 2013). When competition is asymmetric (such that the competitive effect of species A on species B differs widely from the reverse effect of species B on species A), a wide range of outcomes are possible, but the most common effect of increasing the strength of interspecific competition is destabilization of aggregate ecosystem properties (Loreau and de Mazancourt 2013). Instead, a *reduction* in the strength of interspecific competition tends to stabilize ecosystem properties because it generates overyielding, that is, a higher total production and biomass (Chapter 4), which in turn reduces the destabilizing effect of demographic stochasticity on aggregate ecosystem properties (de Mazancourt *et al.* 2013; Loreau and de Mazancourt 2013).

Additional mechanisms can further contribute to ecosystem stability. One such mechanism is differences in the speed at which species respond to perturbations, which can also generate asynchronous population dynamics and thereby promote ecosystem stability (Fowler 2009). This mechanism operates under more restrictive conditions than do differences in species' responses to environmental fluctuations, but, interestingly, it appears to provide the only scenario in which strong interspecific competition can generate compensatory dynamics as envisaged by the competition hypothesis (Loreau and de Mazancourt 2013). Selection effects (Chapter 4) can also affect ecosystem stability. If species that have a higher-than-average invariability tend to dominate multispecies communities, this will generate a positive selection effect of species diversity on ecosystem stability. Conversely, if species that have a lower-than-average invariability tend to dominate multispecies communities, this will generate a negative selection effect on ecosystem stability. Such selection effects are likely to be present, and hence to interact with other factors that affect the relationship between biodiversity and ecosystem stability, under many circumstances. Lastly, observation errors, which arise from the random effects of uncontrolled factors, may significantly inflate observed ecosystem stability as they tend to decorrelate variations in abundance among species (de

Mazancourt *et al.* 2013). Thus, observation errors should be taken into account in interpretating the results of small-scale biodiversity experiments.

This brief overview of the factors known to affect diversity–stability relationships from stochastic Lotka–Volterra competition models shows that asynchrony of species' responses to environmental fluctuations provides a likely general explanation for positive effects of biodiversity on ecosystem stability, but a number of other factors can also come into play and should be considered carefully in the interpretation of experimental or observational data. It is important to realize that the range of potential factors to consider becomes even wider in complex food webs or interaction networks. In particular, one critical additional factor that may affect the stabilizing or destabilizing effect of species diversity when multiple trophic levels are considered is the combined interaction strength of animal consumers, as measured by the total per capita effect of all resources combined on their population growth rate (Ives *et al.* 2000). For species diversity to stabilize ecosystem properties, a trade-off between the niche breadth of animal consumers and their efficiency at exploiting each of their resources is required (Thébault and Loreau 2005; Loreau 2010). Such a trade-off can arise, for instance, when prey diversity forces predators to spend more time on information processing, thereby reducing their consumption efficiency (Kratina *et al.* 2007).

More generally, stochastic competition models have so far assumed that environmental fluctuations affect individuals irrespective of their density or diversity, which is a reasonable assumption as a first approximation. If, however, the effect of environmental fluctuations on per capita population growth rates were to change systematically with diversity because of changes in individual behavior as a result of species interactions, this would obviously alter model predictions. Specifically, if environmental fluctuations were to affect individuals relatively less in species-rich (species-poor) communities, this would act as an additional mechanism contributing to the stabilizing (destabilizing) effect of diversity on aggregate ecosystem properties (Loreau and de Mazancourt 2013). Although there is currently no evidence for such effects of diversity, only careful future experiments and observations will tell us whether existing theory needs to be revisited and expanded in new directions.

7.4. Scaling up diversity–stability relationships

As stability theory develops in new directions that strengthen its connection with empirical data (section 7.2), new areas of theoretical research are emerging. A particularly exciting new research area is exploring how biodiversity, ecosystem stability, and their relationship change with spatial scale (Gonzalez *et al.* 2020).

Theoretical advances in this area have been made possible by the use of temporal invariability as a stability metric. It turns out that variability or its inverse, invariability, can be scaled up in much the same way as species diversity.

Two classic approaches have been used to study species diversity across two or more scales: 1) partitioning gamma (regional) diversity into alpha (local) diversity and beta (between-community) diversity (Chapter 2); and 2) the species–area relationship (SAR), which describes how the number of species increases with study area (Rosenzweig 1999). Recent theoretical work has shown that the same approaches can be used to study ecosystem invariability across scales.

The first approach partitions gamma (regional) variability or invariability into alpha (local) and beta (between-community) components, either multiplicatively or additively, in the same way as gamma diversity is partitioned into its alpha and beta components (Wang and Loreau 2014). This partition predicts that the invariability of aggregate ecosystem properties must necessarily increase as one moves from the local to the regional scale. Interestingly, this partition can be extended to multiple nested scales or hierarchical levels, and doing so offers exciting new insights into the factors that govern ecosystem stability across scales. In particular, this hierarchical approach reveals that beta variability is equivalent to spatial asynchrony between communities, and that the factors that govern alpha invariability and spatial asynchrony are very similar to those that govern population invariability and species asynchrony, respectively, within a local community (Wang and Loreau 2014). Thus, the hierarchical partitioning of invariability provides a powerful integrative framework to understand ecological stability across scales and hierarchical levels.

Since species diversity and ecosystem stability can be studied under the same partitioning framework, the relationship between diversity and stability can also be studied under this framework. Using a dynamical model of competitive metacommunities, theorists have shown that, while alpha diversity increases local ecosystem invariability, beta diversity generally contributes to increasing spatial asynchrony among local ecosystems (Wang and Loreau 2016). Consequently, alpha diversity and beta diversity play complementary roles in stabilizing ecosystem properties at the regional scale: while local diversity provides local insurance for ecosystem functioning by enhancing species asynchrony, beta diversity provides spatial insurance for ecosystem functioning by enhancing spatial asynchrony between ecosystems. Furthermore, the model predicts that the stabilizing effect of biodiversity at the regional scale increases as the correlation between environmental conditions across space increases. This is because when environmental conditions are very heterogeneous across space, the background level of spatial asynchrony is already high and biodiversity cannot enhance it much more; in contrast, when environmental conditions are homogenous across space, the background level of spatial asynchrony is low and biodiversity can increase it dramatically. This theoretical prediction suggests that biodiversity loss may exacerbate the destabilizing effect of the homogenization of environmental conditions across space that is expected to occur under current global environmental changes (Vitousek *et al.* 1997).

The second approach to scaling up ecosystem invariability is the invariability–area relationship (IAR), which describes how ecosystem invariability increases with surface area (Wang *et al.* 2017). The IAR offers a continuous approach to the spatial scaling of ecosystem stability that is complementary to the discrete approach provided by the hierarchical partition. The shape and slope of the IAR are essentially determined by the patterns of spatial synchrony or asynchrony across scales, just as spatial asynchrony (beta variability) is the link that connects local (alpha) and regional (gamma) invariability in the hierarchical partition. In particular, when spatial synchrony decays exponentially with distance, the IAR exhibits three phases, characterized by steeper increases in invariability at both small and large scales (Figure 7.4A). Such a triphasic IAR was observed for primary productivity from plot to continental scales (Figure 7.4B). If spatial synchrony decays as a power law with distance, however, the IAR is nearly linear on a log-log plot (Figure 7.4C). This pattern was observed for the population abundance of North American birds (Figure 7.4D). The IAR provides a new quantitative tool to predict the effects of habitat loss on population and ecosystem stability and to detect possible regime shifts in spatial ecological systems, which are important goals for biodiversity conservation and ecosystem management (Wang *et al.* 2017).

Since species diversity and ecosystem stability show similar relationships with area, the link between the SAR and the IAR can also be studied. A simple theoretical model that simultaneously predicts the SAR and the IAR shows that the link between the two relationships depends strongly on whether the temporal fluctuations of the ecosystem property of interest are more synchronized within than between species (Delsol *et al.* 2018). If fluctuations are synchronized within species but not between species, the IAR is strongly constrained by the SAR. If instead the level of synchrony between individual fluctuations is governed by spatial proximity, the IAR is unrelated to the SAR. These two scenarios of synchrony between species and across space further lead to very different predictions regarding the effects of biodiversity loss and habitat destruction on ecosystem stability (Delsol *et al.* 2018). Thus, a recurrent conclusion from recent theory is that understanding the drivers and patterns of synchrony or asynchrony between species and across space is key to predicting the effects of biodiversity loss and global environmental changes on ecosystem stability.

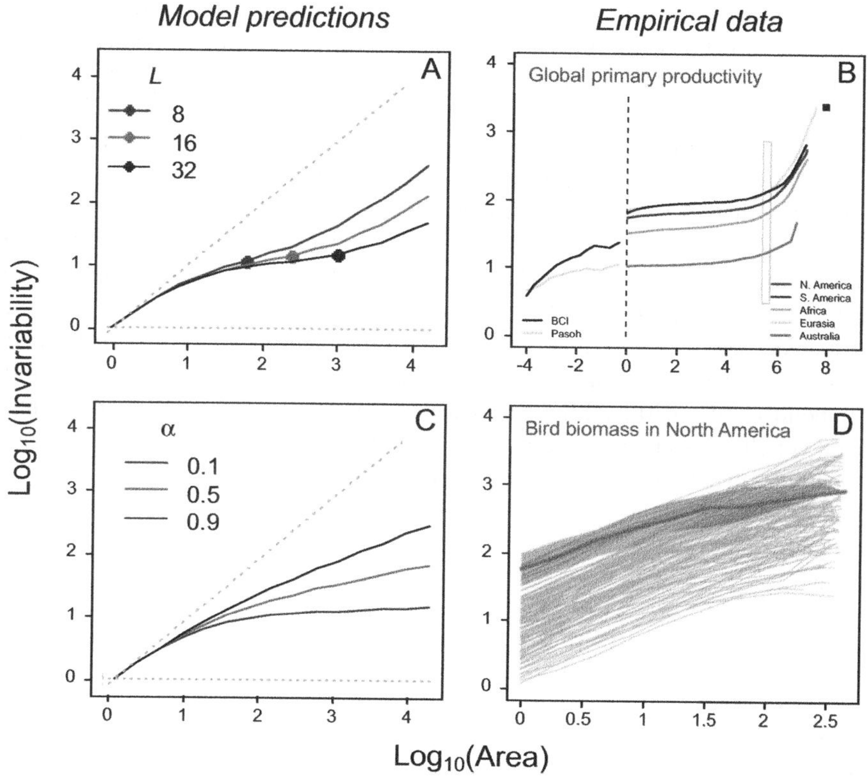

Figure 7.4. *Invariability–area relationships (IARs). Left panels: model predictions. Triphasic IARs are predicted when spatial synchrony decays exponentially with distance (panel A), while near-linear IARs on a log-log plot are predicted when spatial synchrony decays as a power law with distance (panel B). Solid lines show different IARs obtained for different values of either L, the characteristic length of the exponential decay of spatial synchrony with distance, or α, the exponent of the power law. Dotted lines show the two limiting cases where spatial synchrony is either minimum or maximum, irrespective of distance. Right panels: empirical data. Panel B: triphasic IARs observed for primary productivity from plot to continental scales on five continents, where area is measured in km^2. The black filled square on the right shows the invariability of primary productivity for the entire globe. Panel D: near-linear IARs observed for bird biomass in eastern North America, where area is measured by the number of sampling routes in the North American Breeding Bird Survey. The red line shows the median invariability of total bird biomass, and the red shade shows its 25 and 75% quantiles. The gray lines show the IARs of individual species biomass for 121 bird species. After Wang et al. (2017). For a color version of this figure, see www.iste.co.uk/loreau/biodiversity.zip*

Hierarchical partitioning of invariability has recently been extended in two alternative directions to provide consistent measures of asynchrony across hierarchical levels when there is spatial heterogeneity among communities, as is typically the case in empirical data (Wang *et al.* 2019; Hammond *et al.* 2020). The resulting mathematical frameworks have been used to quantify the relative contributions of different sources of asynchrony to large-scale ecosystem stability in several empirical datasets. Local insurance due to alpha diversity was shown to provide stronger stabilizing effects on regional ecosystem functioning than did spatial insurance in a desert grassland ecosystem (Wang *et al.* 2019) and in a kelp forest (Lamy *et al.* 2019). Other studies, however, found that spatial insurance contributed more than did local insurance to the stability of benthic marine fish communities (Thorson *et al.* 2018) and rock-pool invertebrate metacommunities (Hammond *et al.* 2020). Although clearly more work will be necessary to reach general conclusions about the respective roles of different sources of asynchrony in the stability of natural ecosystems, the fact that a complete set of conceptual and mathematical tools is now available to address this issue with empirical data is a remarkable achievement of ecological theory which still seemed out of reach a few years ago.

7.5. Conclusion

Recent studies of diversity–stability relationships have led to the development of a new body of theory that is profoundly changing our views of both ecological stability and its relationships with biodiversity.

This body of theory suggests that stability should be fully embraced as a multidimensional concept, not only because it contains different components that describe different aspects of the response of ecosystems to perturbations, but also because each of these components is governed by different species at different timescales depending on their abundance and the type of perturbation they experience. In particular, temporal invariability is an integrative measure of stability that is affected by the type, amplitude, and direction of perturbations, the response of the system to these perturbations, and the system variable that is observed. Different types of perturbations and different observed variables yield different relationships between community invariability and species abundances, which in turn yield qualitatively different diversity–stability relationships. A fully integrative theory of ecological stability that connects and integrates the various components of stability (resilience, resistance, persistence, invariability) may not yet be available, but this goal does not seem as distant today as it still seemed a few years ago based on recent advances made in that direction. The new body of theory built so far already provides both the elements to resolve the old diversity–stability debate and new perspectives on ecological stability that help reveal the full dynamical richness of ecological systems.

This body of theory has also greatly increased our mechanistic understanding of the ecological processes that generate the positive effects of biodiversity on ecosystem stability observed in small-scale biodiversity experiments. Asynchrony of species' responses to environmental fluctuations appears to provide a likely general explanation for these effects, but a number of other factors can also come into play and should be considered carefully in the interpretation of experimental or observational data.

An exciting extension of this body of theory is the recent development of new approaches to studying the relationship between biodiversity and ecosystem stability across spatial scales using temporal invariability as a stability metric. The hierarchical partitioning of invariability turns out to provide a powerful integrative framework to understand ecological stability across scales and hierarchical levels. Its continuous equivalent, the invariability–area relationship, provides a quantitative tool to predict the effects of biodiversity loss, habitat destruction, and other environmental changes on ecosystem stability – an important goal for biodiversity conservation and ecosystem management. Both approaches highlight the key role played by asynchrony between species and across space in ecosystem stability at large spatial scales. Thus, understanding the drivers and patterns of asynchrony across scales and hierarchical levels appears to be critical. The hierarchical partitioning of invariability precisely provides the necessary conceptual and mathematical tools to detect the main sources of asynchrony in empirical data.

Ecology now has a powerful set of theoretical approaches and predictions that can be connected directly to experimental and observational data across multiple organizational levels and spatial scales. This is a unique strength, which hopefully will open a new area of rigorous quantitative research into ecosystem stability and the consequences of biodiversity loss for ecosystem functioning and ecosystem services at scales relevant to management.

7.6. Acknowledgements

I thank Jean-François Arnoldi, Claire de Mazancourt and Shaopeng Wang for their invaluable assistance in producing Figures 7.2, 7.3 and 7.4, respectively, and Yann Hautier, Forest Isbell, Lin Jiang, Fons van der Plas and Shaopeng Wang for providing helpful comments.

7.7. References

Arnoldi, J.-F., Loreau, M., Haegeman, B. (2016). Resilience, reactivity and variability: A mathematical comparison of ecological stability measures. *Journal of Theoretical Biology*, 389, 47–59.

Arnoldi, J.-F., Bideault, A., Loreau, M., Haegeman, B. (2018). How ecosystems recover from pulse perturbations: A theory of short- to long-term responses. *Journal of Theoretical Biology*, 436, 79–92.

Arnoldi, J.-F., Loreau, M., Haegeman, B. (2019). The inherent multidimensionality of temporal variability: How common and rare species shape stability patterns. *Ecology Letters*, 22, 1557–1567.

Delsol, R., Loreau, M., Haegeman, B. (2018). The relationship between the spatial scaling of biodiversity and ecosystem stability. *Global Ecology and Biogeography*, 27, 439–449.

Domínguez-García, V., Dakos, V., Kéfi, S. (2019). Unveiling dimensions of stability in complex ecological networks. *Proceedings of the National Academy of Sciences of the USA*, 116, 25714–25720.

Donohue, I., Petchey, O.L., Montoya, J.M. *et al.* (2013). On the dimensionality of ecological stability. *Ecology Letters*, 16, 421–429.

Donohue, I., Hillebrand, H., Montoya, J.M. *et al.* (2016). Navigating the complexity of ecological stability. *Ecology Letters*, 19, 1172–1185.

Elton, C.S. (1958). *The Ecology of Invasions by Animals and Plants*. Methuen, London.

Fowler, M.S. (2009). Increasing community size and connectance can increase stability in competitive communities. *Journal of Theoretical Biology*, 258, 179–188.

Gardner, M.R. and Ashby, W.R. (1970). Connectance of large dynamic (cybernetic) systems: Critical values for stability. *Nature*, 228, 784.

Gonzalez, A., Germain, R.M., Srivastava, D.S. *et al.* (2020). Scaling-up biodiversity–ecosystem functioning research. *Ecology Letters*, 23, 757–776.

Grimm, V. and Wissel, C. (1997). Babel, or the ecological stability discussions: An inventory and analysis of terminology and a guide for avoiding confusion. *Oecologia*, 109, 323–334.

Haegeman, B., Arnoldi, J.-F., Wang, S., de Mazancourt, C., Montoya, J.M., Loreau, M. (2016). Resilience, invariability, and ecological stability across levels of organization. *bioRxiv*, 085852.

Hammond, M., Loreau, M., de Mazancourt, C., Kolasa, J. (2020). Disentangling local, metapopulation and cross-community sources of stabilization and asynchrony in metacommunities. *Ecosphere*, 11, e03078.

Harrison, G.W. (1979). Stability under environmental stress: Resistance, resilience, persistence, and variability. *The American Naturalist*, 113, 659–669.

Holling, C.S. (1973). Resilience and stability of ecological systems. *Annual Review of Ecology and Systematics*, 4, 1–23.

Isbell, F., Craven, D., Connolly, J., *et al.* (2015). Biodiversity increases the resistance of ecosystem productivity to climate extremes. *Nature*, 526, 574–577.

Ives, A.R. and Carpenter, S.R. (2007). Stability and diversity of ecosystems. *Science*, 317, 58–62.

Ives, A.R., Gross, K., Klug, J.L. (1999). Stability and variability in competitive communities. *Science*, 286, 542–544.

Ives, A.R., Klug, J.L., Gross, K. (2000). Stability and species richness in complex communities. *Ecology Letters*, 3, 399–411.

Klug, J.L., Fischer, J.M., Ives, A.R., Dennis, B. (2000). Compensatory dynamics in planktonic community responses to pH perturbations. *Ecology*, 81, 387–398.

Kratina, P., Vos, M., Anholt, B.R. (2007). Species diversity modulates predation. *Ecology*, 88, 1917–1923.

Lamy, T., Wang, S., Renard, D., Lafferty, K.D., Reed, D.C., Miller, R.J. (2019). Species insurance trumps spatial insurance in stabilizing biomass of a marine macroalgal metacommunity. *Ecology*, 100, e02719.

Levins, R. (1970). Complex systems. In *Towards a Theoretical Biology,* Waddington, C.H. (ed.). Edinburgh University Press, Edinburgh.

Loreau, M. (1994). Material cycling and the stability of ecosystems. *The American Naturalist*, 143, 508–513.

Loreau, M. (2010). *From Populations to Ecosystems: Theoretical Foundations for a New Ecological Synthesis*. Princeton University Press, Princeton.

Loreau, M. and de Mazancourt, C. (2008). Species synchrony and its drivers: Neutral and nonneutral community dynamics in fluctuating environments. *The American Naturalist*, 172, E48–E66.

Loreau, M. and de Mazancourt, C. (2013). Biodiversity and ecosystem stability: A synthesis of underlying mechanisms. *Ecology Letters*, 16(S1), 106–115.

Loreau, M., Downing, A., Emmerson, M. *et al.* (2002). A new look at the relationship between diversity and stability. In *Biodiversity and Ecosystem Functioning: Synthesis and Perspectives*, Loreau, M., Naeem, S., Inchausti, P. (eds). Oxford University Press, Oxford.

MacArthur, R.H. (1955). Fluctuations of animal populations and a measure of community stability. *Ecology*, 36, 533–535.

May, R.M. (1972). Will a large complex system be stable? *Nature*, 238, 413–414.

de Mazancourt, C., Isbell, F., Larocque, A. *et al.* (2013). Predicting ecosystem stability from community composition and biodiversity. *Ecology Letters*, 16, 617–625.

McCann, K.S. (2000). The diversity–stability debate. *Nature*, 405, 228–233.

McNaughton, S.J. (1977). Diversity and stability of ecological communities: A comment on the role of empiricism in ecology. *The American Naturalist*, 111, 515–525.

Odum, E.P. (1953). *Fundamentals of Ecology*. Saunders, Philadelphia.

Pimm, S.L. (1984). The complexity and stability of ecosystems. *Nature*, 307, 321–326.

Rosenzweig, M.L. (1999). *Species Diversity in Space and Time*. Cambridge University Press, Cambridge.

Scheffer, M., Carpenter, S., Foley, J.A., Folkes, C., Walker, B. (2001). Catastrophic shifts in ecosystems. *Nature*, 413, 591–596.

Thébault, E. and Loreau, M. (2005). Trophic interactions and the relationship between species diversity and ecosystem stability. *The American Naturalist*, 166, E95–E114.

Thorson, J.T., Scheuerell, M.D., Olden, J.D., Schindler, D.E. (2018). Spatial heterogeneity contributes more to portfolio effects than species variability in bottom-associated marine fishes. *Proceedings of the Royal Society B*, 285, 20180915.

Tilman, D. (1996). Biodiversity: Population versus ecosystem stability. *Ecology*, 77, 350–363.

Tilman, D. (1999). The ecological consequences of changes in biodiversity: A search for general principles. *Ecology*, 80, 1455–1474.

Tilman, D., Reich, P.B., Knops, J.M.H. (2006). Biodiversity and ecosystem stability in a decade-long grassland experiment. *Nature*, 441, 629–632.

Tredennick, A.T., de Mazancourt, C., Loreau, M., Adler, P.B. (2017). Environmental responses, not species interactions, determine synchrony of dominant species in semiarid grasslands. *Ecology*, 98, 971–981.

Vitousek, P.M., Mooney, H.A., Lubchenco, J., Melillo, J.M. (1997). Human domination of Earth's ecosystems. *Science*, 277, 494–499.

Wang, S. and Loreau, M. (2014). Ecosystem stability in space: Alpha, beta and gamma variability. *Ecology Letters*, 17, 891–901.

Wang, S. and Loreau, M. (2016). Biodiversity and ecosystem stability across scales in metacommunities. *Ecology Letters*, 19, 510–518.

Wang, S., Loreau, M., Arnoldi, J.-F., *et al.* (2017). An invariability–area relationship sheds new light on the spatial scaling of ecological stability. *Nature Communications*, 8, 15211.

Wang, S., Lamy, T., Hallett, L.M., Loreau, M. (2019). Stability and synchrony across ecological hierarchies in heterogeneous metacommunities: Linking theory to data. *Ecography*, 42, 1200–1211.

Yachi, S. and Loreau, M. (1999). Biodiversity and ecosystem productivity in a fluctuating environment: The insurance hypothesis. *Proceedings of the National Academy of Sciences of the USA*, 96, 1463–1468.

8

What Do Biodiversity Experiments Tell Us About Biodiversity and Ecological Stability Relationships?

Lin JIANG and Qianna XU

School of Biological Sciences, Georgia Institute of Technology, Atlanta, USA

8.1. Introduction

The Earth harbors various ecological communities that differ widely in their biodiversity. Tropical rainforests, for example, may contain hundreds of tree species within one hectare of land, whereas more than 80 herbaceous plant species can be found within a square meter plot of some species-rich grasslands (Wilson *et al.* 2012). By contrast, much lower plant diversity is found in many other ecosystems, particularly human-managed plantations and agricultural fields. Does this difference in biodiversity carry any significance for the stability of these ecosystems? More than 60 years ago, Elton (1958) used six lines of observational evidence, including pest outbreaks being more frequent in species-poor agricultural lands than species-rich tropical rainforests, to argue that more diverse communities are more stable. MacArthur (1955) suggested that generalist predators feeding on multiple prey species should exhibit more stable dynamics than specialist predators feeding on a single prey, based on the simple logic that multiple energy input pathways would buffer generalist predator populations against large fluctuations. Despite the lack of supporting experimental evidence, these early conceptual ideas were influential at the time when observational studies constituted the dominant form of ecological

The Ecological and Societal Consequences of Biodiversity Loss,
coordinated by Michel LOREAU, Andy HECTOR, and Forest ISBELL.

research. Findings of observational studies, however, are prone to alternative interpretations, given the presence of a myriad of potential confounding factors. For example, observed relationships between biodiversity and stability in natural communities may not necessarily demonstrate causality, if biodiversity and stability are both strongly influenced by other factors (e.g. nutrient availability). Only experiments that directly manipulate biodiversity, while controlling for confounding variables, can provide robust tests of diversity-stability linkages. These biodiversity experiments were conducted mostly in the past three decades, in response to the concern that ongoing global biodiversity loss (Pimm *et al.* 2014) may diminish the ability of ecosystems to provide reliable products and services for humanity. In this chapter, we review these biodiversity–stability experiments and summarize their findings.

Before embarking on examining biodiversity–stability experiments, it is important to note that both biodiversity and ecological stability are multi-faceted concepts. Within an ecological community, biodiversity can be quantified in various ways (see Chapter 2), such as species diversity (i.e. species richness and evenness), functional diversity (i.e. the variety of functional attributes or traits possessed by co-occurring species), and phylogenetic diversity (i.e. the diversity component accounting for species evolutionary relationships). Here we will focus on species richness (used interchangeably with species diversity here), as this metric has received the most attention in biodiversity experiments and beyond, and the role of evenness has rarely been explored in the context of ecological stability (but see Isbell *et al.* 2009). We will also touch upon functional and phylogenetic diversity. Ecological stability has even broader meanings (Pimm 1984; Grimm and Wissel 1997; Ives and Carpenter 2007; Donohue *et al.* 2013; also see Chapter 7), and can be studied at different levels of ecological organization (e.g. population vs. ecosystem). Both theory and experiments have shown that different stability components may not necessarily exhibit similar relationships with species diversity (e.g. Ives and Carpenter 2007) and that the same stability components may exhibit different relationships with species diversity at different levels of ecological organization (e.g. Tilman *et al.* 2006). Failure to recognize these issues has contributed to historical debates on diversity–stability relationships (McCann 2000). Our focus will be on temporal stability (also called temporal invariability, often quantified as the inverse of coefficient of variation of ecological properties measured over time), and to a lesser extent, resistance (the ability of an ecological system to resist displacement from disturbance) and resilience (the ability of an ecological system to return to its original state after disturbance) of aggregated ecosystem properties (e.g. community biomass production). Population-level stability will be discussed in relation to ecosystem-level stability responses to changes in diversity.

8.2. Insight from models

Early theoretical explorations of diversity–stability relationships, exemplified by May's (1973) work based on local stability analysis of linear Lotka–Volterra models, suggest that more speciose communities tend to be less resilient. This prediction, translating into lower population temporal stability in more diverse communities (*sensu* Haegeman *et al.* 2016), challenged the earlier belief that more diverse communities are more stable (MacArthur 1955; Elton 1958). This prediction is echoed by more recent models of plant communities (Lehman and Tilman 2000). May's (1973) model, however, was based on the assumption that the strength of species interactions is randomly distributed, whereas natural communities typically exhibit skewed interaction-strength distributions towards weak interactions (Wootton and Emmerson 2005). Considering non-random interaction strength in biologically more realistic nonlinear models, McCann *et al.* (1998) predicted that adding weak predator–prey interactions could dampen large oscillations of strong-interacting predator–prey populations, hence conferring greater population stability in more complex food webs. Stabilizing weak trophic interactions thus provide a plausible, mechanistic explanation for MacArthur's (1955) hypothesis on the greater stability of generalist predator populations and the long-term existence of many diverse natural communities. Models also show that other biological attributes, such as realistic predator–prey body mass allometry (Brose *et al.* 2006) and ecological network architecture (Thebault and Fontaine 2010), may help stabilize complex communities.

Compared with the divergent predictions on diversity–population stability relationships, predictions are more congruent for stability at the ecosystem level, suggesting a general positive effect of species diversity on the temporal stability of aggregated ecosystem properties. This positive diversity–ecosystem stability relationship was explicitly proposed by McNaughton (1977) and later substantiated by mathematical models (Ives *et al.* 1999; Yachi and Loreau 1999; Lehman and Tilman 2000). This positive diversity–ecosystem stability relationship is also predicted to be generally robust to trophic interactions (Thebault and Loreau 2005). One major mechanism underlying this positive diversity–ecosystem stability relationship is the increased prevalence of compensatory dynamics, where declines in the abundance of some species are compensated by increases in the abundance of other species (Gonzalez and Loreau 2009), among a greater number of interacting species, resulting in relatively constant ecosystem functions despite possible substantial fluctuations in the populations of individual species (i.e. the insurance hypothesis, Yachi and Loreau 1999). While the stability of aggregated ecosystem properties, by definition, is influenced by the population stability of constituent species, compensatory dynamics mediate the link between population- and ecosystem-level stability and play a particularly important role in stabilizing the properties of more diverse ecosystems in situations where population-level stability may decline with species diversity. Therefore, partitioning ecosystem-level stability into within-species population stability and among-species compensatory dynamics facilitates the understanding of diversity-ecosystem stability relationships (Thibaut and Connolly 2013).

8.3. A brief account of earlier diversity–stability experiments

Prior to the 1990s, few empirical studies examined diversity–stability relationships (Hairston *et al.* 1968; Hurd *et al.* 1971; McNaughton 1977; Luckinbill 1979), despite the call for more empirical work on this topic (McNaughton 1977). In the early 1990s, as ecologists became increasingly concerned about the functional consequences of biodiversity loss, they began systematic investigations of the relationships between biodiversity and ecosystem functioning (BEF), including ecosystem stability, resulting in the emergence of a new paradigm. One of the earliest experiments addressing diversity–stability relationships during this period was conducted by Tilman and Downing (1994), who altered grassland plant species diversity via different rates of nitrogen addition and found that biomass production in more diverse grasslands was more resistant to drought and more resilient after the drought. This work, however, was criticized by Huston (1997), who pointed out the difficulty of separating effects of species diversity and nitrogen fertilization and the necessity for independent manipulation of biodiversity. Huston (1997) also suggested the need to include communities that differ in species composition for each level of diversity, in order to disentangle the effects of species diversity and composition. Many contemporary biodiversity experiments have adopted these suggestions, by experimentally assembling communities differing in diversity via random draws from a general species pool, with multiple different species compositions nested within each diversity level. When this is not possible (e.g. for highly diverse microbial communities), a dilution to extinction approach, where high-diversity samples from the natural environment are serially diluted to create lower-diversity treatments, has often been used (e.g. Roger *et al.* 2016). Our review here will focus on these contemporary diversity–stability experiments.

8.4. The relationships between biodiversity and temporal stability

8.4.1. *Grassland biodiversity experiments*

Grassland experiments make up the bulk of experimental BEF research, due to the relative ease of manipulating plant species diversity and composition, replicating experimental units, and quantifying species, community, and ecosystem properties in grasslands. These experiments have played indispensable roles in advancing our empirical knowledge on BEF patterns and mechanisms, including those related to stability.

One of the longest-running biodiversity experiments, established in Cedar Creek Ecosystem Science Reserve in Minnesota, USA, in 1994, provided strong empirical evidence that plant species diversity influences ecosystem and population temporal stability (Tilman *et al.* 2006). A diversity gradient of 1, 2, 4, 8, and 16 species was set up from a pool of 18 perennial grassland species, and peak plant biomass was measured in mid-August each year. Data collected over a 10-year period (1996–2005) showed that as species diversity increased, the temporal stability of community biomass increased (Tilman *et al.* 2006), supporting theoretical predictions of the positive relationships between the two. On the other hand, the temporal stability of population-level biomass, averaged across species, declined as species diversity increased (Tilman *et al.* 2006), supporting the prediction of May (1973).

Similar findings have been reported in a large number of other grassland biodiversity experiments. In the Jena Experiment, another long-term (since 2002) biodiversity experiment established in Jena, Germany, many species in the experimental plots exhibited lower population temporal stability in more diverse communities characterized by greater total biomass temporal stability (Roscher *et al.* 2011). This opposite effect of species diversity on the ecosystem and population temporal stability was also reported for the BIODEPTH experiment, which adopted the same experimental design at eight field sites across Europe (Hector *et al.* 2010). The consistency in the findings from multiple sites of the same experiment suggests that these findings were not driven by idiosyncratic characteristics of the experimental sites or plant communities, providing some of the strongest experimental evidence that species diversity stabilizes aggregated ecosystem properties but destabilizes population dynamics in grassland ecosystems.

These contrasting diversity–stability relationships at population and ecosystem levels suggest that compensatory dynamics must increase with diversity to overcome the decline in population stability to confer greater ecosystem stability for more diverse grasslands. Nevertheless, using summed covariances of species abundances as the metric for species compensatory dynamics, several grassland experiments have found that the degree of compensatory dynamics did not increase with species diversity and thus could not explain the observed positive diversity–ecosystem stability relationships (e.g. Tilman *et al.* 2006; van Ruijven and Berendse 2007). However, there are mathematical constraints on the lower limits of summed covariances, which makes it difficult to distinguish communities differing in compensatory dynamics based on summed covariance values (Loreau and de Mazancourt 2008). When compensatory dynamics were quantified using more appropriate metrics (e.g. community-wide asynchrony index, Loreau and de Mazancourt 2008), positive relationships between compensatory dynamics and species diversity were frequently found in experimental grassland communities (e.g. Isbell *et al.* 2009; Roscher *et al.* 2011). Nevertheless, the relative importance of different mechanisms (e.g. interspecific competition, species responses to environmental variation) underlying compensatory dynamics in ecological communities has rarely been explored (Gross *et al.* 2014; Tredennick *et al.* 2017) and remains poorly understood.

While the observed lower population temporal stability in more diverse communities is consistent with their lower resilience as predicted from theoretical models (May 1973), alternative explanations exist. For example, the scaling of the temporal mean and variance of species abundances, known as Taylor's power law, may strongly influence how average population stability scales with diversity (Thibaut and Connolly 2013). There is increasing evidence that in natural plant communities the temporal standard deviation (the square root of variance) of species abundance often does not increase as fast as its mean, such that more abundant species tend to be temporally more stable (Leps 2004; Grman *et al.* 2010; Roscher *et al.* 2011). Under this circumstance, higher species diversity would translate into an increased proportion of subordinate and rare species characterized by lower temporal stability, resulting in a decline in overall population stability (Roscher *et al.* 2011; Thibaut and Connolly 2013). Thus, understanding species abundance distribution and mean-variance scaling patterns would help understand diversity–temporal stability relationships. Note that recent theory has shown that community resilience is largely determined by the population temporal stability of the least stable species (Haegeman *et al.* 2016), suggesting that the above explanation and May's (1973) prediction may not be mutually exclusive.

8.4.2. *Forest biodiversity experiments*

In contrast with the large number of biodiversity experiments conducted in grasslands, biodiversity experiments in forest ecosystems are still rare (Ewel *et al.* 2015; Huang *et al.* 2018; van de Peer *et al.* 2018). The major reasons for the paucity of forest biodiversity experiments are the logistic difficulty of establishing experimental tree plantations and the long generation times of trees, which make it difficult to draw robust conclusions on BEF relationships, particularly diversity–stability relationships, based on immature plantations. The Sarninilla biodiversity experiment, the oldest biodiversity experiment in the tropics (established in Sardinilla, Panama, in 2001), provided a rare opportunity to experimentally test forest diversity–stability relationships. This experiment established a gradient of 1, 2, 3, and 5 tree species in 24 45 × 45 m plots that have experienced strong inter-annual variation in climatic conditions (including wet La Niña and dry El Niño events). Using the 2006–2016 data of the Sarninilla experiment, Schnabel *et al.* (2019) found that increasing tree diversity increased the temporal stability of ecosystem productivity, which was largely driven by compensatory dynamics among the planted species. While population-level stability was not examined in this study, these ecosystem-level results bear a strong similarity to those of grassland experiments.

8.4.3. *Aquatic biodiversity experiments*

Aquatic ecosystems differ from terrestrial ecosystems in many aspects, in part driven by the differences in the size and stoichiometry of primary producers (plants vs. phytoplankton) between the two habitats (Shurin *et al.* 2006). One notable difference, for example, is that the interactions between herbivores and primary producers tend to be stronger in water than on land, resulting in stronger top-down control of producer biomass in aquatic habitats (Shurin *et al.* 2002, 2006). Could these differences translate into differences in the biodiversity–stability relationships between the two habitats?

Much experimental evidence for diversity–stability relationships in aquatic habitats has come from studies of multi-trophic rock pool communities (e.g. Vogt *et al.* 2006, Romanuk *et al.* 2006, 2010). These experimental explorations followed observational studies suggesting that population- and ecosystem-level temporal stability were both greater in more diverse rock pool communities (Kolasa and Li 2003; Romanuk and Kolasa 2004). A variety of invertebrates, composed of primarily crustaceans, insect larvae, and worms, inhabit these rock pools. Natural pools vary substantially in the diversity of invertebrate communities that they support because they differ in size, salinity, and other physical and chemical conditions, as well as disturbance regimes (Kolasa and Li 2003; Romanuk and Kolasa 2004), making it difficult to separate the effect of diversity from that of environmental variables. Experimental manipulations of invertebrate diversity, via the dilution-to-extinction approach, indicate that diversity indeed stabilizes both population and ecosystem dynamics, even under the same environmental conditions (Vogt *et al.* 2006; Romanuk *et al.* 2006). Vogt *et al.* (2006), for example, found that seven out of the nine invertebrate species in their experiment showed increased temporal stability in more diverse communities. Furthermore, these studies identified increased population temporal stability as the main reason for ecosystem-level stability to increase with diversity (Vogt *et al.* 2006; Romanuk *et al.* 2006), in sharp contrast to increased compensatory dynamics being more important for stabilizing more diverse grassland ecosystems. However, why population-level stability increased with diversity, which is at odds with the negative effect of species diversity on population temporal stability in grasslands, was largely unexplained.

These stabilizing effects of species diversity on population and ecosystem dynamics were also reported in other experimental studies of aquatic communities. For example, by directly varying the number of species in four trophic groups of experimentally assembled freshwater food webs, Steiner *et al.* (2005) found that the temporal stability of community biomass was significantly greater in more diverse communities. In addition, they also found that the temporal stability of population-level biomass showed positive, albeit weaker, responses to increases in diversity (Steiner *et al.* 2005). Downing *et al.* (2014) manipulated zooplankton species diversity in experimental freshwater ponds by varying the number of dominant species as well as by controlling species dispersal from external source communities and found that increasing diversity resulted in significant increases in the temporal stability of zooplankton biomass at both population and ecosystem levels. Downing *et al.* (2014) also showed that increased asynchrony and population-level stability both contributed to the positive relationships between zooplankton diversity and ecosystem-level stability. Nevertheless, mechanisms underlying the positive diversity effect on population-level stability in these studies remain elusive.

What could cause species diversity to stabilize population dynamics in aquatic systems but destabilize population dynamics in grasslands? This is an important question, as the contrasting pattern points to potentially different stability-regulating mechanisms between the two types of ecosystems. The answer possibly lies in the distribution pattern of trophic interaction strength in ecological communities, which has important implications for the stability of their constituent populations. The distribution of the strength of trophic interactions in ecological communities is neither random nor uniform. Rather, communities tend to be characterized by a few strong interactions, but many weak interactions (Wootton and Emmerson 2005). This skewed distribution of interaction strength is expected to be particularly pronounced for aquatic communities with average stronger trophic interactions than terrestrial communities. Under these circumstances, increasing diversity could allow weak trophic interactions, which may mitigate the dynamical instability of strong trophic interactions (McCann *et al.* 1998), and stabilize more diverse communities, resulting in positive diversity–population stability relationships in aquatic communities. In contrast, this stabilizing effect of weak interactions is expected to be less effective in terrestrial communities, where weak trophic interactions are prevalent and top-down control is generally weak (Shurin *et al.* 2002, 2006).

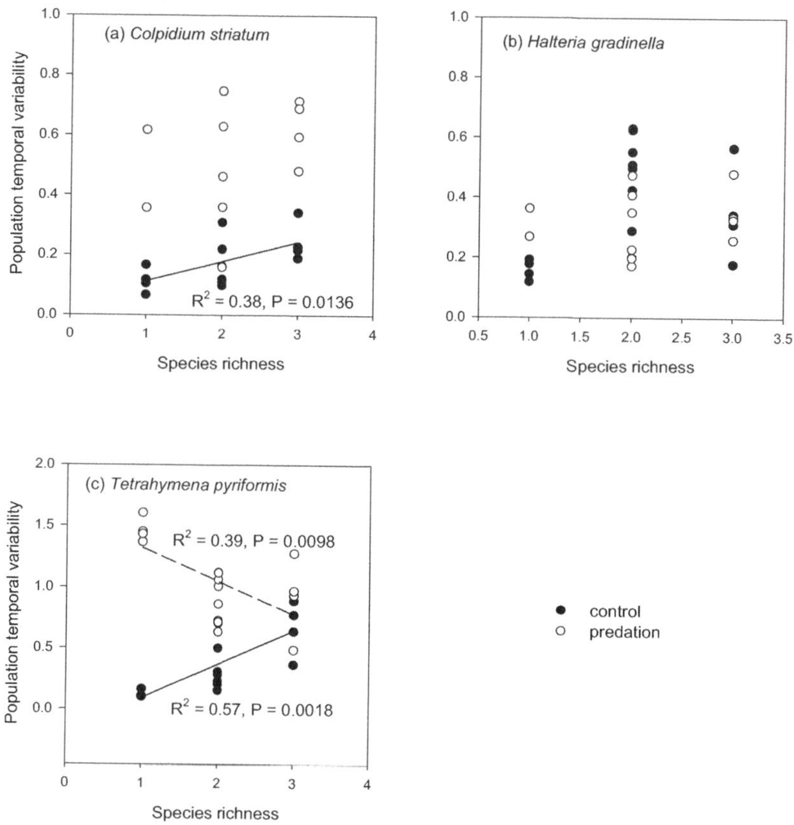

Figure 8.1. *Population temporal variability (the inverse of temporal stability) as a function of species richness of bacterivorous ciliated protist communities in a 23-day laboratory microcosm experiment. The experiment manipulated the number of bacterivorous ciliated protist species (1, 2, and 3 species) from a pool of three species: a)* Colpidium striatum, *b)* Halteria grandinella, *and c)* Tetrahymena pyriformis, *and the absence and presence of the ciliate predator* Lacrymaria olor. *The population variability of* T. pyriformis, *which interacted most strongly with the predator, increased with diversity in the absence of the predator, but decreased with diversity in the presence of the predator, due to the addition of relatively weak interactions between the other two prey species and the predator. Solid lines are linear regression lines for the controls (filled circles), and dashed lines regression lines for the predation treatment (open circles). Only statistically significant regression lines are shown. Temporal variability was quantified as the standard deviation of log-transformed population biovolume; measures based on the coefficient of variation of population biovolume yielded qualitatively similar results. Biovolume was measured as μm³/mL. Adapted from Jiang et al. (2009)*

The above hypothesis suggests that it may be the difference in trophic complexity between the aquatic and terrestrial communities that drives the contrasting relationships between species diversity and population-level stability. In other words, the diversity–population stability relationships should differ between single-trophic (i.e. trophic interactions are weak or absent) and multi-trophic communities, even within the same type of habitats. To test this idea, Jiang *et al.* (2009) assembled freshwater bacterivorous ciliate communities that differed in species diversity (one to three species) and the presence/absence of a ciliated predator, with the three prey species differing widely in the strength of their interactions with the predator. In the absence of the predator, two of the three prey species exhibited lower population temporal stability at higher species diversity (Figure 8.1a, b, c). However, in the presence of the predator, the introduction of weak trophic interactions stabilized the dynamics of the prey strongly interacting with the predator, especially at higher prey diversity, resulting in a positive diversity–population stability relationship for the strong interacting prey species (Figure 8.1c). Weak trophic interactions also helped stabilize the dynamics of prey community biomass, resulting in a positive diversity–community biomass stability relationship that did not exist when the predator was absent (Jiang *et al.* 2009). Likewise, O'Gorman and Emmerson (2009) reported that weak trophic interactions helped stabilize ecosystem production in complex marine food webs, such that the removal of weak interacting species led to a decline in the temporal stability of primary and secondary production. Together, these findings suggest that weak trophic interactions may operate as an important stabilizing mechanism driving positive diversity–ecosystem stability relationships in multi-trophic systems.

8.4.4. *Microbial biodiversity experiments*

Natural microbial communities are enormously diverse and hence it is not possible to experimentally assemble microbial communities that replicate their diversity in nature. As a result, the dilution-to-extinction approach has often been used to establish microbial diversity gradients. So far, microbial BEF studies using this approach have found little relationship between the bacterial diversity and temporal stability of bacterial abundances, which parallels the inconsistent, generally weak bacterial diversity–function (e.g. microbial biomass production, organic carbon degradation) relationships reported by this type of study (Roger *et al.* 2016). These results likely arose because high functional redundancy among bacteria allows lower-diversity bacterial communities, which still contain a substantial number of species, to attain similar constancy in their functions as higher-diversity communities. Antagonistic interactions between bacteria, such as toxin production, are known to affect their functions, especially at higher diversity where such interactions tend to be more frequent (Becker *et al.* 2012). However, how much antagonistic interactions have contributed to the lack of diversity–stability relationships observed in complex microbial communities remains an open question.

Experimentally assembled microbial communities are generally of much lower complexity compared to natural microbial communities. However, the simplicity of these synthetic microbial communities offers the advantage of being better able to gain a mechanistic understanding of microbial BEF relationships. Nevertheless, few BEF studies have examined the relationship between microbial diversity and temporal stability in synthetic microbial communities. Several studies, however, have examined spatial variability (i.e. variation among communities subject to different environmental conditions) in bacterial productivity in relation to bacterial species diversity. These studies showed that increasing bacterial diversity reduced spatial variability in bacterial productivity, which was attributed to the difference in species responses to spatial variation in environmental conditions (Eisenhauer *et al.* 2012; Awasthi *et al.* 2014). A corollary of these results is that compensatory dynamics, driven by non-synchronous species responses to temporal variation in environmental conditions, could conceivably drive positive relationships between microbial diversity and temporal stability of microbial community productivity, although it is likely that such effects are only strong at low diversity levels where there may not be much functional redundancy among species. This hypothesis, of course, would need to be directly tested by experiments.

8.4.5. *How general are the effects of species diversity on temporal stability?*

Meta-analyses, which provide statistical syntheses of results from multiple studies, have proven valuable for integrating the findings of biodiversity experiments (e.g. Cardinale *et al.* 2006). Two meta-analyses have examined the relationships between species diversity and temporal stability. Jiang and Pu (2009) found that across single-trophic (i.e. trophic interactions are absent or insignificant, including mostly terrestrial experiments) experiments species diversity did not affect the temporal stability of community biomass production, presumably due to the limited number of studies included in their analysis. The later meta-analysis of Campbell *et al.* (2011) did find a positive effect of species diversity on the temporal stability of community biomass. Both studies reported a general negative effect of species diversity on population-level temporal stability across single-trophic experiments. A reanalysis of multiple grassland diversity–stability experiments, while not using the tool of meta-analysis, also supported the positive effect of species diversity on community biomass stability and the negative effect of species diversity on population stability (Gross *et al.* 2014). Our updated meta-analysis, which includes diversity–stability experiments conducted after these studies and hence has greater statistical power, indicates that these patterns still hold (Figure 8.2a). Current experimental evidence, therefore, suggests that species diversity stabilizes aggregated ecosystem properties but destabilizes population dynamics in single-trophic communities.

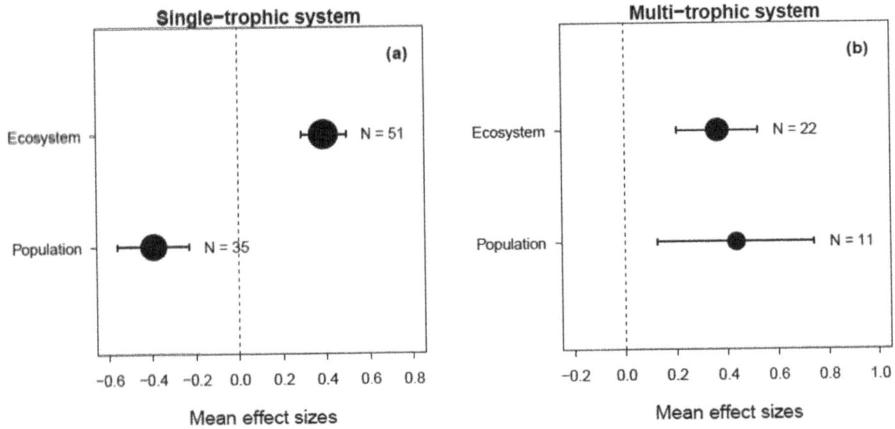

Figure 8.2. *Mean effect sizes (± 95% bootstrap confidence intervals) of the relationship between species richness and temporal stability at the ecosystem and population levels in a) single-trophic and b) multi-trophic experiments. Ecosystem temporal stability increased with species diversity, whereas population temporal stability decreased with species diversity in single-trophic experiments. Both ecosystem- and population-level temporal stability increased with species diversity in multi-trophic experiments. Effect sizes are Fisher's z transforms of correlation coefficients between species richness and temporal stability. N represents the sample size (i.e. the number of studies included in each analysis). The vertical dotted line indicates the case of effect size = 0*

The meta-analyses of Jiang and Pu (2009) and Campbell *et al.* (2011) also found that across multi-trophic (i.e. trophic interactions are significant, including mostly aquatic experiments) experiments, increasing diversity tended to increase temporal stability at both population and ecosystem levels. Our updated meta-analysis indicates that these patterns still hold (Figure 8.2b). Current experimental evidence, therefore, suggests that species diversity stabilizes both population and ecosystem dynamics in multi-trophic systems.

8.4.6. *Other dimensions of biodiversity*

In addition to species diversity, other biodiversity dimensions, such as functional and phylogenetic diversity, have received much recent attention in the BEF field. This trend follows the recognition that the functioning of an ecosystem is largely governed by the distribution of functional traits among its constituent species (Diaz and Cabido 2001). Functional diversity can account for functional trait distribution among species, but the explanatory power hinges on the inclusion of traits that are important for the function of interest (Petchey *et al.* 2004). Phylogenetic diversity, typically measured as the total length of phylogenetic branches connecting a given set of taxa (Faith 1992), accounts for species' evolutionary histories. When traits are phylogenetically conserved (i.e. the tendency for closely related species to possess similar traits), phylogenetic diversity may serve as a reasonable proxy of functional trait diversity, especially when key functional traits are not identified or data on these traits are unavailable. Species diversity can complement functional and phylogenetic diversity as predictors of ecosystem functioning, by capturing distributions of traits that are not considered by the functional diversity component or not phylogenetically conserved. Therefore, considering species, functional, and phylogenetic diversity together may help to better understand the role of biodiversity for ecosystem functioning.

So far, most biodiversity experiments have involved direct manipulations of species diversity, which translate into parallel changes in functional and phylogenetic diversity. The effects of these different dimensions of biodiversity are thus often difficult to separate. This is even true for experiments that have made efforts to manipulate different biodiversity dimensions independently (e.g. species and functional group richness in the Jena experiment). In the Jena experiment, although more asynchronous population dynamics were detected for plant species belonging to different functional groups, functional group richness failed to explain population- or ecosystem-level temporal stability after accounting for species diversity (Roscher *et al.* 2011). Studies comparing the relative importance of plant species and phylogenetic diversity for ecosystem-level temporal stability have reported mixed results (Cadotte *et al.* 2012; Venail *et al.* 2015). At the same time, there is growing evidence that key traits related to plant growth strategies may strongly influence plant population stability, such that slow-growing species with conservative strategies (e.g. those with high leaf dry matter content) tend to exhibit more temporally stable population dynamics (Majekova *et al.* 2014). When considering plant functional diversity based on these growth strategy traits, phylogenetic diversity, and species diversity together in an analysis of 39 grassland biodiversity experiments, Craven *et al.* (2018) found that the three biodiversity components all contributed to reducing community biomass variability over time, thus playing complementary roles in stabilizing grassland biomass production. The use of structural equation modelling allowed Craven *et al.* (2018) to tease apart the different biodiversity effects and the direct and indirect pathways via which biodiversity influenced community biomass stability. Nevertheless, future biodiversity experiments should consider manipulating functional and phylogenetic diversity directly, while fixing species diversity, in order to better assess their roles in stabilizing aggregated ecosystem properties. Data from existing biodiversity experiments in which multiple communities differing in functional and phylogenetic diversity existed within the same diversity levels may also be reanalyzed for the same purpose.

8.5. The relationships between biodiversity and resistance/resilience

The abilities of an ecosystem to resist displacement from disturbances (i.e. resistance) and return to its original state after disturbances (i.e. resilience) are two important facets of ecological stability (Pimm 1984; Ives and Carpenter 2007). Understanding how biodiversity affects ecosystem resistance and resilience is thus essential for obtaining a comprehensive understanding of the consequences of biodiversity loss for ecological stability. The resistance and resilience of an ecosystem are also important determinants of its temporal stability (Pimm 1984). Therefore, understanding how resistance and resilience change with biodiversity may also help in understanding the relationships between biodiversity and ecosystem temporal stability.

Unlike the fairly general positive relationships between species diversity and the temporal stability of aggregated ecosystem properties reported by biodiversity experiments, experimental results on species diversity and resistance/resilience relationships are rather mixed. For example, Pfisterer and Schmid (2002) subjected the experimental grasslands at the Swiss site of the BIODEPTH experiment to experimentally simulated summer drought and found that both resistance and resilience of plant biomass production, measured as the absolute change from pre- to mid-drought biomass and post- to pre-drought biomass ratio, respectively, declined with increasing plant species diversity. However, they also noted that resistance, if quantified as proportional biomass loss, did not change with plant diversity. Using metrics based on the proportional deviation in biomass in dry or wet years from normal precipitation years, Isbell *et al.* (2015) found that ecosystem resistance, but not resilience, increased with plant species diversity across 46 grassland experiments, leading them to conclude that increased biomass temporal stability in more diverse grasslands was primarily due to increased resistance. By contrast, comparing mid- and post-drought biomass to that in the control treatment (i.e.

without drought) in a multi-site grassland mesocosm experiment, Kreyling *et al.* (2017) showed that plant species diversity increased resilience, but not resistance. Similar incongruent results have been reported by other studies (e.g. van Ruijven and Berendse 2010; Wright *et al.* 2015; Wagg *et al.* 2017).

Aside from differences in the characteristics of study systems, several other factors have also contributed to these incongruent results. First, the shifting definition of components of ecological stability, particularly resilience (e.g. Holling 1973; Pimm 1984; Hodgson *et al.* 2015), in the literature has led to confusion among ecologists. Second, widely different metrics have been used for both resistance and resilience (Matos *et al.* 2020), which can result in different diversity–stability relationships even for the same resistance/resilience concepts (e.g. Pfisterer and Schmid 2002). Third, the type and intensity of disturbance may also influence resistance/resilience measures. Disturbances that increase the availability of resources (e.g. flood, Wright *et al.* 2015) may impose different effects on ecological communities than stress-based disturbance (e.g. drought). Extreme disturbances may induce greater ecosystem biomass loss (i.e. lower resistance) than disturbances of lower intensity (Ruppert *et al.* 2014), and can diminish compensatory dynamics among coexisting species (Muraina *et al.* 2021). Potentially different diversity and resistance/resilience relationships can thus emerge under different types and intensities of disturbance. Standardizing stability concepts and metrics and disturbance regimes is thus essential for future biodiversity experiments if we are to draw general conclusions on resistance/resilience and their linkages to biodiversity.

8.6. The relevance of biodiversity experiments to real-world ecosystems

Here we have focused on stability-related findings of experiments in which communities varying in biodiversity have been assembled by experimenters. These biodiversity experiments have proven instrumental in advancing our understanding of diversity–stability and, more generally, BEF patterns and mechanisms. Principles derived from these experiments have been applied to understand the role of crop diversity in stabilizing the yield of agricultural systems (Renard and Tilman 2019; Egli *et al.* 2020), which are also intensively managed by humans. Natural assembly processes (e.g. species colonization and extinction) are certainly non-random, and as a result, the structure of experimentally and naturally assembled communities may not be the same (Wardle 2016). Nevertheless, BEF mechanisms identified in biodiversity experiments are also known to operate in natural communities (see Chapter 9), and findings from biodiversity experiments have facilitated our understanding of the functional significance of biodiversity in natural communities

(Eisenhauer *et al.* 2016; van der Plas 2019; Jochum *et al.* 2020). Comparative studies of biodiversity and stability patterns between the two types of communities would be particularly useful for directly evaluating the relevance of findings of biodiversity experiments to real-world ecosystems (e.g. Kohli *et al.* 2019).

8.7. Conclusion

Consistent with theoretical predictions, experiments conducted over the last three decades have shown that increasing species diversity tends to increase the temporal stability of aggregate ecosystem properties, with increased compensatory dynamics identified as an important stabilizing mechanism. However, these experiments also revealed the dependency of the relationship between species diversity and population temporal stability on trophic complexity, such that species diversity destabilizes population dynamics in single-trophic systems, but stabilizes population dynamics in multi-trophic systems. This contrasting pattern may be driven by weak trophic interactions (and possibly other mechanisms) stabilizing population and community dynamics in multi-trophic systems.

There are, however, several significant gaps in our knowledge. First, although both theory and experiments have identified species compensatory dynamics as a key mechanism stabilizing ecosystem properties, the relative importance of "low-level" ecological processes (e.g. competition, different species responses to environmental fluctuation) that generate compensatory dynamics in ecological communities remains poorly understood. Carefully designed experiments are needed to separate different sources of compensatory dynamics. Second, while here we focus on reviewing experimental findings of biodiversity effects on temporal stability, a host of abiotic environmental factors, such as nutrient availability, are known to influence both biodiversity and stability (see Chapter 9). Future experiments that independently manipulate biodiversity and environmental factors are needed to better understand their independent and interactive effects on ecological stability. Third, our chapter focuses on diversity–stability relationships at local spatial scales, whereas recent theories have extended diversity–stability research to regional scales (Wang and Loreau 2016, Chapter 7). These theories on the diversity–stability relationships across multiple spatial scales have been tested by several observational studies (see Chapter 9) but would need to be evaluated more rigorously via experimentation. Finally, despite much research, a general understanding of how biodiversity influences ecosystem resistance and resilience is still lacking. To streamline research on this topic, future studies should begin adopting the same stability concepts and comparable resistance/resilience metrics. Moreover, other components of ecological stability, in addition to temporal stability, resilience, and resistance, should also be more extensively explored for a more comprehensive understanding of biodiversity and stability relationships.

8.8. Acknowledgements

We thank Yann Hautier, Michel Loreau, and Shaopeng Wang for constructive comments that significantly improved this chapter and the US National Science Foundation (DEB-1342754, DEB-1856318, and CBET-1833988) for supporting our research.

8.9. References

Awasthi, A., Singh, M., Soni, S.K., Singh, R., Kalra, A. (2014). Biodiversity acts as insurance of productivity of bacterial communities under abiotic perturbations. *The ISME Journal*, 8(12), 2445–2452.

Becker, J., Eisenhauer, N., Scheu, S., Jousset, A. (2012). Increasing antagonistic interactions cause bacterial communities to collapse at high diversity. *Ecology Letters*, 15(5), 468–474.

Brose, U., Williams, R.J., Martinez, N.D. (2006). Allometric scaling enhances stability in complex food webs. *Ecology Letters*, 9(11), 1228–1236.

Cadotte, M.W., Dinnage, R., Tilman, D. (2012). Phylogenetic diversity promotes ecosystem stability. *Ecology*, 93(8), S223–S233.

Campbell, V., Murphy, G., Romanuk, T.N. (2011). Experimental design and the outcome and interpretation of diversity-stability relations. *Oikos*, 120(3), 399–408.

Cardinale, B.J., Srivastava, D.S., Duffy, J.E., Wright, J.P., Downing, A.L., Sankaran, M. (2006). Effects of biodiversity on the functioning of trophic groups and ecosystems. *Nature*, 443(7114), 989–992.

Craven, D., Eisenhauer, N., Pearse, W.D. *et al.* (2018). Multiple facets of biodiversity drive the diversity–stability relationship. *Nature Ecology and Evolution*, 2(10), 1579–1587.

Diaz, S. and Cabido, M. (2001). Vive la difference: Plant functional diversity matters to ecosystem processes. *Trends in Ecology and Evolution*, 16(11), 646–655.

Donohue, I., Petchey, O.L., Montoya, J.M. *et al.* (2013). On the dimensionality of ecological stability. *Ecology Letters*, 16(4), 421–429.

Downing, A.L., Brown, B.L., Leibold, M.A. (2014). Multiple diversity–stability mechanisms enhance population and community stability in aquatic food webs. *Ecology*, 95(1), 173–184.

Egli, L., Schroter, M., Scherber, C., Tscharntke, T., Seppelt, R. (2020). Crop asynchrony stabilizes food production. *Nature*, 588(7837), E7–E12.

Eisenhauer, N., Scheu, S., Jousset, A. (2012). Bacterial diversity stabilizes community productivity. *PLoS ONE*, 7(3), e34517.

Eisenhauer, N., Barnes, A.D., Cesarz, S. *et al.* (2016). Biodiversity–ecosystem function experiments reveal the mechanisms underlying the consequences of biodiversity change in real world ecosystems. *Journal of Vegetation Science*, 27(5), 1061–1070.

Elton, C.S. (1958). *The Ecology of Invasions by Animals and Plants.* University of Chicago Press, Chicago.

Emmerson, M.C. and Raffaelli, D. (2004). Predator–prey body size, interaction strength and the stability of a real food web. *Journal of Animal Ecology*, 73(3), 399–409.

Ewel, J.J., Celis, G., Schreeg, L. (2015). Steeply increasing growth differential between mixture and monocultures of tropical trees. *Biotropica*, 47(2), 162–171.

Faith, D.P. (1992). Conservation evaluation and phylogenetic diversity. *Biological Conservation*, 61(1), 1–10.

Gonzalez, A. and Loreau, M. (2009). The causes and consequences of compensatory dynamics in ecological communities. *Annual Review of Ecology Evolution and Systematics*, 40, 393–414.

Grimm, V. and Wissel, C. (1997). Babel, or the ecological stability discussions: An inventory and analysis of terminology and a guide for avoiding confusion. *Oecologia*, 109(3), 323–334.

Grman, E., Lau, J.A., Schoolmaster, D.R., Gross, K.L. (2010). Mechanisms contributing to stability in ecosystem function depend on the environmental context. *Ecology Letters*, 13(11), 1400–1410.

Gross, K., Cardinale, B.J., Fox, J.W. *et al.* (2014). Species richness and the temporal stability of biomass production: A new analysis of recent biodiversity experiments. *American Naturalist*, 183(1), 1–12.

Haegeman, B., Arnodi, J.-F., Wang, S.-P., de Mazancourt, C., Montoya, J.M., Loreau, M. (2016). Resilience, invariability, and ecological stability across levels of organization. *bioRxiv*.

Hairston, N.G., Allan, J.D., Colwell, R.K. *et al.* (1968). The relationship between species diversity and stability: An experimental approach with protozoa and bacteria. *Ecology*, 49(6), 1091–1101.

Hector, A., Hautier, Y., Saner, P. *et al.* (2010). General stabilizing effects of plant diversity on grassland productivity through population asynchrony and overyielding. *Ecology*, 91(8), 2213–2220.

Hodgson, D., McDonald, J.L., Hosken, D.J. (2015). What do you mean, "resilient"? *Trends in Ecology and Evolution*, 30(9), 503–506.

Holling, C.S. (1973). Resilience and stability of ecological systems. *Annual Review of Ecology and Systematics*, 4(1), 1–23.

Huang, Y., Chen, Y., Castro-Izaguirre, N. *et al.* (2018). Impacts of species richness on productivity in a large-scale subtropical forest experiment. *Science*, 362(6410), 80–83.

Hurd, L.E., Mellinger, M.V., Wolf, L.L., McNaughton, S.J. (1971). Stability and diversity at three trophic levels in terrestrial successional ecosystems. *Science*, 173(4002), 1134–1136.

Huston, M.A. (1997). Hidden treatments in ecological experiments: Re-evaluating the ecosystem function of biodiversity. *Oecologia*, 110(4), 449–460.

Isbell, F.I., Polley, H.W., Wilsey, B.J. (2009). Biodiversity, productivity and the temporal stability of productivity: Patterns and processes. *Ecology Letters*, 12(5), 443–451.

Isbell, F.I., Craven, D., Connolly, J. *et al.* (2015). Biodiversity increases the resistance of ecosystem productivity to climate extremes. *Nature*, 526(7574), 574–577.

Ives, A.R. and Carpenter, S.R. (2007). Stability and diversity of ecosystems. *Science*, 317(5834), 58–62.

Ives, A.R., Gross, K., Klug, J.L. (1999). Stability and variability in competitive communities. *Science*, 286(5439), 542–544.

Jiang, L. and Pu, Z.C. (2009). Different effects of species diversity on temporal stability in single-trophic and multitrophic communities. *American Naturalist*, 174(5), 651–659.

Jiang, L., Joshi, H., Patel, S.N. (2009). Predation alters relationships between biodiversity and temporal stability. *American Naturalist*, 173(3), 389–399.

Jochum, M., Fischer, M., Isbell, F.I. *et al.* (2020). The results of biodiversity–ecosystem functioning experiments are realistic. *Nature Ecology and Evolution*, 4(11), 1485–1494.

Kohli, M., Borer, E.T., Kinkel, L., Seabloom, E.W. (2019). Stability of grassland production is robust to changes in the consumer food web. *Ecology Letters*, 22(4), 707–716.

Kolasa, J. and Li, B.L. (2003). Removing the confounding effect of habitat specialization reveals the stabilizing contribution of diversity to species variability. *Proceedings of the Royal Society B – Biological Sciences*, 270(Supplement 2), S198–S201.

Kreyling, J., Dengler, J., Walter, J. *et al.* (2017). Species richness effects on grassland recovery from drought depend on community productivity in a multisite experiment. *Ecology Letters*, 20(11), 1405–1413.

Lehman, C.L. and Tilman, D. (2000). Biodiversity, stability, and productivity in competitive communities. *American Naturalist*, 156(5), 534–552.

Leps, J. (2004). Variability in population and community biomass in a grassland community affected by environmental productivity and diversity. *Oikos*, 107(1), 64–71.

Loreau, M. and de Mazancourt, C. (2008). Species synchrony and its drivers: Neutral and nonneutral community dynamics in fluctuating environments. *American Naturalist*, 172(2), E48–E66.

Luckinbill, L.S. (1979). Regulation, stability, and diversity in a model experimental microcosm. *Ecology*, 60(6), 1098–1102.

MacArthur, R. (1955). Fluctuations of animal populations and a measure of community stability. *Ecology*, 36(3), 533–536.

Majekova, M., de Bello, F., Dolezal, J., Leps, J. (2014). Plant functional traits as determinants of population stability. *Ecology*, 95(9), 2369–2374.

Matos, I.S., Menor, I.O., Rifai, S.W., Rosado, B.H.P. (2020). Deciphering the stability of grassland productivity in response to rainfall manipulation experiments. *Global Ecology and Biogeography*, 29(3), 558–572.

May, R.M. (1973). *Stability and Complexity in Model Ecosystems*. Princeton University Press, Princeton.

McCann, K.S. (2000). The diversity–stability debate. *Nature*, 405(6783), 228–233.

McCann, K.S., Hastings, A., Huxel, G.R. (1998). Weak trophic interactions and the balance of nature. *Nature*, 395(6704), 794–798.

McNaughton, S.J. (1977). Diversity and stability of ecological communities: A comment on the role of empiricism in ecology. *American Naturalist*, 111(979), 515–525.

Muraina, T.O., Xu, C., Yu, Q. (2021). Species asynchrony stabilises productivity under extreme drought across Northern China grasslands. *Journal of Ecology*, 109(4), 1665–1675.

O'Gorman, E.J. and Emmerson, M.C. (2009). Perturbations to trophic interactions and the stability of complex food webs. *Proceedings of the National Academy of Sciences of the United States of America*, 106(32), 13393–13398.

Petchey, O.L., Hector, A., Gaston, K.J. (2004). How do different measures of functional diversity perform? *Ecology*, 85(3), 847–857.

van de Peer, T., Verheyen, K., Ponette, Q., Setiawan, N.N., Muys, B. (2018). Overyielding in young tree plantations is driven by local complementarity and selection effects related to shade tolerance. *Journal of Ecology*, 106(3), 1096–1105.

Pfisterer, A.B. and Schmid, B. (2002). Diversity-dependent production can decrease the stability of ecosystem functioning. *Nature*, 416(6876), 84–86.

Pimm, S.L. (1984). The complexity and stability of ecosystems. *Nature*, 307(5949), 321–326.

Pimm, S.L., Jenkins, C.N., Abell, R. *et al.* (2014). The biodiversity of species and their rates of extinction, distribution, and protection. *Science*, 344(6187), 1246752.

van der Plas, F. (2019). Biodiversity and ecosystem functioning in naturally assembled communities. *Biological Reviews*, 94(4), 1220–1245.

Renard, D. and Tilman, D. (2019). National food production stabilized by crop diversity. *Nature*, 571(7764), 257–260.

Roger, F., Bertilsson, S., Langenheder, S., Osman, O.A., Gamfeldt, L. (2016). Effects of multiple dimensions of bacterial diversity on functioning, stability and multifunctionality. *Ecology*, 97(10), 2716–2728.

Romanuk, T.N. and Kolasa, J. (2004). Population variability is lower in diverse rock pools when the obscuring effects of local processes are removed. *Ecoscience*, 11(4), 455–462.

Romanuk, T.N., Vogt, R.J., Kolasa, J. (2006). Nutrient enrichment weakens the stabilizing effect of species richness. *Oikos*, 114(2), 291–302.

Romanuk, T.N., Vogt, R.J., Young, A., Tuck, C., Carscallen, M.W. (2010). Maintenance of positive diversity–stability relations along a gradient of environmental stress. *PLoS ONE*, 5(4), e10378.

Roscher, C., Weigelt, A., Proulx, R. *et al.* (2011). Identifying population- and community-level mechanisms of diversity–stability relationships in experimental grasslands. *Journal of Ecology*, 99(6), 1460–1469.

van Ruijven, J. and Berendse, F. (2007). Contrasting effects of diversity on the temporal stability of plant populations. *Oikos*, 116(8), 1323–1330.

van Ruijven, J. and Berendse, F. (2010). Diversity enhances community recovery, but not resistance, after drought. *Journal of Ecology*, 98(1), 81–86.

Ruppert, J.C., Harmoney, K., Henkin, Z. *et al.* (2015). Quantifying drylands' drought resistance and recovery: The importance of drought intensity, dominant life history and grazing regime. *Global Change Biology*, 21(3), 1258–1270.

Schnabel, F., Schwarz, J.A., Danescu, A. *et al.* (2019). Drivers of productivity and its temporal stability in a tropical tree diversity experiment. *Global Change Biology*, 25(12), 4257–4272.

Shurin, J.B., Borer, E.T., Seabloom, E.W. *et al.* (2002). A cross-ecosystem comparison of the strength of trophic cascades. *Ecology Letters*, 5(6), 785–791.

Shurin, J.B., Gruner, D.S., Hillebrand, H. (2006). All wet or dried up? Real differences between aquatic and terrestrial food webs. *Proceedings of the Royal Society B: Biological Sciences*, 273(1582), 1–9.

Steiner, C.F., Long, Z.T., Krumins, J.A., Morin, P.J. (2005). Temporal stability of aquatic food webs: Partitioning the effects of species diversity, species composition and enrichment. *Ecology Letters*, 8(8), 819–828.

Thebault, E. and Fontaine, C. (2010). Stability of ecological communities and the architecture of mutualistic and trophic networks. *Science*, 329(5993), 853–856.

Thebault, E. and Loreau, M. (2005). Trophic interactions and the relationship between species diversity and ecosystem stability. *American Naturalist*, 166(4), E95–E114.

Thibaut, L.M. and Connolly, S.R. (2013). Understanding diversity–stability relationships: Towards a unified model of portfolio effects. *Ecology Letters*, 16(2), 140–150.

Tilman, D. and Downing, J.A. (1994). Biodiversity and stability in grasslands. *Nature*, 367(6461), 363–365.

Tilman, D., Reich, P.B., Knops, J.M.H. (2006). Biodiversity and ecosystem stability in a decade-long grassland experiment. *Nature*, 441(7093), 629–632.

Tredennick, A.T., de Mazancourt, C., Loreau, M., Adler, P.B. (2017). Environmental responses, not species interactions, determine synchrony of dominant species in semiarid grasslands. *Ecology*, 98(4), 971–981.

Venail, P., Gross, K., Oakley, T.H. *et al.* (2015). Species richness, but not phylogenetic diversity, influences community biomass production and temporal stability in a re-examination of 16 grassland biodiversity studies. *Functional Ecology*, 29(5), 615–626.

Vogt, R.J., Romanuk, T.N., Kolasa, J. (2006). Species richness–variability relationships in multi-trophic aquatic microcosms. *Oikos*, 113(1), 55–66.

Wagg, C., O'Brien, M.J., Vogel, A. *et al.* (2017). Plant diversity maintains long-term ecosystem productivity under frequent drought by increasing short-term variation. *Ecology*, 98(11), 2952–2961.

Wang, S.P. and Loreau, M. (2016). Biodiversity and ecosystem stability across scales in metacommunities. *Ecology Letters*, 19(5), 510–518.

Wardle, D.A. (2016). Do experiments exploring plant diversity–ecosystem functioning relationships inform how biodiversity loss impacts natural ecosystems? *Journal of Vegetation Science*, 27(3), 646–653.

Wilson, J.B., Peet, R.K., Dengler, J., Partel, M. (2012). Plant species richness: The world records. *Journal of Vegetation Science*, 23(4), 796–802.

Wootton, J.T. and Emmerson, M. (2005). Measurement of interaction strength in nature. *Annual Review of Ecology Evolution and Systematics*, 36, 419–444.

Wright, A.J., Ebeling, A., de Kroon, H. *et al.* (2015). Flooding disturbances increase resource availability and productivity but reduce stability in diverse plant communities. *Nature Communications*, 6, 6092.

Yachi, S. and Loreau, M. (1999). Biodiversity and ecosystem productivity in a fluctuating environment: The insurance hypothesis. *Proceedings of the National Academy of Sciences USA*, 96(04), 1463–1468.

9

Biodiversity and Temporal Stability of Naturally Assembled Ecosystems Across Spatial Scales in a Changing World

Yann HAUTIER[1] and Fons VAN DER PLAS[2]

[1] *Ecology and Biodiversity Group, Utrecht University, The Netherlands*

[2] *Plant Ecology and Nature Conservation Group, Wageningen University and Research, The Netherlands*

9.1. Introduction

Both theory and experiments have demonstrated the positive effect of biodiversity on the temporal stability of aggregated ecosystem properties (Chapters 7 and 8). Hereafter, by temporal stability, we mean the temporal invariability of ecosystem properties (e.g. primary productivity) measured as the inverse of the coefficient of variation. While biodiversity experiments have established the causal effects of biodiversity on functional stability, the transferability of these results to the management of real-world ecosystems has been questioned for several reasons (Wardle 2016).

First, most experiments simulate a random loss of diversity from a local species pool and minimize variability in abiotic conditions. However, in natural ecosystems, species loss is not random but the result of multiple factors such as nutrient availability, climatic conditions, and land use (Selmants *et al.* 2012). For example, nutrient

The Ecological and Societal Consequences of Biodiversity Loss, coordinated by Michel LOREAU, Andy HECTOR, and Forest ISBELL.

enrichment usually leads to the dominance of a few fast-growing or taller species that exclude slow-growing or smaller species due to increased competition for light (Hautier *et al.* 2009). As a result, in natural systems, dominant and rare species have an unequal probability of being lost, with rare species being more at risk (Gaston 2008). Additionally, biodiversity is likely not the only driver of ecosystem function responses and abiotic conditions could outweigh biodiversity effects (Diaz *et al.* 2007).

Second, most experiments prevent immigration by continuously removing non-target species. This limits the role played by species dispersal and species sorting in maintaining both biodiversity and ecosystem functioning across temporal and spatial scales (Leibold *et al.* 2017). For example, dispersal helps species with different environmental optima to effectively track spatial changes in local environmental conditions, promoting species persistence and increasing ecosystem functioning (Loreau *et al.* 2003).

Third, experimental studies have primarily focused on plant responses at relatively small spatial scales (i.e. within plots with a median size of 3 m^2) (Cardinale *et al.* 2012). This hinders our ability to predict the extent to which biodiversity will maintain ecosystem services at broader spatial scales most relevant for policy, nature management, and biodiversity conservation (Isbell *et al.* 2017). Understanding whether biodiversity safeguards ecosystem functioning against environmental fluctuations in natural ecosystems at larger spatial scales has thus become a major challenge of modern ecology (Isbell *et al.* 2017; Gonzalez *et al.* 2020). This is of particular importance given rapid biodiversity changes at multiple spatial scales due to anthropogenic activities, including habitat loss and fragmentation, climate change, pollution, overexploitation, and species introductions (Chase *et al.* 2019; McGill *et al.* 2015).

Although earlier examples exist (Dodd *et al.* 1994; McNaughton 1978), a new generation of research quantifying biodiversity–stability relationships in natural and semi-natural ecosystems has emerged. These studies can be classified into two types. First, there are observational studies linking natural gradients of biodiversity with the temporal stability of ecosystem functioning (Path 1 of Figure 9.1). Second, there are global change experiments assessing how environmental drivers influence temporal stability directly (Path 3 of Figure 9.1) or indirectly by jointly changing biodiversity and stability (Paths 2 and 3 of Figure 9.1) or by changing the biodiversity–stability relationship (Path 4 of Figure 9.1).

Since biodiversity and ecosystem functioning are shaped by environmental drivers and global change factors, the link between biodiversity and stability in these studies is correlational and inference relies upon statistical control of covariates (Duffy *et al.* 2017). However, confident inference of causal links is limited due to the high probability of missing important variables that can confound the relationship between them (Ferraro *et al.* 2019). Despite this limitation, studies of natural ecosystems are a crucial step towards transferring knowledge from controlled experiments with high internal validity (confidence in the result) to natural settings with high external validity (generality of the result).

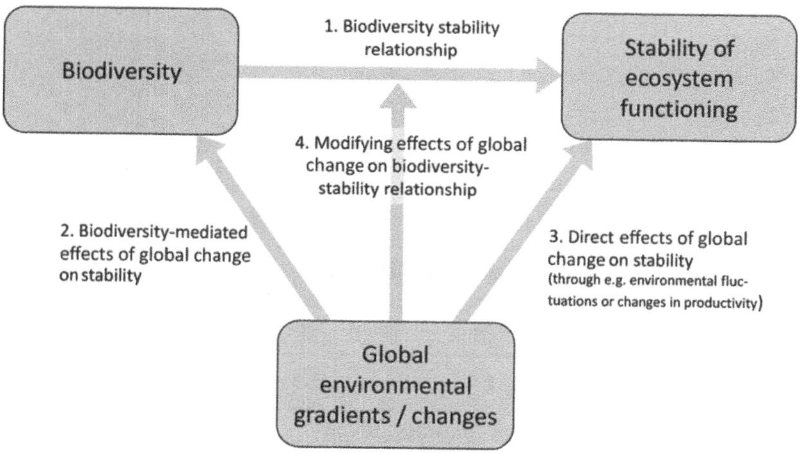

Figure 9.1. *Conceptual framework illustrating how global change drivers (e.g. fertilization, increased livestock densities, and climate change) can affect biodiversity, stability, and their relationships*

Concurrently to the emergence of these empirical studies in natural ecosystems, new theoretical developments have contributed to clarifying the mechanisms by which biodiversity can stabilize functioning at different spatial scales (Chapters 4 and 7). Local species diversity (α-diversity) can provide *local insurance* effects to enhance community stability (α-stability) because different species with different functional traits exhibit asynchronous temporal responses to their shared local environment (species asynchrony) (Figure 9.2). Similarly, variation in species composition among local communities (β-diversity) can provide *spatial insurance* effects to enhance stability at the larger spatial scale (γ-stability) because communities with different species compositions exhibit asynchronous responses to a spatially correlated environment (spatial asynchrony) (Figure 9.2). Hence, global change drivers that reduce biodiversity in local communities or homogenize community composition across space should reduce the local or spatial insurance effects of alpha or beta diversity respectively (Path 2 of Figure 9.1). Additionally, global change drivers may alter the stabilizing effects of biodiversity, leading to a decoupling of biodiversity and stability in systems highly altered by global change drivers (Path 4 of Figure 9.1) (Hautier *et al.* 2020; Hautier *et al.* 2014).

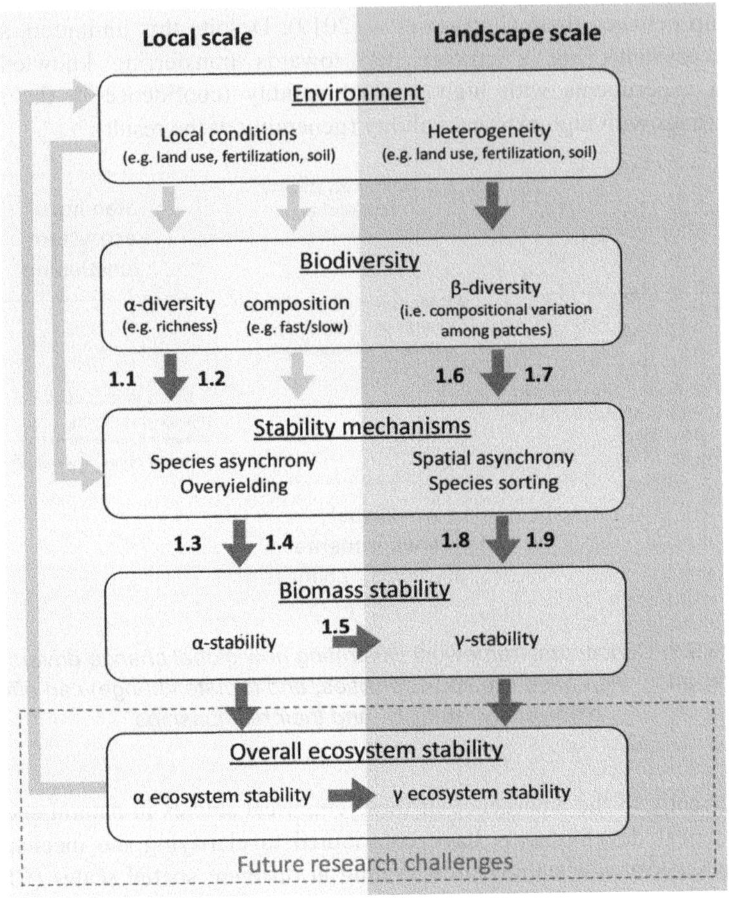

Figure 9.2. *Mechanisms by which environmental conditions (including global change drivers) influence biodiversity, and by which biodiversity influences stability in biomass production, at both local and landscape scales. Eventually, "overall ecosystem stability", that is, the stability of multiple ecosystem functions (not exclusively biomass production), may be influenced by biodiversity, which is generally an avenue for future research. Paths depict hypothesized, causal relationships and are green when expected to be positive and yellow when expected to be context dependent. Numbered paths depict relationships that receive special attention in this review. Path 1.1: effect of α-diversity on species asynchrony. Path 1.2: effect of α-diversity on overyielding. Path 1.3: effect of species asynchrony on α-stability. Path 1.4: effect of overyielding on species asynchrony. Path 1.5: effect of α-stability on γ-stability. Path 1.6: effect of β-diversity on spatial asynchrony. Path 1.7: effect of β-diversity on species sorting. Path 1.8: effect of spatial asynchrony on γ-stability. Path 1.9: effect of species sorting on γ-stability. For a color version of this figure, see www.iste.co.uk/loreau/biodiversity.zip*

Here, we first review the literature and assess the balance of evidence regarding the direction of biodiversity–stability relationships and underlying mechanisms in (semi-)naturally assembled communities at the local and larger spatial scales. Studies include both observational studies linking natural gradients of biodiversity with the temporal stability of ecosystem functioning as well as global change experiments assessing how environmental drivers influence temporal stability, either by jointly changing biodiversity (thus altering biodiversity–stability relationships *across* gradients created by the global change driver), or by changing the biodiversity–stability relationship *within* global change contexts. Next, we discuss the contributions of dominant and rare species to functional stability. Finally, we identify knowledge gaps and opportunities for future research.

9.2. Biodiversity–stability relationships along natural gradients

We found 39 publications assessing biodiversity–stability relationships along natural gradients (see the online supplementary material[1]). Most were carried out in North America, Europe, and Asia (Figure 9.3a), especially in drylands, temperate grasslands, and temperate forests (Figure 9.3b). Some other ecosystem types, such as tropical rainforests and the open ocean, have not been studied at all, despite their widespread global distribution. Furthermore, most studies focused on relationships between plant diversity and the stability of primary production, rather than on higher trophic levels (Figure 9.3c).

Across the 39 publications, 63 biodiversity–stability relationships were tested. The majority (44 out of 63) showed positive relationships between α-diversity and α-stability along natural gradients (Figure 9.4) (e.g. Blüthgen *et al.* 2016). Negative relationships were found in some contexts in two papers only (Polley *et al.* 2007; Jourdan *et al.* 2020). Thus, along natural gradients, more biodiverse communities are generally more stable, and this pattern applies to plants, as well as to higher trophic levels such as birds and bats (Blüthgen *et al.* 2016), fishes (Franssen *et al.* 2011), and invertebrates (Blüthgen *et al.* 2016).

In line with theory (Chapter 7), higher α-diversity was associated with a higher species asynchrony in 21 out of 28 studies (e.g. Gilbert *et al.* 2020), and higher species asynchrony was associated with higher stability in 32 out of 33 studies (e.g. Zhang *et al.* 2018). Various mechanisms can cause asynchrony in the fluctuations of co-occurring species, including stochasticity, species interactions and responses to environmental fluctuations (Chapter 7). Fluctuations in precipitation and temperature throughout years may play a key role in driving relationships between α-diversity

1 Available at: ww.iste.co.uk/loreau/biodiversity.zip.

and species asynchrony, so that asynchronous responses of plants to weather conditions may promote stability in more diverse communities (Gilbert *et al.* 2020). In contrast, Lamy *et al.* (2019) suggested that resource competition, rather than responses to environmental fluctuations, was most important in driving a relationship between α-diversity and α-stability. However, with so few studies assessing which mechanism is most important, this remains an unresolved issue.

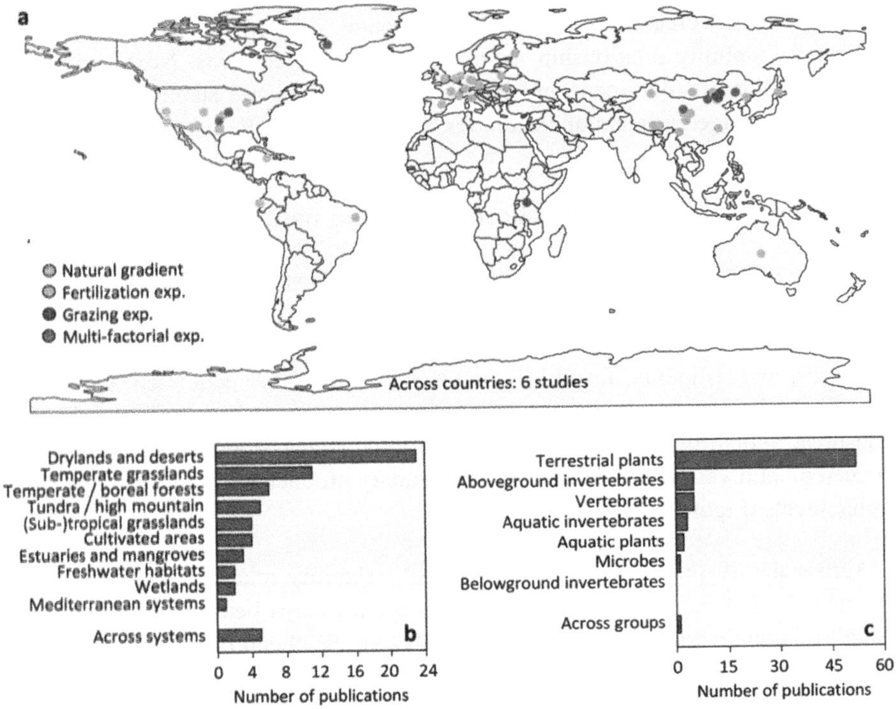

Figure 9.3. *Overview of studies assessing relationships between biodiversity and stability, along natural gradients or in global change experiments. A) Map with locations of studies. Colors indicate whether the study was along a gradient or based on a fertilization experiment, grazing experiment, or a multi-factorial global change experiment. B) Main ecosystem types in which studies were carried out. C) Taxonomic groups for which biodiversity was studied. For a color version of this figure, see www.iste.co.uk/loreau/biodiversity.zip*

Theory suggests that biodiversity may also promote biomass stability through overyielding, that is, higher biomass in mixtures than expected based on monocultures (Chapter 7). While many studies along natural gradients found that diverse communities produce on average more biomass than species-poor communities (e.g. Zhang *et al.* 2018, Chapter 6), evidence that overyielding promotes stability was rather mixed (Figure 9.4), with some studies finding positive relationships (e.g. Dolezal *et al.* 2020), but many others finding neutral relationships (e.g. Gilbert *et al.* 2020). Our results indicate that at the local scale, species asynchrony, rather than overyielding, plays the most important role in underlying positive biodiversity–stability relationships.

Response	Driver	Path	*N*	Results
Spec. asynchrony	α-diversity	1.1	28	
Overyielding	α-diversity	1.2	17	
α-stability	Spec. asynchrony	1.3	33	
α-stability	overyielding	1.4	8	
α-stability	α-diversity	1.1–1.4	63	
γ-stability	α-stability	1.5	5	
γ-stability	α-diversity	1.1–1.5	3	
Spat. asynchrony	β-diversity	1.6	8	
Species sorting	β-diversity	1.7	0	
γ-stability	Spat. asynchrony	1.8	5	
γ-stability	Species sorting	1.9	0	
γ-stability	β-diversity	1.6–1.9	2	

Figure 9.4. *Balance of evidence regarding the different mechanisms by which biodiversity may affect biomass stability. Response: the variable whose response to the driver is assessed. Path: the corresponding path of the assessed relationship in Figure 9.2. N: the number of studies that assessed the relationship. The bars indicate the proportion of positive (green), neutral (yellow), and negative (red) relationships reported in studies, while the arrows indicate whether relationships are, across studies, generally positive (green, up) or unresolved (white, horizontal). For a color version of this figure, see www.iste.co.uk/loreau/biodiversity.zip*

Larger scale γ-stability can simply arise from α-stability (and hence α-diversity) or from processes taking place at larger spatial scales, such as species sorting and spatial insurance effects (Chapter 7). Few studies (e.g. Hautier *et al.* 2020; Wilcox *et al.* 2017) have investigated how α-stability relates to γ-stability, but their outcomes are consistent and show that γ-stability generally increases with α-stability (Figure 9.4). Thus, local scale mechanisms, such as local insurance effects, driving positive biodiversity–stability relationships can propagate to larger spatial scales. Only two studies (Zhang *et al.* 2019; Hautier *et al.* 2020) assessed the overall relationship between β-diversity and γ-stability, which was found to be positive (Figure 9.4). This indicates that larger scale processes may also contribute to positive relationships between biodiversity and stability.

Some studies tested the idea that spatial asynchrony can drive relationships between β-diversity and γ-stability. Although studies are only starting to emerge, most show that β-diversity is positively related to spatial asynchrony (e.g. Hautier *et al.* 2020; Wilcox *et al.* 2017) (Figure 9.4) and thereby to γ-stability (e.g. Catano *et al.* 2020; Hautier *et al.* 2020).

Positive relationships between β-diversity and γ-stability may also occur through species sorting, that is, where patches with different abiotic conditions are occupied by different species, so that each species is present in the environment where it grows best (Loreau *et al.* 2003). However, the importance of this biomass-enhancing mechanism in driving relationships between β-diversity and γ-stability has not been directly assessed along natural gradients (Figure 9.4). That said, there is indirect evidence that this mechanism may be important, as different species maximize biomass production in different environments (Isbell *et al.* 2011) and because β-diversity can be positively related to biomass production (Grman *et al.* 2018). Thus, it is possible that β-diversity can promote γ-stability through species sorting processes, although this merits further study.

9.3. Global change drivers and biodiversity–stability relationships

We found 27 studies experimentally assessing how single (e.g. McNaughton 1985) or multiple (e.g. Ma *et al.* 2017) global change drivers influence relationships between biodiversity and stability (see the online supplementary material). Effects of fertilization were most often assessed (19 studies), while effects of grazing (7 studies), warming (6 studies), and changes in precipitation (4 studies) or other global change drivers (e.g. fire, increased CO_2, tilling and mowing, in single studies only) were less frequently assessed (Figures 9.3 and 9.5). Many of the studies assessed the joint effects of the given global change driver on biodiversity and stability directly (Paths 2 and 3 in Figure 9.1), while modifying effects on biodiversity–stability relationships (Path 4 in Figure 9.1) were less frequently assessed. Almost all studies focused on plant diversity and its relationship to the stability of primary production, except for Wagg *et al.* (2018), who studied how microbial diversity related to their biomass stability in experimentally disturbed soils.

Response	Driver	Path	*N*	Results
Biodiversity and stability	Fertilization	2+3	19	
Biodiv-stability relationship	Fertilization	4	2	
Biodiversity and stability	Grazing	2+3	7	
Biodiv-stability relationship	Grazing	4	2	
Biodiversity and stability	Warming	2+3	6	
Biodiv-stability relationship	Warming	4	1	
Biodiversity and stability	Drought	2+3	5	
Biodiv-stability relationship	Drought	4	1	

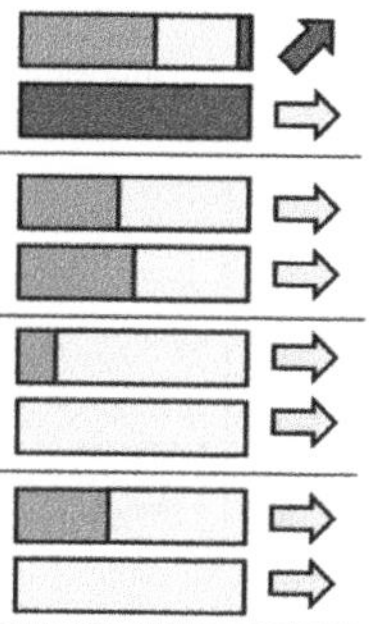

Figure 9.5. *Balance of evidence regarding how global change drivers affect biodiversity, biomass stability, and their relationship, either through joint responses of biodiversity and biomass stability or through direct effects on their relationship. Response: the variable whose response to the driver is assessed. Path: the corresponding path of the assessed relationship in Figure 9.1. N: the number of studies that assessed the relationship. For the response "biodiversity and stability", the bars indicate the proportion of cases where the variables responded qualitatively in the same direction (either both positive or both negative; Figure 9.6a–c) to the global change driver (green), in opposite directions (biodiversity positive and stability negative, or vice versa; red; Figure 6d–f), or whether responses were unrelated, because at least one of the variables did not respond significantly to the global change driver (Figure 9.6g–i). The bars on lines assessing the "biodiversity–stability relationship" as the response indicate whether this relationship was strengthened (more positive or less negative; green), not affected, or weakened (less positive or more negative; red) by the global change driver. The arrows indicate whether relationships were, across studies, generally positive (green, up), negative (red, down), or unresolved (white, horizontal). For a color version of this figure, see www.iste.co.uk/loreau/biodiversity.zip*

Most studies showed that biodiversity and stability respond similarly to fertilization (Figure 9.5, Hautier *et al.* 2015), which is in line with existing meta-analyses (Midolo *et al.* 2019; Avolio *et al.* 2020). The joint negative responses suggest that along gradients in fertilization, biodiversity and stability should be positively related to each other (right panels in Figure 9.6), as found in most studies (e.g. Zhang *et al.* 2017). However, studies assessing the impact of fertilization on the relationship between biodiversity and stability found weaker relationships under fertilized compared to unfertilized conditions (e.g. Hautier *et al.* 2020), as conceptually illustrated in Figure 9.6c.

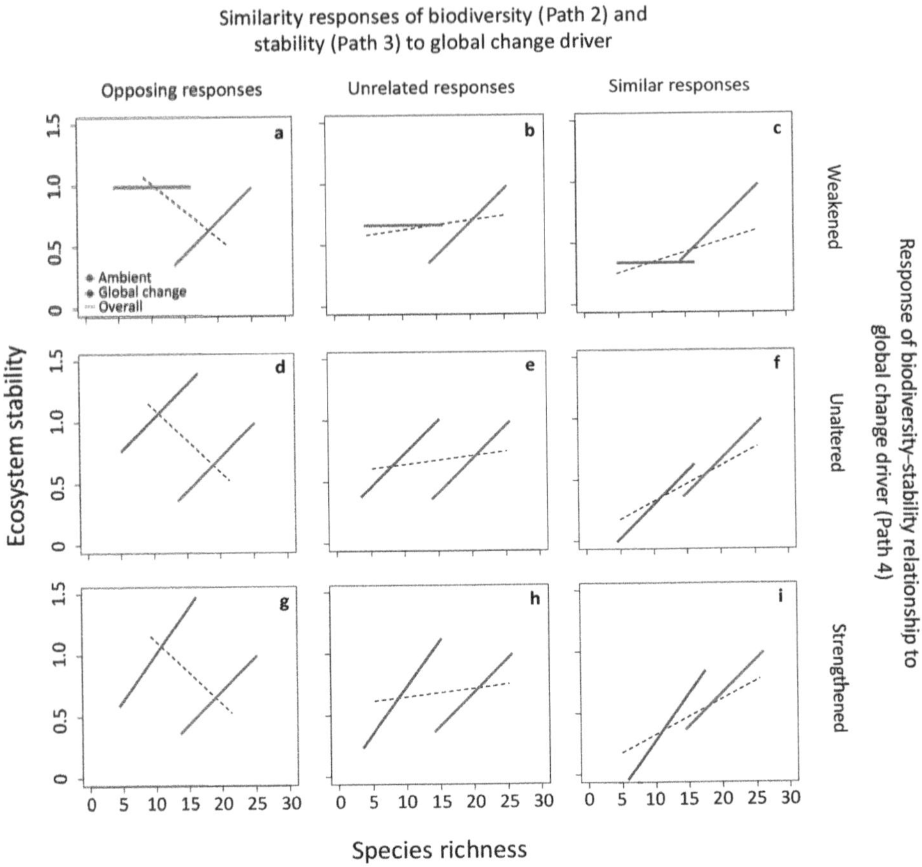

Figure 9.6. *Possible scenarios of how biodiversity, stability, and relationships between them respond to global change drivers. a, b, c: scenarios where the relationship between biodiversity and stability is weakened by the global change driver (compare the blue and red relationships). d, e, f: scenarios where the relationship between biodiversity and stability is unaltered by the global change driver. g, h, i: scenarios where the relationship between biodiversity and stability is strengthened by the global change driver. a, d, g: biodiversity and stability have opposing responses (positive and negative, respectively) to the global change driver, causing a negative overall relationship (dashed black line) between them. b, e, h: biodiversity and stability have unrelated responses (neutral and negative, respectively) to the global change driver, thereby having a limited effect on the overall relationship between them. c, f, i: biodiversity and stability have similar responses (both negative, respectively) to the global change driver, strengthening the overall relationship between them. For a color version of this figure, see www.iste.co.uk/loreau/biodiversity.zip*

Seven studies on biodiversity–stability relationships assessed the effects of grazing on biodiversity and stability. Three studies reported that biodiversity and stability responded similarly to grazing, although one of these showed joint positive responses (Post 2013), while the others showed joint negative responses (Qin *et al.* 2019; Liang *et al.* 2020) (right panels in Figure 9.6). Four other studies showed qualitatively unrelated responses of biodiversity and stability (e.g. Hautier *et al.* 2015; Xu *et al.* 2020). Thus, while gradients in grazing intensity can cause positive relationships between biodiversity and stability, in some cases they do not. Only two studies investigated whether grazing altered the strength of the relationship between biodiversity and stability. One found a strengthening effect (Post 2013), while the other found a neutral effect (McNaughton 1985). Thus, it is too early to draw conclusions concerning an overall grazing effect on the strength of biodiversity–stability relationships.

Six studies on biodiversity–stability relationships reported on the effects of experimental warming. While one study found joint negative effects of warming on biodiversity and stability (Ma *et al.* 2017), other studies showed unrelated responses (e.g. Yang *et al.* 2020). Only one study (Post 2013) investigated the effect of warming on the strength of the biodiversity–stability relationship, and it found a neutral response (Figure 9.5). This suggests that biodiversity should be equally strongly related to stability in future, warmer climates as in present-day conditions.

Studies on the joint effects of drought on biodiversity and stability are still relatively rare. Two studies showed joint negative responses (Muraina *et al.* 2020) and three other studies showed unrelated responses (e.g. Hautier *et al.* 2015). Only one study assessed the impact of drought on the strength of the biodiversity–stability relationship, and it found no significant response (Muraina *et al.* 2021). Thus, more studies are needed to draw general conclusions on how climate change may alter the relationship between biodiversity and stability.

In summary, global changes can alter biodiversity and stability in many ways. Fertilization usually decreases both plant diversity and stability, leading to positive biodiversity–stability relationships. At the same time, relationships between biodiversity and stability are weakened in fertilized areas (Figure 9.6c). Other global change drivers, such as grazing management, global warming, and precipitation, can strengthen or weaken relationships between biodiversity and stability, although it is still too early to say which scenario is most common. However, our findings suggest that in many scenarios of global change, relationships between biodiversity and stability may weaken. The predominantly positive relationships along natural gradients we observed when reviewing the literature contrast somewhat with the less frequent positive relationships found in other meta-analyses (e.g. Valencia *et al.*

2020), which lumped studies along natural gradients with global change experiments. Possibly, the inclusion of various global change contexts may have weakened the strength of relationships between biodiversity and stability. On the other hand, truly "natural" gradients are extremely rare, and almost all observational studies include sites that vary at least to some extent regarding global change drivers such as land use. Thus, more studies on how global change drivers affect biodiversity, stability, and the relationships between them, are urgently needed.

9.4. Contribution of dominant and rare species to stability

High dominance in natural ecosystems can lead to a stronger contribution of the dominant species to the stability of aggregate properties relative to rare species (Grime 1998), thereby diminishing the role of biodiversity per se (Loreau and de Mazancourt 2013). This would lead to a positive selection effect when the dominant species have lower variance than expected (compared to monocultures or to ambient conditions) and a negative selection effect when the dominant species have higher variance than expected.

Supporting this idea, many real-world studies have shown that the temporal stability of the dominant species disproportionately contributes to the temporal stability of community productivity (e.g. Xu *et al.* 2015). As predicted by theory (Haegeman *et al.* 2016), most of these studies found a positive selection effect where dominant species were more stable than expected. Another way dominance could disproportionately contribute to stability and override diversity effects is through higher species asynchrony; that is, when species asynchrony is higher in communities with low evenness that are dominated by a few species compared with communities with high evenness. This could be the case when species-rich communities contain many species with similar ecological attributes that respond similarly to environmental fluctuations, as found in some studies (Song *et al.* 2020; Valencia *et al.* 2020).

Accordingly, most of the studies discussed above found a stronger contribution of dominant species stability or dominant species asynchrony to α-stability compared to species richness, and a high frequency of neutral or negative biodiversity–stability relationships (Valencia *et al.* 2020). These results contrast with theoretical predictions (de Mazancourt *et al.* 2013), as well as with results from along natural gradients of biodiversity (Figure 9.4) (Houlahan *et al.* 2018; Hautier *et al.* 2020), and from synthetic ecosystems (Cardinale *et al.* 2012), which all show a strong association between biodiversity and temporal stability. This discrepancy can be explained by two reasons.

First, neutral or negative richness–stability relationships found in real-world ecosystems are almost exclusively based on single site studies in which biodiversity and stability respond to global change treatments (but see Polley *et al.* 2007; Jourdan *et al.* 2020). However, the simultaneous effect of global change drivers on both biodiversity and functioning is likely to confound the relationship between biodiversity and functional stability (Huston 1997). For example, a simultaneous increase or decrease in both diversity and stability in response to global changes would lead to a positive diversity–stability relationship (Figure 9.6c,f,i; e.g. Tilman 1996; Ma *et al.* 2020). In contrast, a simultaneous decrease (increase) in diversity and increase (decrease) in stability would lead to a negative diversity–stability relationship (Figure 9.6a,d,g; e.g. Polley *et al.* 2007; Yu *et al.* 2020). Additionally, global changes that simultaneously increase or decrease stability (diversity) but have no effect on diversity (stability; as often happens: see Figure 9.5) would lead to a neutral or very weak diversity–stability relationship (Figure 9.6 b,e,h; e.g. Yang *et al.* 2017).

Second, in synthetic experiments, communities are usually initiated with even relative-abundance distribution and the random loss of species gives equal probability to rare and abundant species of being lost. While patterns of dominance can quickly develop (Hector *et al.* 2010), ecosystem responses in experiments may, at least in the short-term, largely be influenced by an initially very even abundance distribution and depend on the identity of the species being added or excluded from the community (Huston 1997). However, the only experiment, to our knowledge, that manipulated species diversity together with species abundance found no evidence that stabilizing effects of diversity are influenced by abundance distribution (Isbell *et al.* 2009). This could be because experimentally imposed differences in evenness quickly dissipated. Additionally, the range of the diversity gradient may be relatively small in observational studies, and particularly in global change experiments, as compared to that of synthetic experiments (Hautier *et al.* 2015). This may limit diversity effects when dominant species effects are strong.

In addition to the strong contribution of dominant species, there is growing evidence that rare species may significantly and disproportionately contribute to ecosystem functioning (Dee *et al.* 2019) and functional stability (Xiong *et al.* 2020). New theoretical advances have clarified the role of species abundances in shaping the diversity–stability relationships in response to perturbations (Chapter 7). In particular, perturbations that predominantly affect the many rare, highly unstable species lead to a negative diversity–stability relationship. In contrast, perturbations that predominantly affect the dominant, stable species lead to a positive diversity–stability relationship.

Finally, the type of perturbation that dominates depends on the spatial scale considered and thus may determine the diversity–stability relationships at multiple spatial scales (Arnoldi *et al.* 2019). For example, at the small spatial scale, communities are expected to be driven by demographic stochasticity and thus to depend more on the gains and losses of species compared to environmental perturbations. This suggests a stronger role of the rare species at small spatial scales. In contrast, at larger spatial scales, the role of environmental perturbations is expected to become more important compared to demographic stochasticity and dominant species could play a stronger role.

9.5. Future directions

While it is clear that the diversity of various taxonomic groups is related to stability in natural settings, various questions remain. Some of the questions we addressed here remain partly unresolved: for example, to what extent and how biodiversity at larger spatial scales contributes to stability, and to what extent global change drivers alter biodiversity–stability relationships. Biodiversity loss occurs at multiple spatial scales and many communities are homogenizing in their composition (β-diversity loss) (McGill *et al.* 2015). An emerging insight is that these larger scale biodiversity losses also hamper stability, but whether the consequences are less, equally, or more detrimental than local biodiversity loss is unresolved. Furthermore, it is likely that several global change drivers will become more important in the future. For example, continued and increasing global warming is almost inevitable in the coming decades, but to what extent this will strengthen or weaken links between biodiversity and stability is unknown. Thus, increased efforts into the effects of climate change on biodiversity–stability relationships are much needed.

Furthermore, existing biodiversity–stability studies are highly biased towards terrestrial, temperate systems and towards primary producers (Figure 9.3). However, in various understudied ecosystems, stability is also of great importance. For example, the oceans provide a great source of food for humans in the form of fish and seafood. While it is known that a high fish marine diversity is generally associated with high fish biomass (e.g. Lefcheck *et al.* 2019), to what extent this biomass is also more stable over time is unknown. Similarly, tropical forests harbor the most aboveground carbon of all terrestrial systems (Crowther *et al.* 2019). However, relationships between tree diversity and aboveground carbon sequestration are generally weaker in tropical forests than in temperate forests (van der Plas 2019). Whether this is also the case for stability in carbon sequestration is unknown. Thus, we urgently need to investigate the role of biodiversity for stability among higher trophic levels and in both marine and tropical systems.

Future studies could also focus more on how organisms' traits drive relationships between biodiversity and stability. Under optimal conditions (e.g. resource rich environments), primary production is often maximized by plant species that have "fast" traits that maximize photosynthetic rates, such as high specific leaf area (e.g. Grigulis *et al.* 2013). On the other hand, in stressful situations, more conservative traits enable a species to conserve resources and maintain moderate growth (Ruiz-Benito *et al.* 2014). Therefore, it has been proposed, and shown in experiments, that plant communities containing both "slow" and "fast" species are most stable in their biomass production when environmental conditions change over time (Craven *et al.* 2018). Similarly, across drylands, natural gradients in functional diversity are a stronger predictor of stability than species richness per se (Garcia-Palacios *et al.* 2018). It is likely that, for higher trophic levels also, a high functional diversity in traits related to the "speed of life" (i.e. traits related to growth rates) may be crucial for biodiversity–stability relationships.

Another question still hardly addressed is to what extent biodiversity regulates the stability of ecosystem functions other than biomass production. While primary productivity is only one property of ecosystems, ecosystems consist of multiple organisms across multiple trophic levels, which are intricately related to each other. Therefore, a high primary productivity may have propagating effects on higher trophic levels by also promoting their biomass stocks and process rates (e.g. Barnes *et al.* 2018) and may thereby also affect ecosystem functions, such as pollination or soil carbon storage, and their stability. On the other hand, levels of different ecosystem functions can also trade off (Lavorel and Grigulis 2012). The main questions include how 1) biodiversity regulates the stability of other ecosystem functions and 2) whether biodiversity can promote "multifunctional stability", that is, the stability of multiple ecosystem functions simultaneously (Figure 9.2). Orford *et al.* (2016) found that a high pollinator diversity is associated with a higher temporal stability in flower visitation and hence likely with a higher stability in pollination services. Similarly, Wagg *et al.* (2021) found that microbial diversity is positively related to the stability of multiple functions, including plant biomass production, litter decomposition, and carbon assimilation. On the other hand, Sasaki *et al.* (2019) show that plant diversity had an overall negative effect on multifunctional stability. So, while in theory one would expect biodiversity to promote "multifunctional stability", the few studies so far (mostly performed in experimental settings) offer rather mixed evidence. Hence, one of the main open questions is whether, along natural gradients, positive or negative relationships between biodiversity and multifunctional stability are most common.

A key challenge is also to apply our current understanding of biodiversity–stability relationships for the stabilization of ecosystem services that we depend on. One promising study in this direction showed that countries with a higher diversity of crop species also had a higher temporal stability of agricultural production (Renard and Tilman 2019). Famines are typically caused by years in which local agricultural production is lower than in normal years, rather than by overall low productivity across years. Thus, to avoid famines it is important that food production is stable, and the study by Renard and Tilman (2019) suggests that diversifying cropping systems at national levels is a promising way to do so.

In summary, in line with theory and experimental findings, the biodiversity of multiple trophic levels is typically associated with a high stability in biomass production. While these relationships have only been explored for a subset of ecosystem types and organisms, patterns to date are generally consistent, and indicate that the conservation and restoration of biodiversity in natural systems has several benefits for the stability of biomass production. It is possible that this is also true for other types of ecosystem functions, such as soil carbon storage, pollination services, and nutrient cycling. Knowledge on biodiversity and stability in natural systems is starting to have applied benefits for the design of natural and semi-natural systems, and these benefits may become greater with an increased understanding yielded by future studies.

9.6. References

Arnoldi, J.F., Loreau, M., Haegeman, B. (2019). The inherent multidimensionality of temporal variability: How common and rare species shape stability patterns. *Ecology Letters,* 22(10), 1557–1567.

Avolio, M. *et al.* (2020). Temporal variability in production is not consistently affected by global change drivers across herbaceous-dominated ecosystems. *Oecologia,* 194(194), 735–744.

Barnes, A.D. *et al.* (2018). Energy flux: The link between multitrophic biodiversity and ecosystem functioning. *Trends in Ecology and Evolution,* 33(3), 186–197.

Blüthgen, N. *et al.* (2016). Land use imperils plant and animal community stability through changes in asynchrony rather than diversity. *Nature Communications,* 7(10697).

Cardinale, B.J. *et al.* (2012). Biodiversity loss and its impact on humanity. *Nature,* 486(7401), 59–67.

Catano, C.P., Fristoe, T.S., Lamanna, J.A., Myers, J.A. (2020). Local species diversity, β-diversity and climate influence the regional stability of bird biomass across North America. *Proceedings of the Royal Society B-Biological Sciences,* 287(20192520).

Chase, J.M. *et al.* (2019). Species richness change across spatial scales. *Oikos,* 128(8), 1079–1091.

Craven, D. *et al.* (2018). Multiple facets of biodiversity drive the diversity-stability relationship. *Nature Ecology and Evolution,* 2(10), 1579–1587.

Crowther, T.W. *et al.* (2019). The global soil community and its influence on biogeochemistry. *Science,* 365(6455), 772–782.

Dee, L.E., Cowles, J., Isbell, F., Pau, S., Gaines, S.D., Reich, P.B. (2019). When do ecosystem services depend on rare species? *Trends in Ecology and Evolution,* 34(8), 746–758.

Diaz, S., Lavorel, S., de Bello, F., Quetier, F., Grigulis, K., Robson, M. (2007). Incorporating plant functional diversity effects in ecosystem service assessments. *Proceedings of the National Academy of Sciences of the United States of America,* 104(52), 20684–20689.

Dodd, M.E., Silvertown, J., Mcconway, K., Potts, J., Crawley, M. (1994). Stability in the plant communities of the Park Grass experiment – the relationships between species richness, soil pH and biomass variability. *Philosophical Transactions of the Royal Society of London Series B – Biological Sciences,* 346(1316), 185–193.

Dolezal, J. *et al.* (2020). Determinants of ecosystem stability in a diverse temperate forest. *Oikos,* 129(11), 1692–1703.

Duffy, J.E., Godwin, C.M., Cardinale, B.J. (2017). Biodiversity effects in the wild are common and as strong as key drivers of productivity. *Nature,* 549(7671), 261–264.

Ferraro, P.J., Sanchirico, J.N., Smith, M.D. (2019). Causal inference in coupled human and natural systems. *Proceedings of the National Academy of Sciences of the United States of America,* 116(12), 5311–5318.

Franssen, N.R., Tobler, M., Gido, K.B. (2011). Annual variation of community biomass is lower in more diverse stream fish communities. *Oikos,* 120(4), 582–590.

Garcia-Palacios, P., Gross, N., Gaitan, J., Maestre, F.T. (2018). Climate mediates the biodiversity-ecosystem stability relationship globally. *Proceedings of the National Academy of Sciences of the United States of America,* 115(33), 8400–8405.

Gaston, K.J. (2008). Bliodiversity and extinction: The importance of being common. *Progress in Physical Geography-Earth and Environment,* 32(1), 73–79.

Gilbert, B. *et al.* (2020). Climate and local environment structure asynchrony and the stability of primary production in grasslands. *Global Ecology and Biogeography*, 29, 1177–1188.

Gonzalez, A. *et al.* (2020). Scaling-up biodiversity-ecosystem functioning research. *Ecology Letters,* 23(4), 757–776.

Grigulis, K. *et al.* (2013). Relative contributions of plant traits and soil microbial properties to mountain grassland ecosystem services. *Journal of Ecology,* 101(1), 47–57.

Grime, J.P. (1998). Benefits of plant diversity to ecosystems: Immediate, filter and founder effects. *Journal of Ecology,* 86(6), 902–910.

Grman, E., Zirbel, C.R., Bassett, T., Brudvig, L.A. (2018). Ecosystem multifunctionality increases with beta diversity in restored prairies. *Oecologia,* 188(3), 837–848.

Haegeman, B., Arnoldi, J., Wang, S., de Mazancourt, C., Montoya, J.M., Loreau, M. (2016). Resilience, invariability, and ecological stability across levels of organization. *bioRxiv.*

Hautier, Y., Niklaus, P.A., Hector, A. (2009). Competition for light causes plant biodiversity loss after eutrophication. *Science,* 324(5927), 636–638.

Hautier, Y. *et al.* (2014). Eutrophication weakens stabilizing effects of diversity in natural grasslands. *Nature,* 508(7497), 521–525.

Hautier, Y., Tilman, D., Isbell, F., Seabloom, E.W., Borer, E.T., Reich, P.B. (2015). Anthropogenic environmental changes affect ecosystem stability via biodiversity. *Science,* 348(6232), 336–340.

Hautier, Y. *et al.* (2020). General destabilizing effects of eutrophication on grassland productivity at multiple spatial scales. *Nature Communications*, 11, 5375–5384.

Hector, A. *et al.* (2010). General stabilizing effects of plant diversity on grassland productivity through population asynchrony and overyielding. *Ecology,* 91(8), 2213–2220.

Houlahan, J.E. *et al.* (2018). Negative relationships between species richness and temporal variability are common but weak in natural systems. *Ecology,* 99(11), 2592–2604.

Huston, M.A. (1997). Hidden treatments in ecological experiments: Re-evaluating the ecosystem function of biodiversity. *Oecologia,* 110(4), 449–460.

Isbell, F.I., Polley, H.W., Wilsey, B.J. (2009). Biodiversity, productivity and the temporal stability of productivity: Patterns and processes. *Ecology Letters,* 12(5), 443–451.

Isbell, F.I. *et al.* (2011). High plant diversity is needed to maintain ecosystem services. *Nature,* 477(7363), 199–202.

Isbell, F.I. *et al.* (2017). Linking the influence and dependence of people on biodiversity across scales. *Nature,* 546(7656), 65–72.

Jourdan, M., Piedallu, C., Baudry, J., Defossez, E., Morin, X. (2020). Tree diversity and the temporal stability of mountain forest productivity: Testing the effect of species composition, through asynchrony and overyielding. *European Journal of Forest Research*, 140 (273–286).

Lamy, T., Wang, S.P., Renard, D., Lafferty, K.D., Reed, D.C., Miller, R.J. (2019). Species insurance trumps spatial insurance in stabilizing biomass of a marine macroalgal metacommunity. *Ecology,* 100(7).

Lavorel, S. and Grigulis, K. (2012). How fundamental plant functional trait relationships scale-up to trade-offs and synergies in ecosystem services. *Journal of Ecology,* 100(1), 128–140.

Lefcheck, J.S., Innes-Gold, A.A., Brandl, S.J., Steneck, R.S., Torres, R.E., Rasher, D.B. (2019). Tropical fish diversity enhances coral reef functioning across multiple scales. *Science Advances,* 5(3).

Leibold, M.A., Chase, J.M., Ernest, S.K.M. (2017). Community assembly and the functioning of ecosystems: How metacommunity processes alter ecosystems attributes. *Ecology,* 98(4), 909–919.

Liang, M., Liang, C., Hautier, Y., Wilcox, K.R., Wang, S. (2020). Grazing-induced biodiversity loss impairs grassland ecosystem stability at multiple scales. *Authorea.*

Loreau, M. and de Mazancourt, C. (2013). Biodiversity and ecosystem stability: A synthesis of underlying mechanisms. *Ecology Letters,* 16, 106–115).

Loreau, M., Mouquet, N., Gonzalez, A. (2003). Biodiversity as spatial insurance in heterogeneous landscapes. *Proceedings of the National Academy of Sciences of the United States of America,* 100, 12765–12770.

Ma, Z.Y. *et al.* (2017). Climate warming reduces the temporal stability of plant community biomass production. *Nature Communications,* 8.

Ma, F.F. *et al.* (2020). Common species stability and species asynchrony rather than richness determine ecosystem stability under nitrogen enrichment. *Ecosystems*, 24(3), 686–698.

de Mazancourt, C. *et al.* (2013). Predicting ecosystem stability from community composition and biodiversity. *Ecology Letters,* 16(5), 617–625.

McGill, B.J., Dornelas, M., Gotelli, N.J., Magurran, A.E. (2015). Fifteen forms of biodiversity trend in the Anthropocene. *Trends in Ecology and Evolution,* 30(2), 104–113.

McNaughton, S.J. (1978). Stability and diversity of ecological communities. *Nature,* 274(5668), 251–253.

McNaughton, S.J. (1985). Ecology of a grazing ecosystem: The Serengeti. *Ecological Monographs,* 55(3), 259–294.

Midolo, G., Alkemade, R., Schipper, A.M., Benitez-Lopez, A., Perring, M.P., de Vries, W. (2019). Impacts of nitrogen addition on plant species richness and abundance: A global meta-analysis. *Global Ecology and Biogeography,* 28(3), 398–413.

Muraina, T.O. *et al.* (2021). Species asynchrony stabilizes productivity under extreme drought across Northern China grasslands. *Journal of Ecology,* 109(4), 1665–1675.

Orford, K.A., Murray, P.J., Vaughan, I.P., Memmott, J. (2016). Modest enhancements to conventional grassland diversity improve the provision of pollination services. *Journal of Applied Ecology,* 53(3), 906–915.

van der Plas, F. (2019). Biodiversity and ecosystem functioning in naturally assembled communities. *Biological Reviews,* 94(4), 1220–1245.

Polley, H.W., Wilsey, B.J., Derner, J.D. (2007). Dominant species constrain effects of species diversity on temporal variability in biomass production of tallgrass prairie. *Oikos,* 116(12), 2044–2052.

Post, E. (2013). Erosion of community diversity and stability by herbivore removal under warming. *Proceedings of the Royal Society B – Biological Sciences,* 280(1757).

Qin, J. *et al.* (2019). Grazing reduces the temporal stability of temperate grasslands in northern China. *Flora,* 259.

Renard, D. and Tilman, D. (2019). National food production stabilized by crop diversity. *Nature,* 571(7764), 257–260.

Ruiz-Benito, P., Gomez-Aparicio, L., Paquette, A., Messier, C., Kattge, J., Zavala, M.A. (2014). Diversity increases carbon storage and tree productivity in Spanish forests. *Global Ecology and Biogeography,* 23(3), 311–322.

Sasaki, T., Lu, X.M., Hirota, M., Bai, Y.F. (2019). Species asynchrony and response diversity determine multifunctional stability of natural grasslands. *Journal of Ecology,* 107(4), 1862–1875.

Selmants, P.C., Zavaleta, E.S., Pasari, J.R., Hernandez, D.L. (2012). Realistic plant species losses reduce invasion resistance in a California serpentine grassland. *Journal of Ecology,* 100(3), 723–731.

Song, M.H., Chen, J., Xu, X.L., Li, Y.K., Gao, J.Q., Ouyang, H. (2020). Grazing offsets nitrogen enrichment effects on species richness by promoting the random colonization of local species in an alpine grassland. *Ecosystems,* 23(2), 278–291.

Tilman, D. (1996). Biodiversity: Population versus ecosystem stability. *Ecology,* 77(2), 350–353.

Valencia, E. *et al.* (2020). Synchrony matters more than species richness in plant community stability at a global scale. *Proceedings of the National Academy of Sciences of the United States of America,* 117(39), 24345–24351.

Wagg, C., Dudenhoffer, J.H., Widmer, F., van der Heijden, M.G.A. (2018). Linking diversity, synchrony and stability in soil microbial communities. *Functional Ecology,* 32(5), 1280–1292.

Wagg, C., Hautier, Y., Pellkofer, S., Banerjee, S., Schmid, B., van der Heijden, M.G.A. (2021). Diversity and asynchrony in soil microbial communities stabilizes ecosystem functioning. *Elife,* 10.

Wardle, D.A. (2016). Do experiments exploring plant diversity-ecosystem functioning relationships inform how biodiversity loss impacts natural ecosystems? *Journal of Vegetation Science,* 27(3), 646–653.

Wilcox, K.R. *et al.* (2017). Asynchrony among local communities stabilises ecosystem function of metacommunities. *Ecology Letters,* 20(12), 1534–1545.

Xiong, C. *et al.* (2020). Rare taxa maintain the stability of crop mycobiomes and ecosystem functions. *Environmental Microbiology,* 23(4), 1907–1924.

Xu, Z.W. *et al.* (2015). Environmental changes drive the temporal stability of semi-arid natural grasslands through altering species asynchrony. *Journal of Ecology,* 103(5), 1308–1316.

Xu, F.W. *et al.* (2020). Resource enrichment combined with biomass removal maintains plant diversity and community stability in a long-term grazed grassland. *Journal of Plant Ecology,* 13(5), 611–620.

Yang, Z.L. *et al.* (2017). Daytime warming lowers community temporal stability by reducing the abundance of dominant, stable species. *Global Change Biology,* 23(1), 154–163.

Yang, X., Mariotte, P., Guo, J., Hautier, Y., Zhanga, T. (2020). Suppression of arbuscular mycorrhizal fungi decreases the temporal stability of community productivity under elevated temperature and nitrogen addition in a temperate meadow. *Science of the Total Environment,* 762.

Yu, Q.S. *et al.* (2020). Species dominance rather than species asynchrony determines the temporal stability of productivity in four subtropical forests along 30 years of restoration. *Forest Ecology and Management,* 457.

Zhang, Y., Loreau, M., He, N., Zhang, G., Han, X. (2017). Mowing exacerbates the loss of ecosystem stability under nitrogen enrichment in a temperate grassland. *Functional Ecology,* 31(8), 1637–1646.

Zhang, Y. *et al.* (2018). Climate variability decreases species richness and community stability in a temperate grassland. *Oecologia,* 188(1), 183–192.

Zhang, Y., Feng, J.C., Loreau, M., He, N.P., Han, X.G., Jiang, L. (2019). Nitrogen addition does not reduce the role of spatial asynchrony in stabilising grassland communities. *Ecology Letters,* 22(4), 563–571.

PART 4

How Biodiversity Affects Human Societies

10

Biodiversity and Ecosystem Services in Managed Ecosystems

Bernhard SCHMID[1] **and Christian SCHÖB**[2]
[1] *University of Zurich, Switzerland*
[2] *Institute of Agricultural Sciences, ETH Zurich, Switzerland*

10.1. A brief history of the role of biodiversity in managed ecosystems

Humans have influenced ecosystems for a very long time. However, only with the deliberate management of ecosystems for production, combined with breeding, could human population density increase to ever higher levels. This agricultural revolution led to the development of a large number of crops in various regions of the world, which still form the basis of modern-day food production systems. It can be assumed that managed ecosystems in prehistoric times were small-scale and diverse among and within field plots and over time. During much of history, crop rotation was practiced to allow fields to recover from resource extraction or pathogen accumulation. Traditional milpa agriculture in Central America is based on cleared pieces of forest where crops are cultivated on various fields, including various species and varieties, for two or three years before fields are abandoned again (Teran and Rasmussen 1995). A common milpa crop production system combines maize with squash and bean. These so-called "three sisters" work together such that squash covers the ground, beans fix atmospheric nitrogen, and maize provides physical support for the climbing beans. A high diversity of cultured plants can also be found in traditional fruit gardens in Southeast Asia. Although some of the biodiversity in early managed ecosystems may have been due to the inefficiency

The Ecological and Societal Consequences of Biodiversity Loss,
coordinated by Michel LOREAU, Andy HECTOR, and Forest ISBELL.

of weed control or lack of homogenous management of large plots, there must have been an appreciation of benefits of biodiversity for productivity, as for example noted by Darwin and Wallace (1858). They refer to an early experiment that had shown increased productivity of plots sown with more than two or three species or genera of grasses.

To find good mechanistic explanations of beneficial effects of unintended or deliberate management of biodiversity in space and time may have been difficult for farmers at the time. Biodiversity in managed ecosystems often served multiple purposes, such as silvopastoral systems on the Iberian Peninsula (dehesas; Moreno *et al.* 2016), or at least it did not focus on the production of a single crop, such as in the case of unfertilized meadows. Easily observable problems when using single crops for a prolonged time on the same plot include pest outbreaks and less specific phenomena such as soil sickness. However, once the problem of pathogens could be treated with chemicals, what would be the benefit of mixtures compared with monocultures, the latter offering many advantages for production systems due to crop homogeneity? Indeed, after the start of the industrial revolution in Europe, attempts to intensify agriculture by reducing biodiversity in managed ecosystems became mainstream. Inspired by the work of Liebig (1840), viewing production largely as a chemical process of converting nutrients into biomass with the help of light, the effects of different fertilization regimes on productivity were tested. The industrialization of agriculture led to a focus on optimizing environmental conditions by management for high monoculture yields. Under such conditions, it was claimed, there would always be a best monoculture that outperforms mixtures with regard to yields (Harper 1977). Modern agriculture and forestry were consequently focused on monocultures, and it became a common belief that in intensively managed agri- and silvicultural ecosystems there is always a trade-off between biodiversity and productivity.

10.2. Biodiversity as the basis for a new green revolution

Despite the mainstream focus on monocultures, some research on and use of species or genotype mixtures in managed ecosystems were still present during the development of modern agriculture as described above. Thus, the replacement-series approach introduced by de Wit (1960) showed that, in particular in mixtures involving legumes, component species could have greater relative yield than in monocultures, thus together producing a greater so-called "relative yield total" than expected based on their monoculture yields. In this case, the positive mixture effect was explained by the ability of the legumes to fix atmospheric nitrogen, which then also benefitted the other species or at least reduced competition for soil nitrogen. These and other beneficial effects of species mixtures can best be realized in managed ecosystems where not a single crop is to be harvested; for example in hay meadows or pastures. Experimental work (summarized in Sanderson *et al.* 2004) and theoretical considerations (e.g. Vandermeer 1981) in this context provided early incentives for a deeper look into biodiversity–productivity relationships in agriculture; however, they remained largely unrecognized. It was rather ecologists who started the field of biodiversity–ecosystem functioning (BEF) research, at the same time as the United Nations Earth Summit was held in Rio de Janeiro in 1992 and concerns about environmental problems reached the global society at large. The early experiments, often done in grassland ecosystems that mimicked hay meadows or pastures, overall showed a general trend in which biomass production increased linearly with every doubling of plant species richness (Balvanera *et al.* 2006). Many other ecosystem functions, such as soil fertility and pest and weed suppression, also increased with biodiversity (Isbell *et al.* 2017). Furthermore, production stability over the years or in the face of perturbations was generally increased with species richness (Isbell *et al.* 2015, Chapter 7).

While positive effects of biodiversity in the face of environmental variation in time or space can be explained by insurance-type mechanisms (Chapter 7) where different species become important under different conditions, it is more challenging to explain effects of biodiversity under constant environmental conditions (see the online supplementary material[1]). In these cases, it is rather because of the inability of a single species to exploit all resources within that constant environment – for example, because of limits or costs of plasticity – that monocultures may have low performance. Unexploited resources can then be used by other species, which by doing so can contribute to greater total resource extraction by the community and thus also greater ecosystem performance. This mechanism of resource complementarity between species can be particularly important if biotope space is large or if resource environments are complex (Jousset *et al.* 2013). As resource use is related to traits, for example rooting depth in plants, positive BEF relationships under constant environmental conditions should be related to functional trait diversity (Loreau 2000). However, identifying single or even multivariate trait differences among species that are responsible for diversity effects remains a major challenge (van der Plas *et al.* 2020). One reason for this could be that complementarity between species can also involve different "enemy niches", allowing, for example, plants in mixtures to hide from pests or herbivores that may reach high abundance in monocultures, or the reduction of dominant species in mixtures (for the latter, see e.g. Grime *et al.* 1987). In addition to complementarity, there can also be facilitation between species such as the above-mentioned example of legumes fixing nitrogen, which then becomes available for other plants, or maize

1 Available at: ww.iste.co.uk/loreau/biodiversity.zip.

providing physical support for beans. Further examples include direct facilitation among plants by hydraulic lift or indirect facilitation, where beneficial associated species such as soil microbes are enhanced by one plant species (Brooker *et al.* 2021).

The findings of early BEF research have initially been met with skepticism and thought to apply only for newly established ecosystems or ecosystems that cannot be intensively managed (e.g. Huston *et al.* 2000; Thompson *et al.* 2005). However, as experiments matured, it was found that positive BEF relationships increased with time (Meyer *et al.* 2018) and that they were even stronger under intensive than under extensive management (Weigelt *et al.* 2009). This is increasingly being recognized as a potential for ecological intensification in agriculture (Gurr *et al.* 2016). Increasing plant biodiversity in managed ecosystems can also offer solutions to problems caused by high chemical inputs in intensive agriculture, which can be reduced if, for example, nutrients are provided by cover crops or mixed cropping (Li *et al.* 2020) or if pathogen loads are reduced and natural enemy control is increased by adding species, perhaps even weeds, to simplified low-diversity systems. In addition, more diverse managed ecosystems may not only provide fewer disservices, such as greenhouse gas emissions and environmental pollution, but also multiple services in addition to biomass production (Sanderson *et al.* 2004; Isbell *et al.* 2017). This can go as far as the esthetic services of species-rich meadows (Lindemann-Matthies *et al.* 2009) and recreational services of species-rich forests (Lindemann-Matthies *et al.* 2014).

Overall, there is an interesting temporal coincidence between problems caused by conventional production systems focused on maximizing monoculture yields and the arrival of solutions indicated by results of biodiversity research. These not only apply to effects of species richness in multi-species managed ecosystems but also to genotypic richness and variety mixtures in single-species cropping systems. Tremendous yield increases have been achieved by breeding for high monoculture yields in all major crops in the past century, yet these increases have stalled recently or begin to even give way to reductions (Ray *et al.* 2012). Of particular concern are outbreaks of pathogens and pests that could hit varieties over large areas. This has made it clear for some time already that maintaining genetic resources to respond to such challenges is of paramount importance as an insurance strategy but, additionally, that genetic diversity within crop fields can create yield benefits (Østergård *et al.* 2009). In the following we present examples for the use of biodiversity to increase ecosystem services in managed ecosystems, in particular the sustainable provision of production services in agro- and forestry ecosystems. We do this for reasons of focus and brevity. Many of the principles apply to other managed systems, including wetlands (e.g. water purification; Geng *et al.* 2019), freshwater and marine systems (e.g. fisheries or nutritional diversity; Worm *et al.* 2006; Bernhardt and O'Connor 2021), and to multiple other ecosystem services (Balvanera *et al.* 2006), as already indicated above.

10.3. Biodiversity in agriculture

10.3.1. *Crop genetic diversity*

Breeding for high monoculture yields in crops has focused on particular varieties that combine all traits necessary for high performance under particular environmental conditions and management regimes. Advantages of genetically uniform crops are simultaneous maturation and homogenous harvest products, as well as their fulfilment of the need to meet regulatory requirements. In this case, any diversity that might be desirable for increased performance, such as defense against multiple pathogens, has to be stacked into a single genotype, potentially causing costs of plasticity that would be avoided if diversity could be applied at the population level, for example by mixing varieties of different pathogen resistance within a field (Djidjou-Demasse *et al.* 2017).

While the example of within-crop genetic diversity as insurance strategy against pests has been recognized for a long time (Østergård *et al.* 2009), the possibility that resource-use complementarity or facilitation between varieties might further increase yield quantity and quality has so far received little consideration (e.g. Montazeaud *et al.* 2018). Empirical evidence that this "physical basis" contributes almost as much to the performance of multi-variety cropping systems as the "disease basis" has been compiled in a recent meta-analysis of 91 studies involving several crop species (Reiss and Drinkwater 2018). It is very likely that combining varieties specialized in extracting resources at particular times or places or in particular forms and with particular microbial partners would allow further yield increases that cannot be realized with any single variety (Barot *et al.* 2017). In Switzerland, three different wheat cultivar mixtures are currently available (www.ipsuisse.ch). Each cultivar mixture is marketed under a brand (e.g. Isuela), and is composed of two cultivars with complementary characteristics. In the case of Isuela for example, the mixture is composed of a high-quality cultivar with high protein content and a high-yielding cultivar, with the aim of achieving high-quantity yields of top quality. In cases where varieties within a crop field vary with regard to the harvestable product, the varieties may, for example, be planted in alternate rows. This was done with mixtures of two rice varieties, one used for confections and one for staple, in Yunnan, China, where farmers harvested the fields manually (Zhu *et al.* 2000). The more valuable variety in monoculture is highly susceptible to a fungus causing blast disease. The combination of the two varieties at the population level reduced fungal infections to a level where no fungicides were required anymore and in addition the reduced infections allowed for 89% increased yields of the more valuable variety in the mixture compared to in monoculture.

Issues that have to be resolved include the maintenance of varietal characteristics, which is easier in largely self-pollinated crops such as wheat; the acceptance of product heterogeneity; and the question of which traits of varieties are most likely to increase performance via complementarity or facilitation. In a proof-of-principle experiment, Wuest and Niklaus (2018) demonstrated with mixtures of *Arabidopsis thaliana* recombinant inbred lines how overyielding of variety mixtures can be mapped to genomic differences, thus directly identifying the genes involved in complementarity and at the same time making it possible to identify the traits being involved. A further question, which likely will become irrelevant due to the economic benefit of variety mixtures, is their monitoring in cases in which farmers are paid incentives for their use.

10.3.2. *Species diversity in grasslands and intercropping*

Opportunities to manage species diversity in agro-ecosystems arise both in grasslands and with arable crops. The advantage of using biodiversity to increase ecosystem functioning in grassland agro-ecosystems is that mixtures of plant species can directly be used as animal forage and that other ecosystem services such as soil fertility, carbon storage, the biodiversity of other trophic groups and esthetics can simultaneously be increased (Lindemann-Matthies *et al.* 2009; Isbell *et al.* 2017; Lange *et al.* 2021). Beyond the already achieved benefits of grassland mixtures, biodiversity experiments indicate several directions in which biodiversity benefits in managed grassland could be further enhanced. One is to consider the functional traits of species to be combined, such that complementarity and facilitation between species can be enhanced and competition reduced. Another possibility is to select for genotypes within species that increase complementarity and facilitation between species. For example, the Jena Experiment demonstrated evolutionarily enhanced levels of complementarity and facilitation over an eight-year period (Zuppinger-Dingley *et al.* 2014; Schöb *et al.* 2018). Thus, using such selected plant material when sowing biodiverse grassland promises to increase productivity and also resistance in the face of extreme climatic events, as shown by two recent studies (van Moorsel *et al.* 2018, 2021).

Compared with grasslands, annual mixed-species cropping systems are easier to maintain because of the shorter duration. Recent meta-analyses have shown that, in such cases, mixtures generally produced the same yield on 77% of the area that the sum of the monocultures would have required; and this applied to the same degree to mixtures with or without legumes (Martin-Guay *et al.* 2018; Li *et al.* 2020). Importantly, consistent with results from BEF experiments (Craven *et al.* 2016), there were no trade-offs between mixed-cropping benefits and management intensity (e.g. fertilizer inputs), and the observed yield increases could also be converted into economic benefits. There are many possibilities for designing mixed-cropping systems by varying sowing patterns or times (Li *et al.* 2020). However, because most crops have a long history of breeding for high monoculture performance, it is conceivable that even higher benefits could be achieved by mixed cropping if breeding for high mixture yields were implemented at large. It may be necessary to first bring back the genomic resources from older varieties into modern crops before such breeding can be done. Thus, in a recent experiment comparing mixtures of crops with mixtures of their wild relatives, selection for increased biodiversity benefits in productivity was possible for the latter but disrupted by domestication in the former (Chacon-Labella *et al.* 2019).

The above examples are focused on using plant diversity to increase ecosystem services in agriculture, as plants are themselves the first harvested product. However, biodiversity at other trophic levels may also increase the performance of managed ecosystems. For example, it has been shown that mixed-cattle grazing can promote grassland multidiversity and multifunctionality (Wang *et al.* 2019). Furthermore, soil microbial diversity can increase soil fertility and plant production (Wagg *et al.* 2014) and may be managed by avoiding its destruction and potentially increasing it via enhanced plant diversity (Weisser *et al.* 2017; Dietrich *et al.* 2020). Thus, a recent study in Switzerland revealed strong reductions in soil microbial diversity due to the use of more than 50 different chemicals in conventional agriculture, but at the same time it demonstrated a slow reduction of their residues and microbial recovery during decades of organic farming (Riedo *et al.* 2021). Despite the many benefits of biodiversity for ecosystem functioning and services demonstrated in ecological research and novel applications in agriculture, there are still many obstacles to its large-scale use. First of all, we have focused for so long on monocultures that it takes time to change production systems from field management to harvest equipment to newly-bred varieties. There are still incentives for farmers to grow monocultures, including subsidies for inputs, requests for product homogeneity, and the availability of equipment and seed material (Mamine and Farès 2020). Furthermore, there may still be a belief, even at agricultural extension stations, that there is a trade-off between biodiversity and productivity under high-intensity agricultural management. This may be true for a transition period during which additional positive incentives might have to be provided, although these add the burden of monitoring farmer compliance.

10.3.3. *Farm-scale diversity*

Perhaps the easiest way to provide biodiversity-dependent ecosystem services in agriculture is to increase diversity between fields within farms or at even larger scales. A recent study relating remotely-sensed ecosystem functions such as plant productivity and albedo to diversity of land-cover types across Switzerland found similarly positive landscape diversity–landscape functioning relationships as those reported for BEF relationships from experimental studies (Oehri *et al.* 2020). There are many ways in which farm-scale diversity may provide ecological and economic benefits (Pretty 2018; Rosa-Schleich *et al.* 2019). First of all, mixtures of single-crop fields in space and time can serve as insurance in case of environmental fluctuations favoring or disfavoring a particular crop under a particular set of conditions. Furthermore, diversified field margins (Wan *et al.* 2018) or ecological compensation areas, such as flower-rich meadows or weed strips, may provide pollinators or predators of pests to adjacent crop fields, avoiding the costs and detrimental environmental effects of using chemicals (Dainese *et al.* 2019; Sirami 2019; Wan *et al.* 2020).

Farm-scale diversity can become the central principle for sustainable intensification of agriculture, reducing negative impacts of inputs and undesired subsidies while at the same time increasing productivity and economic benefits for farmers, but policymakers, manufacturers, and consumers must also support this major redesign effort (Pretty 2018). This should become easier as the detrimental effects of conventional agriculture on biodiversity and its provisioning of ecosystem services become more apparent (see e.g. Riedo *et al.* 2021) and because yields of conventional agriculture are stagnating (Ray *et al.* 2012). Farm-scale diversity also increases the independence of farmers and consumers from large-scale regional or even global suppliers, thus reducing transport losses of input and output goods and the uniformity of production systems elsewhere, together with their risks of environmental destruction and production failures. Finally, together with the increased physical landscape functioning mentioned above, mixtures of, for example, crop fields, grassland plots, forest patches, and other landscape elements can be preferred by people enjoying the landscape as residents or tourists, as shown in a survey with accordingly manipulated photographs (Junge *et al.* 2015). In some countries, such ecosystem services are often very highly valued, as indicated by survey respondents stating that they would want spending for biodiversity protection to be ranked after retirement schemes and public transportation, but before relations with foreign countries, order and security, and culture and leisure in the expenses of the state (Meinard *et al.* 2015).

10.4. Biodiversity in forestry

10.4.1. *Evidence for positive biodiversity effects on forest ecosystem services*

Forest ecosystem services are not as heavily linked to a single product as, for example, crop production in agricultural systems, but monoculture plantations are also very common in silviculture. Even in cases where multiple ecosystem services, such as production of timber, fruits, carbon storage, and the biodiversity of associated organisms, are the target, such as in the world's largest reforestation program, the Chinese Grain-for-Green Program, most plantations are monocultures (Hua *et al.* 2016). The potential of increasing productivity, carbon storage, and other ecosystem services with mixed plantings was perhaps even less expected for forest than for grassland ecosystems, where the first BEF experiments were carried out (Ammer 2019). This was in part because it had been suggested that the coexistence of different species, especially in highly diverse subtropical or tropical forests, may represent a neutral equilibrium among population dynamics of multiple community members, all with similar niches and similar competitive strengths (Hubbell 2006).

However, sample surveys over the past 15 years with random plot selection and comparative studies, where plots of different tree species richness levels were selected deliberately, often found positive BEF relationships in managed and naturally established forests, especially in the northern hemisphere (Kelty 2006; Baruffol *et al.* 2013; Gamfeldt *et al.* 2013; Liang *et al.* 2016; Chamagne *et al.* 2017; Liu *et al.* 2018). In one comparative study, where several carbon-related ecosystem functions were assessed, a doubling of total ecosystem carbon storage was observed along a richness gradient from 3–20 tree species per plot of 1/15 ha in Chinese subtropical forest (Liu *et al.* 2018). This biodiversity effect was related to both increased tree individual numbers and sizes (Baruffol *et al.* 2013). Scaling up biodiversity effects on forest productivity from a large sample survey, Liang *et al.* (2016) estimated the economic value of biodiversity for maintaining commercial global forest productivity at 166–490 billion (10^9) US dollars per year, an amount that would be further increased if multiple ecosystem services were considered.

In the meantime, several forest biodiversity experiments have been established, using similar designs as those used in grassland biodiversity experiments. Even though these forest plots are still young, early results show similarly positive BEF relationships as found in grassland experiments or in the above-mentioned observational studies carried out in older forests (Sapijanskas *et al.* 2014; Verheyen *et al.* 2016; Williams *et al.* 2017; Huang *et al.* 2018). The similarities concern both mechanisms underpinning biodiversity effects as well as multiple ecosystem functions being enhanced by tree species richness. During the earliest phases of forest establishment, community-average traits and effects of particular species, related to the so-called mass-ratio hypothesis, may dominate (Verheyen *et al.* 2016; Luo *et al.* 2020); but, over time, effects of functional trait diversity and species complementarity, related to the so-called diversity hypothesis, seem to become more important (Huang *et al.* 2018), just as in grassland biodiversity experiments (Meyer *et al.* 2016). Complementarity between tree species may include physical crown complementarity (Williams *et al.* 2017), facilitation regarding water use (Salmon *et al.* 2018), microbe-related nitrogen partitioning (Luo *et al.* 2018), and other mechanisms.

Although most studies about BEF relationships in forests focused on demonstrating effects of tree species richness or other measures of tree diversity on tree growth and primary productivity, other ecosystem functions, often related to primary productivity, were also shown to benefit from increased tree diversity, including above- and belowground carbon storage (Verheyen *et al.* 2016; Huang *et al.* 2018; Liu *et al.* 2018; Li *et al.* 2019), pest resistance (Verheyen *et al.* 2016), maintenance of tree genetic diversity within species (Ang *et al.* 2016), and the biodiversity of associated species (Schuldt *et al.* 2018). These ecosystem functions relate to ecosystem services other than goods production for the direct uses mentioned above (Liang *et al.* 2016). They contribute to regulation of climate, erosion control, soil fertility, pests, and animal habitats and are highly appreciated and demanded by society at large (Carnol *et al.* 2014; Lindemann-Matthies *et al.* 2014). In the latter survey, participants from different sectors expressed a clear preference for diverse forests over monocultures, which was related to expected benefits for recreation and other ecosystem services.

As for agro-ecosystems, there still seems to be a general assumption that production services of forests may show a negative trade-off with biodiversity and the other ecosystem services it enhances. Therefore, in the largest forest biodiversity experiment, BEF-China, plots with the presumably most productive monocultures were planted for comparison. While the two species tested had similar productivity as the most diverse mixtures after 10 years (Huang *et al.* 2018), one of them is now affected by a pest outbreak and may have to be abandoned altogether (Xiaojuan Liu, Institute of Botany, Chinese Academy of Sciences, personal communication). Besides the skepticism regarding positive biodiversity effects, especially for productive ecosystem services, technical difficulties, such as production of seedlings for multi-species forest plantations, management to maintain tree species mixtures, and the availability of harvesting and wood processing equipment, may be further causes for the slow implementation of biodiversity principles in forestry (Ammer 2019). Payment for regulatory ecosystem services, which do not only benefit forest owners, may come from governments or firms that need to compensate their CO_2 output. However, benefits of planting diverse as opposed to monoculture forests, shown to be in the order of a doubling of services (Liang *et al.* 2016; Huang *et al.* 2018; Liu *et al.* 2018), are not being implemented as yet (Di Saccco *et al.* 2021). Indeed, international programs such as REDD+ still consider biodiversity more of an outcome than an input variable of forest management and focus on monitoring carbon storage in existing forests rather than designing new forests for carbon storage (Palomo *et al.* 2019). While the first is important to prevent deforestation, the second would be important to encourage afforestation.

10.4.2. *Ecosystem services provided by agroforestry*

The combination of forestry and agriculture (agroforestry) has a long tradition in many parts of the world (Nair 1993). This includes using forests to feed cattle by collecting forage or direct pasture (silvopasture) or to provide a habitat for crop production (agrosilviculture), in particular fruit trees such as coffee or cocoa. Here, biodiversity is part of management by definition, yet increases in tree species richness may provide benefits in addition to their mere presence, for example, by increasing pollinator diversity and fruit set in the shade of coffee (Klein *et al.* 2003) or providing insurance by allowing tree species compositions to shift in response to climate change, thus allowing the continued use of plots for coffee and cocoa production (Sousa *et al.* 2019). The Mediterranean oak savannahs in the Iberian Peninsula and in California combine a uniform low-density distribution of trees – predominantly oak – with shrubs and highly diverse pasture grasslands or occasionally arable crops. The trees in these ecosystems act as ecosystem engineers that improve the water and nutrient balance of the understory vegetation through shading, hydraulic lift, trapping of dust, or attracting animals and their droppings (Marañon *et al.* 2009). These multiple benefits of trees for the understory pasture or crop are complemented by the use of the tree fruits (e.g. acorns in the case of oaks) as animal feed and the pruned wood as energy source for the heating of houses during winter. This traditionally managed agroforestry system therefore provides a multitude of ecosystem services thanks to its diverse species composition.

Negative environmental effects of intensive agriculture with monoculture cropping systems and the increased need to combine food production in managed ecosystems with co-benefits such as carbon storage and other ecosystem services have helped agroforestry to make a comeback (Kay *et al.* 2019). Recent meta-analyses confirm that conversion from agriculture to agroforestry increases soil carbon by more than 20% and has beneficial effects on other soil ecosystem services – including 50% erosion reduction and increased nutrient availability – and human well-being (de Stefano and Jacobson 2018; Mucane *et al.* 2020).

10.5. Outlook

10.5.1. *Potential of biodiversity to support the next green revolution*

More sustainable production of food and assuring a healthy diet for an increasing human population are major challenges for agriculture. In this chapter, we have shown the potential of biodiversity to address this challenge. Research on the diversification of agroecosystems at the genotype, species, and landscape levels has provided the proof-of-concept that diversification works. The next steps require the validation of these methods at the farm scale and their implementation into mainstream agricultural production. This is a long way to go and will require fundamental adjustments in legislation, education, and thinking, but opportunities will arise and the societal pressure is there to achieve zero hunger, good health and well-being, sustainable cities and communities, responsible consumption and production, climate action, life on land, and partnerships for the goals (Griggs *et al.* 2013). Beyond the practical implementation of the new knowledge about the potential benefits of agroecosystem diversification, we also outlined a number of opportunities for further improvement of diverse agroecosystems, including the optimization of intercropping systems.

10.5.2. *Obstacles*

The profound changes required for the implementation of diversification strategies into mainstream agriculture and forestry suffer in the first instance from the inertia of all the stakeholders involved. The primary production sector is very well established economically and changes to this system require idealism for some and incentives for many others. Therefore, the proposed changes towards more diversified agriculture and forestry might initially require adaptations of the currently applied policy that force producers and consumers to implement these changes for the good of society. For example, many countries currently have policies that subsidize a small number of crop species or production irrespective of sustainability trade-offs. These are an enormous obstacle to diversifying croplands. Thus, changes will likely be slow and will have to deal with resistance from influential stakeholders, depending on their economic interests or conceptual belief systems. We have named numerous challenges throughout this chapter and more have been identified elsewhere (IPES Food 2016). Ongoing and future research and innovation actions, together with political willingness, will allow the required changes to happen in the future.

10.5.3. *Solutions*

Participatory research, where scientists develop research questions and activities together with stakeholders, has proven useful to change a system. Diversification of agriculture and forestry require a systemic transformation. Therefore, transdisciplinary research is crucial and should be at the core of future activities on the topic. Furthermore, the interdisciplinary context of agriculture, spanning from the biological and physical sciences to socioeconomic and cultural aspects, requires novel research initiatives that address the complexity of the transformation towards diversified agriculture from multiple viewpoints. For example, participatory research together with breeders, farmers, and downstream industries might identify promising intercropping systems for the sustainable production of food for healthy diets in large quantities able to fulfill global market needs. For such research to be implemented at a large scale, it should be accompanied by the development of industry products that facilitate its production and policy amendments that incentivize the uptake, but also by extension services and education centers that transfer this knowledge on to practitioners of today and tomorrow. In parallel to supporting diversified agriculture and forestry for production benefits, accounting for the currently under-costed ecological disservices of current-day intensive agriculture – e.g. greenhouse gas emissions or run-offs causing water pollution – might indirectly facilitate the agroecological transformation.

10.6. Acknowledgements

We thank Andy Hector and Forest Isbell for the many helpful comments on the manuscript.

10.7. References

Ammer. C. (2019). Diversity and forest productivity in a changing climate. *New Phytologist*, 221, 50–66.

Ang, C.C., O' Brien, M., Siong Ng, K.K. *et al.* (2017). Genetic diversity of two tropical trees (Dipterocarpaceae) following logging and restoration in Borneo: High genetic diversity in plots with high species diversity. *Plant Ecology and Diversity*, 9, 459–469.

Balvanera, P., Pfisterer, A.B., Buchmann, N. *et al.* (2006). Quantifying the evidence for biodiversity effects on ecosystem functioning and services. *Ecology Letters*, 9, 1146–1156.

Barot, S., Allard, V., Cantarel, E. *et al.* (2017). Designing mixtures of varieties for multifunctional agriculture with the help of ecology. A review. *Agron. Sustain. Dev.*, 37, 13.

Baruffol, M., Schmid, B., Bruelheide, H. *et al.* (2013). Biodiversity promotes tree growth during succession in subtropical forest. *PLoS ONE*, 8, e81246.

Bernhardt, J.R. and O'Connor, M.I. (2021). Aquatic biodiversity enhances multiple nutritional benefits to humans. *Proceedings of the National Academy of Sciences USA*, 118 (15), e1917487118.

Brooker, R.W., George, T.S., Homulle, Z. *et al.* (2021). Facilitation and biodiversity–ecosystem function relationships in crop production systems and their role in sustainable farming. *Journal of Ecology*, 1–14.

Carnol, M., Baeten, L., Branquart, E. *et al.* (2014). Ecosystem services of mixed species forest stands and monocultures: Comparing practitioners' and scientists' perceptions with formal scientific knowledge. *Forestry*, 87, 639–653.

Chacon-Labella, J., Garcia Palacios, P., Matesanz, S., Schöb, C., Milla, R. (2019). Plant domestication disrupts biodiversity effects across major crop types. *Ecology Letters*, 22, 1472–1482.

Chamagne, J., Tanadini, M., Frank, D. *et al.* (2017). Forest diversity promotes individual tree growth in central European forest stands. *Journal of Applied Ecology*, 54, 71–79.

Craven, C., Isbell, F., Manning, P. *et al.* (2016). Plant diversity effects on grassland productivity are robust to both nutrient enrichment and drought. *Philosophical Transactions of the Royal Society B*, 371, 20150277.

Dainese, M., Martin, E.A., Aizen, M.A. *et al.* (2019). A global synthesis reveals biodiversity-mediated benefits for crop production. *Science Advances*, 5.

Darwin, C.R., Wallace, A.R. (1858). On the tendency of species to form varieties; and on the perpetuation of varieties and species by natural means of selection. *Journal of the Proceedings of the Linnean Society of London, Zoology*, 3, 45–62.

Di Sacco, A., Hardwick, K.A., Blakesley, D. *et al.* (2021). Ten golden rules for reforestation to optimize carbon sequestration, biodiversity recovery and livelihood benefits. *Global Change Biology*, 27, 1328–1348.

Dietrich, P., Roscher, C., Clark, T.A., Eisenhauer, N., Schmid, B., Wagg, C. (2020). Diverse plant mixtures sustain a greater arbuscular mycorrhizal fungi spore viability than monocultures after 12 years. *Journal of Plant Ecology*, 13, 478–488.

Djidjou-Demasse, R., Moury, B., Fabre, F. (2017). Mosaics often outperform pyramids: Insights from a model comparing strategies for the deployment of plant resistance genes against viruses in agricultural landscapes. *New Phytologist*, 216, 239–253.

Gamfeldt, L., Snäll, T., Bagchi, R. *et al.* (2013). Higher levels of multiple ecosystem services are found in forests with more tree species. *Nature Communications*, 4, 1340.

Geng, Y., Ge, Y., Luo, B. *et al.* (2019). Plant diversity became more important in N removal when multiple, rather than single N processes were considered in a constructed wetland. *Ecological Applications*, 29, e01965.

Griggs, D., Stafford-Smith, M., Gaffney, O., *et al.* (2013). Sustainable development goals for people and planet. *Nature*, 495, 305–307.

Grime, P., Mackey, J.M.L., Hillier, S.H., Read, D.J. (1987). Floristic diversity in a model system using experimental microcosms. *Nature*, 328, 420–422.

Gurr, G., Lu, Z., Zheng, X. *et al.* (2016). Multi-country evidence that crop diversification promotes ecological intensification of agriculture. *Nature Plants*, 2, 16014.

Harper, J.L. (1977). *Population Biology of Plants*. Academic Press, London.

Hua, F., Wang, X., Zheng, X. *et al.* (2016). Opportunities for biodiversity gains under the world's largest reforestation programme. *Nature Communications*, 7, 12717.

Huang, Y., Chen, Y., Castro-Izaguirre, N. *et al.* (2018). Impacts of species richness on productivity in a large-scale subtropical forest experiment. *Science*, 362, 80–83.

Hubbell, S.P. (2006). Neutral theory and the evolution of ecological equivalence. *Ecology*, 87, 1387–1398.

Huston, M.A., Aarssen, L.W., Austin, M.P. *et al.* (2000). No consistent effect of plant diversity on productivity. *Science*, 289, 1255.

IPES Food (2016). From Uniformity to Diversity: A Paradigm Shift from Industrial Agriculture to Diversified Agroecological Systems. International Panel of Experts on Sustainable Food Systems.

Isbell, F., Craven, D., Connolly, J. *et al.* (2015). Biodiversity increases the resistance of ecosystem productivity to climate extremes. *Nature*, 526, 574–577.

Isbell, F., Adler, P.R., Eisenhauer, N. *et al.* (2017). Benefits of increasing plant diversity in sustainable agroecosystems. *Journal of Ecology*, 105, 871–879.

Jousset, A., Schmid, B., Scheu, S., Eisenhauer, N. (2011). Genotypic richness and dissimilarity opposingly affect ecosystem functioning. *Ecology Letters*, 14, 537–545.

Junge, X., Schüpbach, B., Walter, T., Schmid, B., Lindemann-Matthies, P. (2015). The aesthetic quality of agricultural landscape elements in different seasonal stages in Switzerland. *Landscape and Urban Planning*, 133, 67–77.

Kay, S., Graves, A., Palma, J.H.N. *et al.* (2019). Agroforestry is paying off – Economic evaluation of ecosystem services in European landscapes with and without agroforestry systems. *Ecosystem Services*, 36, 100896.

Kelty, M.J. (2006). The role of species mixtures in plantation forestry. *Forest Ecology and Management*, 233, 195–204.

Klein, A.-M., Steffan-Dewenter, I., Tscharntke, T. (2003). Fruit set of highland coffee increases with the diversity of pollinating bees. *Proceedings of the Royal Society London B*, 270, 955–961.

Lange, M., Roth, V.-N., Eisenhauer, N. *et al.* (2021). Plant diversity enhances production and downward transport of biodegradable dissolved organic matter. *Journal of Ecology*, 109, 1284–1297.

Li, L., Li, S.-M., Sun, J.-H. *et al.* (2007) Diversity enhances agricultural productivity via rhizosphere phosphorus facilitation on phosphorus-deficient soils. *Proceedings of the National Academy of Sciences USA*, 104, 11192–11196.

Li, Y., Bruelheide, H., Scholten, T. *et al.* (2019). Early positive effects of tree species richness on soil organic carbon accumulation in a large-scale forest biodiversity experiment. *Journal of Plant Ecology*, 12, 882–893.

Li, C., Hoffland, E., Kuyper, T.W. *et al.* (2020). Syndromes of production in intercropping impact yield gains. *Nature Plants*, 6, 653–660.

Liang, J., Crowther, T.W., Picard, N. *et al.* (2016). Positive biodiversity–productivity relationships predominant in global forests. *Science*, 354.

Liebig, J. (1840). *Die Chemie in ihrer Anwendung auf Agricultur und Physiologie*. Verlag Vieweg, Braunschweig.

Lindemann-Matthies, P., Junge, X., Matthies, D. (2009). The influence of plant diversity on people's perception and aesthetic appreciation of grassland vegetation. *Biological Conservation*, 143, 195–202.

Lindemann-Matthies, P., Keller, D., Li, X., Schmid, B. (2014). Attitudes towards forest diversity and forest ecosystem services – A cross-cultural comparison between China and Switzerland. *Journal of Plant Ecology*, 7, 1–9.

Liu, X., Trogisch, S., He, J.-S. *et al.* (2018). Tree species richness increases ecosystem carbon storage in subtropical forests. *Proceedings of the Royal Society London B*, 285, 20181240.

Loreau, M. (2000). Biodiversity and ecosystem functioning: Recent theoretical advances. *Oikos*, 91, 3–17.

Luo, S., Schmid, B., de Deyn, G., Yu, S. (2018). Soil microbes promote complementarity effects among co-existing trees through soil nitrogen partitioning. *Functional Ecology*, 32, 1879–1889.

Luo, S., Wagg, C., Schmid, B. *et al.* (2020). Community-wide trait means and variations affect biomass in a biodiversity experiment with tree seedlings. *Oikos*, 129, 799–810.

Mamine, F. and Farès, M. (2020). Barriers and levers to developing wheat–pea intercropping in Europe: A review. *Sustainability*, 12, 6962.

Marañon, T., Pugnaire, F.I., Callaway, R.M. (2009). Mediterranean-climate oak savannas: The interplay between abiotic environment and species interactions. *Web Ecology*, 9, 30–43.

Martin-Guay, M.-O., Paquette, A., Dupras, J., Rivest, D. (2018). The new Green Revolution: Sustainable intensification of agriculture by intercropping. *Science of the Total Environment*, 615, 767–772.

Meinard, Y., Remy, A., Schmid, B. (2017). Measuring impartial preference for biodiversity. *Ecological Economics*, 132, 45–54.

Meyer, S.T., Ebeling, A., Eisenhauer, N. *et al.* (2016). Effects of biodiversity strengthen over time as ecosystem functioning declines at low and increases at high biodiversity. *Ecosphere*, 7, e01619.

Montazeaud, G., Violle, C., Fréville, H. *et al.* (2018). Crop mixtures: Does niche complementarity hold for belowground resources? An experimental test using rice genotypic pairs. *Plant Soil*, 424, 187–202.

van Moorsel, S.J., Hahl, T., Wagg, C. *et al.* (2018). Community evolution increases plant productivity at low diversity. *Ecology Letters*, 21, 128–137.

van Moorsel, S.J., Hahl, T., Petchey, O.L. *et al.* (2021). Co-occurrence history increases ecosystem stability and resilience in experimental plant communities. *Ecology*, 102, e03205.

Moreno, G., Gonzalez-Bornay, G., Pulido, F. *et al.* (2016). Exploring the causes of high biodiversity of Iberian dehesas: The importance of wood pastures and marginal habitats. *Agroforestry Systems*, 90, 87–105.

Mucane, M.N., Sileshi, G.W., Gripenberg, S., Jonsson, M., Pumariño, L., Barriosa, E. (2020). Agroforestry boosts soil health in the humid and sub-humid tropics: A meta-analysis. *Agriculture, Ecosystems and Environment*, 295, 106899.

Nair, P.K.R. (1993). *An Introduction to Agroforestry*. Kluwer Academic Publishers, Dordrecht.

Oehri, J., Schmid, B., Schaepman-Strub, G., Niklaus, P.A. (2020). Terrestrial land-cover type richness is positively linked to landscape-level functioning. *Nature Communications*, 11, 154.

Østergård, H., Finckh, M.R., Fontaine, L. *et al.* (2009). Time for a shift in crop production: Embracing complexity through diversity at all levels. *Journal of the Science of Food and Agriculture*, 89, 1439–1445.

Palomo, I., Dujardin, Y., Midler, E., Robin, M., Sanza, M.J., Pascuala, U. (2019). Modeling trade-offs across carbon sequestration, biodiversity conservation, and equity in the distribution of global REDD+ funds. *Proceedings of the National Academy of Sciences*, 116, 22645–22650.

van der Plas, F., Schröder-Georgi, T., Weigelt, A. *et al.* (2020). Plant traits alone are poor predictors of ecosystem properties and long-term ecosystem functioning. *Nature Ecology and Evolution*, 4, 1602–1611.

Pretty, J. (2018). Intensification for redesigned and sustainable agricultural systems. *Science*, 362, eaav0294.

Ray, D.K., Ramankutty, N., Mueller, N.D., West, P.C., Foley, J.A. (2012). Recent patterns of crop yield growth and stagnation. *Nature Communications*, 3, 1293.

Reiss, E.R. and Drinkwater, L.E. (2018). Cultivar mixtures: A meta-analysis of the effect of intraspecific diversity on crop yield. *Ecology*, 28, 62–77.

Riedo, J., Wettstein, F.E., Rösch, A. *et al.* (2021). Widespread occurrence of pesticides in organically managed agricultural soils – The ghost of a conventional agricultural past? *Environmental Science and Technology*, 55(5), 2919–2928.

Rosa-Schleich, J., Loos, J., Mußhoff, O., Tscharntke, T. (2019). Ecological–economic trade-offs of diversified farming systems – A review. *Ecological Economics*, 160, 251–263.

Salmon, Y., Li, X., Yang, B., Ma, K., Siegwolf, R., Schmid B. (2018). Surrounding species diversity improves sub-tropical seedlings' carbon dynamics. *Ecology and Evolution*, 8, 7055–7067.

Sanderson, M.A., Skinner, R.H., Barker, D.J., Edwards, G.R., Tracy, B.F., Wedin, D.A. (2004). Plant species diversity and management of temperate forage and grazing land ecosystems. *Crop Science*, 44, 1132–1144.

Sapijanskas, J., Paquette, A., Potvin, C., Kunert, N., Loreau, M. (2014). Tropical tree diversity enhances light capture through crown plasticity and spatial and temporal niche differences. *Ecology*, 95, 2479–2492.

Schöb, C., Brooker, R.W., Zuppinger-Dingley, D. (2018). Evolution of facilitation requires diverse communities. *Nature Ecology and Evolution*, 2, 1381–1385.

Schuldt, A., Assmann, T., Brezzi, M. *et al.* (2018). Biodiversity across trophic levels drives multifunctionality in highly diverse forests. *Nature Communications*, 9, 2989.

Sirami, C., Gross, N., Baillod, A.B. *et al.* (2019). Increasing crop heterogeneity enhances multitrophic diversity across agricultural regions. *Proceedings of the National Academy of Sciences*, 116, 16442–16447.

de Sousa, K., van Zonneveld, M., Holmgren, M., Kindt, R., Ordoñez, J.C. (2019). The future of coffee and cocoa agroforestry in a warmer Mesoamerica. *Scientific Reports*, 9, 8828.

de Stefano, A. and Jacobson, M.G. (2018). Soil carbon sequestration in agroforestry systems: A meta-analysis. *Agroforestry Systems*, 92, 285–299.

Teran, S. and Rasmussen, C.H. (1995). Genetic diversity and agricultural strategy in 16th century and present-day Yucatecan Milpa agriculture. *Biodiversity and Conservation*, 4, 363–381.

Thompson, K., Askew, A.P., Grime, J.P., Dunnett, N.P., Willis, A.J. (2005). Biodiversity, ecosystem function and plant traits in mature and immature plant communities. *Functional Ecology*, 19, 355–358.

Vandermeer, J. (1981). The interference production principle: An ecological theory for agriculture. *BioScience*, 31, 361–364.

Vandermeer, J., Lawrence, D., Symstad, A., Hobbie, S. (2002). Effect of biodiversity on ecosystem functioning in managed ecosystems. In *Biodiversity and Ecosystem Functioning: Synthesis and Perspectives*, Loreau, M., Naeem, S., Inchausti, P. (eds). Oxford University Press, Oxford.

Verheyen, K., Vanhellemont, M., Auge, H. *et al.* (2016). Contributions of a global network of tree diversity experiments to sustainable forest plantations. *Ambio*, 45, 29–41.

Wagg, C., Bender, F., Widmer, F., van der Heijden, M.G.A. (2014). Soil biodiversity and soil community composition determine ecosystem multifunctionality. *Proceedings of the National Academy of Sciences*, 111, 5266–5270.

Wan, N.-F., Cai, Y.-M., Shen, Y.-J. *et al.* (2018). Increasing plant diversity with border crops reduces insecticide use and increases crop yield in urban agriculture. *eLife*, 7, e35103.

Wan, N.-F., Zheng, X.-R., Fu, L.-W. *et al.* (2020). Global synthesis of effects of plant species diversity on trophic groups and interactions. *Nature Plants*, 6, 503–510.

Wang, L., Delgado-Baquerizo, M., Wang, D. *et al.* (2019). Diversifying livestock promotes multidiversity and multifunctionality in managed grasslands. *Proceedings of the National Academy of Sciences*, 116, 6187–6192.

Weigelt, A., Weisser, W.W., Buchmann, N., Scherer-Lorenzen, M. (2009). Biodiversity for multifunctional grasslands: Equal productivity in high-diversity low-input and low-diversity high-input systems, *Biogeosciences*, 6, 1695–1706.

Weisser, W.W., Roscher, C., Meyer, S. *et al.* (2017). Biodiversity effects on ecosystem functioning in a 15-year grassland experiment: Patterns, mechanisms, and open questions. *Basic and Applied Ecology*, 23, 1–73.

Williams, L.J., Paquette, A., Cavender-Bares, J., Messier, C., Reich, P.B. (2017). Spatial complementarity in tree crowns explains overyielding in species mixtures. *Nature Ecology and Evolution*, 1, 63.

de Wit, C.T. (1960). *On Competition*. Verslag Landbouwkundig Onderzoek, Wageningen.

Worm, B., Barbier, E.B., Beaumont, N. *et al.* (2006). Impacts of biodiversity loss on ocean ecosystem services. *Science*, 314, 787–790.

Wuest, S.E. and Niklaus, P.A. (2018). A plant biodiversity effect resolved to a single chromosomal region. *Nature Ecology and Evolution*, 2, 1933–1939.

Zhu, Y., Chen, H., Fan, J. *et al.* (2000). Genetic diversity and disease control in rice. *Nature*, 406, 718–722.

Zuppinger-Dingley, D., Schmid, B., Petermann, J.S., Yadav, V., de Deyn, G.B., Flynn, D. (2014). Selection for niche differentiation in plant communities increases biodiversity effects. *Nature*, 515, 108–111.

11

Biodiversity and Human Health: On the Necessity of Combining Ecology and Public Health

Jean-François GUÉGAN[1], Benjamin ROCHE[2], and Serge MORAND[2]

[1] *UMR EPIA, INRAE–VetAgroSup, Saint-Genès-Champanelle, France*

[2] *UMR MIVEGEC, IRD–CNRS–Montpellier University, IRD Montpellier, France*

11.1. Introduction

All human health, and well-being also, largely depends on ecosystem services that derive from biodiversity and their resulting products (Convention on Biological Diversity 2021). The interconnections between biodiversity, ecosystem services, and human health are inherently complex, and they have resulted over time in an extremely limited understanding in modern medicine due to the rise of individual-based, highly technical medicine which would like to free itself from these naturalistic foundations. In resonance, at the end of the 1970s the end of infectious diseases was declared. This optimism was the result of the successes in the fight against diseases due to the development of hygiene, environmental sanitation, the advent of antimicrobials, vaccines, and vaccination programs, including that against smallpox which allowed its eradication, and social progress (Reingold 2000). Unfortunately, the 1980s would toll the bell with the appearance of new infectious disease agents previously unknown to science, problems linked to bacterial resistance to antibiotic drugs, the worrying spread of infections already known, and

The Ecological and Societal Consequences of Biodiversity Loss,
coordinated by Michel LOREAU, Andy HECTOR, and Forest ISBELL.

the emerging understanding of relationships between certain microorganisms and chronic diseases (Desenclos and de Valk 2005).

Less than 20 years after the triumphant speech proclaiming the end of human infectious diseases (wrongly attributed to the US General Surgeon Dr H.W. Stewart), Lederberg and collaborators, in a now seminal book (Lederberg *et al.* 1992) from the US Institute of Medicine on emerging microbial threats to health, posed the problem of the complexity of infectious disease systems, suggesting that they should be studied in their dynamical aspects, taking into account not only host–pathogen interactions and their adaptation, but also social, environmental, economic, technological, and demographic modifications. The following years, and in particular the beginning of the 21st century, would come to confirm these statements, bringing back to light the ideas developed by the French medical microbiologist Charles Nicolle in 1920–1930 (Nicolle 1933). Among the many ideas developed by Nicolle, one in particular revives our attention today: a microbe does not become pathogenic by necessity but by circumstance, precisely those to which we as a society expose it. Apart from recalling the current debates concerning the SARS-CoV-2 virus responsible for COVID-19 and a form of reductionism of thought concerning this pandemic, ecosystem and biodiversity components have always been important determinants to be considered in order to understand human infectious and parasitic diseases.

Human pathogens can be divided into broad categories: they can be facultative or obligate pathogens. Facultative pathogens are microbial forms for which a human host is only one of the ecological niches they can exploit as a resource. They are primarily environmental bacteria and fungi – that is, sapronotic diseases that can occasionally cause infections in human patients – or animal viruses, bacteria, fungi, and parasites that are hosted by wild and domestic animals – that is, zoonotic diseases, which cause important diseases in humans as well. A distinction can be made between facultative and accidental pathogens, with the latter representing pathogens which only occasionally infect weakened or immunocompromised human hosts. As we will see in this chapter, there is no Great Watchmaker in the sense given by the philosopher Voltaire, who would have attributed the first microbial pathogens to the first human populations; this would constitute a fixist vision of the origin of human infectious diseases. All human pathogens have an environmental, animal, or often lesser known soil or plant origin, and through different uses and practices it then acquires these microbes *en route*, causing the past, current and future human infections.

This chapter provides an overview of the biodiversity–health relationships that are either ignored, misunderstood, or misguided by modern infectious disease medicine and biology. In this chapter, we address the challenge by first discussing the pivotal, but unsuspected, role of parasites and microorganisms in ecosystem functioning, stability, and resilience. Applying existing ecological concepts from epidemiology, invasion biology, and population and community ecology, we then discuss the major ecological and epidemiological changes operating worldwide that are strongly affecting the interactions between biodiversity and human health, causing important new emerging risks for humanity. Of particular interest is the possibility of using biodiversity as a functional barrier to stop epidemics and pandemics, and therefore to create synergies between public health and environmental protection strategies. Although this work focuses more on infectious disease, which constitutes only a part of health, we discuss at the end of this chapter the relationships that exist between biodiversity and chronic health. We summarize the most recent research findings; present selected examples at the interfaces between ecosystems, biodiversity, and human infectious diseases; and discuss the absolute need for a refoundation of medicine with more depth, with teaching in medicine and biology that better considers the reality of infectious diseases as complex systems, necessitating integrative and transversal thinking and better collaboration between separated disciplines.

11.2. Microbial biodiversity is a key component of ecosystems

Carl Linnaeus developed a classification system that extended to animals, plants, and rocks only (Achtman and Wagner 2008), and he did not classify microbes in order to refer to bacteria, archaea, microscopic fungi, viruses, and other prions and prions-like forms, some of which cause diseases in plants, animals, and humans. However, since the mid-19th century, binomial Linnaean classification has been used by microbiologists and virologists to designate microbial species and pathogenic agents as well. The Earth contains a huge number of largely uncharacterized microbes, and microbiologists are today struggling to classify them and summarize their genetic diversity, which has resulted in heated debates over the last two decades (Pike *et al.* 2018). The explosion of data during the genomic era has revealed the extraordinary nature of the microbial world, with estimates suggesting that more than 1 trillion (10^{12}) species of bacteria, archea, viruses, and fungi exist (Locey and Kennon 2016; but see Willis (2016) for discussion). It is well-documented that only a very small number of microbial or parasitic species thought to exist today have generated diseases in human populations (Table 11.1). Specifically, about one in a billion microbial species is a human pathogen only (Balloux and van Dorp 2017), and it is therefore interesting to ask the question why so few human pathogens exist in so diverse a microbial universe.

Contrasting with the emerged pathogenic part of the microbial iceberg, the microbial world is apparently endless, with many ecological roles and functions devoted to these microbial species and communities in Earth ecosystems (Sala *et al.* 2008). For instance, parasitism is the most common animal lifestyle for obvious consumers (Combes 2005). Recent years have seen a major reconsideration of the role of microbes in the organization and evolution of visible biodiversity, including human beings, exemplified by biomedical studies on human organ microbiomes' effects on human health (Young 2017; Fan and Pedersen 2021).

Total number of known parasitic and infectious disease agents in humans	>2100	Wardeh *et al.* (2015) Murray *et al.* (2018)
Total number of parasitic and infectious disease agents in humans with good information	1415	Cleaveland *et al.* (2001). See Smith and Guégan (2010) for other values
Number of emerging parasitic and infectious disease agents in humans during the last 60 years (including SARS-CoV-1 and 2 and MERS)	>180	Woolhouse and Gowtage-Sequeria (2005). Re-adjusted in the present work
Number of parasitic and infectious disease agents for which information on their spatial distribution exists	7	Hay *et al.* (2017)

Table 11.1. *Recent tallies of the pathogens known to infect human populations. For emerging pathogens, only new species agents are illustrated and not pathogen strains resistant to drugs. Note that current knowledge about the geographic distribution of human pathogens is extremely limited due to the virtual disappearance of health geography, sacrificed by medicine as a discipline*

One of the most remarkable results concerns the work by Kevin Lafferty and his collaborators on the coastal salt marshes of California, US, and their both micro- and macro-biodiversity. In a series of cogent studies, Lafferty *et al.* (2006) analyzed the role exerted by parasitism on food–web structure and organization in several brackish water ecosystems. They incorporated parasites (e.g. cestodes, trematodes) into the salt marsh food-webs by using subwebs: (1) predator–prey subwebs (these constitute most published food webs), (2) host–parasite subwebs, (3) subwebs that contain links where predators eat parasites (this is the case when parasites use predators as hosts), and, finally, (4) subwebs that consist of parasite–parasite links (Lafferty 2014). They found that the predator–parasite subwebs contain the highest linkage densities of all subwebs, and for the Carpinteria salt marsh, two-thirds of the total links occurred in the parasite subwebs (see Figure 11.1). Including all four subwebs, connectance is three times higher with parasites (i.e. increase by 93%) than without parasites, with parasites having twice the number of hosts as predators have prey. These results show that food webs, and so animal (and plant) communities, are incomplete without parasites and other tiny microbes. As counterintuitive as it sounds, life is made and organized around this invisible microbial world. Most notably, recognition of these microbial links may have important consequences for ecosystem stability and resilience and irremediably poses the question of pathogen eradication, while this disease agent is above all a microbe that originally exercised an essential function in the ecosystem from which it originated. Interestingly, a similar questioning and approach are used today to understand the impact of antimicrobials on human microbiome and individual health.

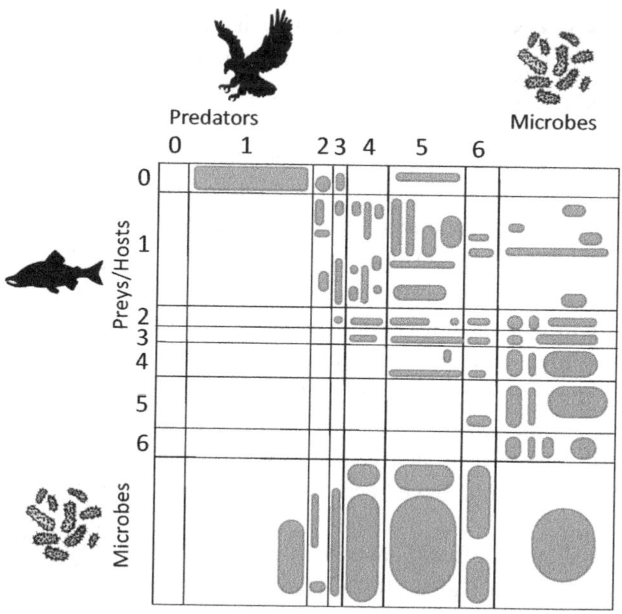

Figure 11.1. *Carpinteria salt marsh food-web divided into four subwebs (see text for explanation). Consumer species are represented as a column, and rows contain the same list of species but considered as prey or hosts. Gray areas indicate the existence of links in the web. Numbers indicates the food-web layer in the food web. Parasites strongly participate in links in the Carpinteria ecosystem as evidenced by this diagram. Redrawn and interpreted from Lafferty et al. (2006). Similar results have been found with other ecosystems*

11.3. The linkages between biodiversity and human infectious diseases

Animal pathogens represent one of the main threats for pandemics, as seen with SARS-CoV-2. Hopefully, few of them will reach the stage of a pandemic, which strongly depends on high human densities, global trade, and travel. These factors facilitate global spread of so-called new emerging pathogens that fulfill the basic requirement of being able to access humans as hosts and show inter-human transmission, directly or indirectly, through anthropophilic arthropod vectors for instance. Nevertheless, the initial event of such catastrophes relies on the circulation of pathogens within wildlife.

Biodiversity can play a prophylactic role against pathogen transmission. Different mechanisms can operate in this context. First, it is worth pointing out that, for a given pathogen, some host species may have high competence (i.e. a high probability of becoming infected and then infectious) while others, in contrast, have low competence. Since pathogens should be adapted to the most abundant species in order to increase their fitness, the most abundant species should be the most competent ones. Therefore, if the less abundant species disappear first, low biodiversity settings would include only highly competent species and high biodiversity settings a high number of host species with a low competence. Such community structuration can (1) decrease contact rates between host species with high competence, (2) "waste" infectious contacts if transmission depends only on the frequency of infectious individuals, and/or (3) decrease the abundance of the host species with high competence through ecological interactions. This "dilution effect", which can be the result of any of these processes, was the focus of a large scientific debate during the last decade (Ostfeld and Keesing 2012). Depending on the metric used, the way of analyzing it, and/or the host–pathogen interaction studied, very different results have been found.

Nevertheless, dilution effects have been observed for West Nile virus (Ezenwa *et al.* 2006; Swaddle and Calos 2008), Hantavirus (Suzán *et al.* 2009; Luis *et al.* 2018), South American cutaneous leishmaniasis (de Thoisy *et al.* 2021), Buruli ulcer (Garcia-Pena *et al.* 2016), and different plant pathogens (Ostfeld and Keesing, 2012). The emblematic example of Lyme disease has been heavily disputed (Randolph *et al.* 2012), while other studies at both small and large scales cast doubts on the generality of such a dilution effect (Salked *et al.* 2012). Notably opposite patterns for an amplification effect have been found for West Nile virus (Levine *et al.* 2017), Hantavirus (Luis *et al.* 2018), cutaneous leishmaniosis (de Thoisy *et al.* 2021), and Buruli ulcer (Garcia-Pena *et al.* 2016), respectively, and this for the same studied disease systems and in the same areas. Today, a potential dilution effect strongly depends on the host–pathogen system. Many pathogens of specialized hosts will become extinct with their hosts, but some may adapt. Therefore, detecting a dilution or an amplification effect also strongly depends on the pathogen detection strategy and the way in which the sampling has been carried out. Furthermore, too much attention has been paid during recent years to looking for a dilution effect as a mandatory result (de Thoisy *et al.* 2021).

The great majority of emerging infectious diseases (EIDs) are zoonoses, which means that the responsible pathogens are transmitted from animals to humans. From a strict academic point of view, a zoonosis corresponds more to the transmission of an infectious disease agent, such as the rabies virus transmitted by mammal species, causing human infection but not inter-human transmission. In this sense, COVID-19 is more pertinent, being a contagious human infectious disease of probable animal origin. In general, pathogens are host-specific and cannot easily overcome species barriers to access new hosts. However, changes in the density of and contact rates between hosts, as well as contact with new hosts, for example, human encroachment in pristine ecosystems, offer opportunities for spillover infections, that is, pathogen transmission from one species to another. Anthropogenic environmental change largely drives ecosystem disturbance, biodiversity loss, and changes in host species communities and, consequently, the emergence of new infectious diseases. Prominent examples include the currently ongoing SARS-CoV-2 pandemic and other emerging pathogens originating from wildlife, such as HIV, Ebola virus, influenza virus, and yellow fever virus.

If the dilution effect is a general phenomenon, it would highlight that the biodiversity erosion observed throughout the world for decades may be a factor contributing to the current era of pathogen emergence. Moreover, it suggests that biodiversity conservation may become an opportunity in public health and open the door to the design of innovative strategies combining environmental protection and public health. Nevertheless, in line with the critiques against the dilution effect, the use of conservation for preventing pathogen emergence has also been the focus of deep attacks (Kilpatrick *et al.* 2017).

First, it is fair to acknowledge that the current knowledge on disease ecology has not reached the stage where we can produce certainties about the public health benefits of biodiversity conservation. Second, it is important to highlight that biodiversity degradation is generally linked with economic activities that support human well-being. It is therefore important to distinguish the relevance of such strategies in low/middle-income and high-income economies.

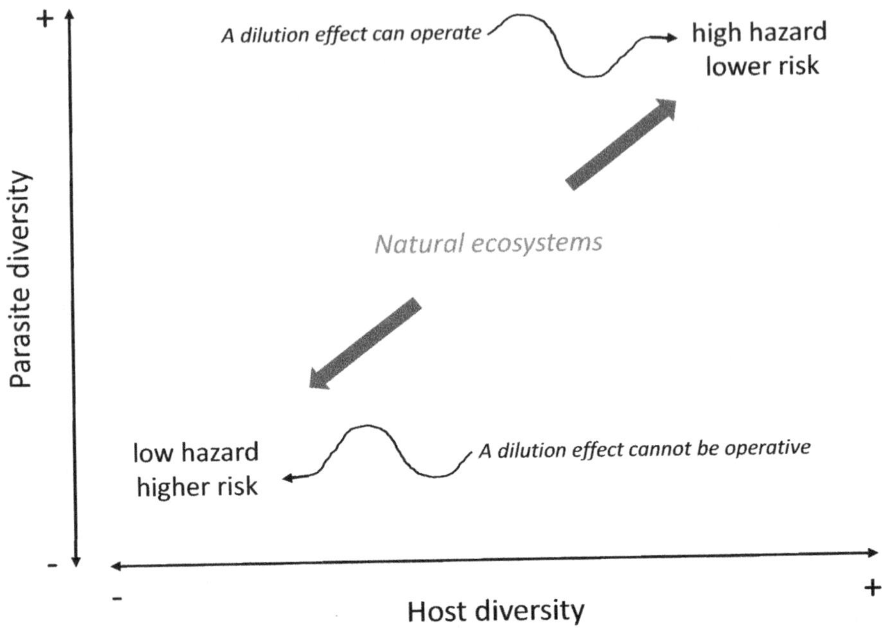

Figure 11.2. *The relationship between parasite diversity and host diversity, and their impact on microbial hazards and pathogen risks. Low host and parasite diversities imply a low hazard (because parasite diversity is weak) but a high disease risk (because the dilution effect cannot work). In contrast high host and parasite diversities yield high microbial hazards (i.e. the diversity of microbes with potential pathogenicity power) but a low pathogen transmission risk thanks to a potential dilution effect. Adapted from Hosseini et al. (2017)*

Finally, it is important to highlight that higher biodiversity in macrofauna also means a higher diversity of microbes or parasites (Figure 11.2; see also previous section). Therefore, such conservation strategies may decrease the risk of pathogen spillover from wildlife (because of lower prevalence) but could increase the microbial hazards resulting from biodiversity (Hosseini *et al.* 2017), since higher biodiversity also means a higher number of microbial species and thus of potential pathogens. To apply such a strategy, it is therefore fundamental to be able to perform hazard management, which is crucially linked to the interface between humans and wildlife, and this interface can be direct or indirect through domestic animals. However, recent evidence suggests that, in human-made ecosystems, disturbance events causing losses of biodiversity drive the presence and abundance of some specific host species and thus also high prevalence rates of pathogens adaptable to humans, which are actually high risk (Halliday *et al.* 2020).

11.4. The evolution of human society is punctuated by epidemiological phases

If life is not quite life – to almost paraphrase the song "Life is Life" by English folk band Noah and the Whale from 2011 – because it consists above all of a myriad of microorganisms and parasites, when then did people contract the first infectious or parasitic disease agents?

Usually, health historians classify three to five periods in the development of infectious and parasitic diseases during the evolution of human society and its organization. Some historians also consider sub-periods, but here we jump to the point. The traditional view is that many human pathogens emerged during the Neolithic revolution sometime around 8,000 to 4,000 BCE (10,000 to 6,000 years ago), even if the starting point of this period is still debated. During that time, humans developed agriculture, learned to raise crops, and kept domestic livestock with a wild origin. The development of the first villages, with several reaching nearly 20,000 inhabitants, then resulted in stable settlements enabled by agricultural subsistence and favoring higher human population growth and density. The proximity of the earliest farmers to livestock, for example sheep and goats, later supplemented by cattle and pigs, and natural elements, for example water and soil when raising barley or wheat, created foci for the development of human pathogens, host reservoirs, and vectors. Permanent dwellings congregating in villages had been achieved by 7,000 BCE in the Tigris and Euphrates river valleys in Mesopotamia, and the first evidence of cultivation and domestication has been dated to roughly 9,500 BCE. This period is called the epidemiological transition of the Neolithic period, and it is accepted that many human pathogens were acquired during this period (see Figure 11.3). Humans are also considered to have come into contact with pathogens long before, but the Neolithic conditions favored the development of the first crowded human infectious diseases (Guégan *et al.* 2007). Other important epidemiological transitions would occur in the era of the development of the great empires and the wars that accompanied them, for example the Roman Empire and the Punic wars, followed by the colonization of other continents by Europeans and the period of land conquests during and after the 15th century. The epidemiological transition of the 19th century constituted a decisive stasis because it led to the development of large cities in the West with the rise of industrialization, for example, in the Victorian period in British history, and massive exoduses from the countryside to cities in search of better life (Smith and Guégan 2010). This transition period is important since it saw the return of human infections that had formerly been controlled, for example cholera fever and malaria, because of greater crowding and significant poverty and the arrival of new infectious disease agents, for example measles (but see below). According to some health historians, the period we have entered, what some call the Anthropocene, unmistakably marks an important epoch in the modern history of human civilization, namely that of the epidemiological transition of new emerging diseases.

The two examples of human measles and pertussis (or whooping cough) can be illustrated here. Human measles is caused by only a few genotypes of a virus that belongs to the genus *Morbillivirus* and the family Paramyxovirus. Human measles virus (MeV) is close to the Rinderpest virus (RPV), which is a pathogen of ruminants, and classical knowledge has traditionally said that MeV originated with the earliest livestock in the Neolithic period. Recent findings based on progress in sequencing technology and phylogenetic reconstruction and dating has shown that MeV diverged from RPV around the 11th and 12th centuries only (Furuse *et al.* 2010), which is much more recent than an appearance in the Neolithic period. Even if the earliest human communities in the ancient Middle East region could have been infected by a progenitor form of the current MeV, the virus probably had different characteristics from the current MeV. This also indicates that the emergence process requires many jump trials, many subsequent failures, and some rarer successes to establish and spread. The second example concerns the childhood disease whooping cough, the causative agent of which is the bacterium *Bordetella pertussis*. Just like MeV, historians state that the human whooping cough appeared in the Neolithic time, but this hypothesis is now under attack. *B. pertussis* infecting humans diverged from an ancestor, *B. bronchiseptica*, hosted by a wide range of wild and domestic animals, for example pigs, rabbits, and goats, around the 16th to 17th centuries (Bjørnstad and Harvill 2005). This coincides well with the dating of the first description of the first cases of whooping cough in humans by medical doctors in Persia. *B. bronchiseptica* is also the sister group of *B. parapertussis*, which causes respiratory disease in sheep. As an aside, during the mad cow disease crisis in France at the beginning of the 1990s, people no longer wanted to consume beef and it was decided to develop rabbit farms for consumption. This resulted in an epidemic recrudescence of asymptomatic pertussis cases in the population due to *B. bronchiseptica*. This anecdote alone reveals the close links that exist between biodiversity, agriculture, and infectious diseases.

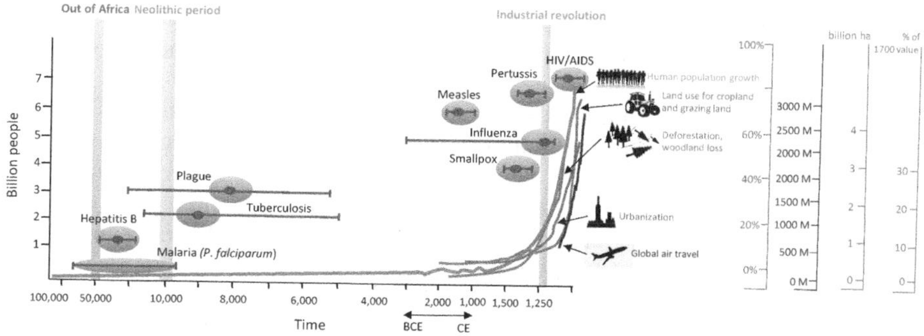

Figure 11.3. *Time of emergence for various human infectious and parasitic diseases based on published information. Horizontal red bars indicate confidence intervals for date of possible appearance. The gray exponential curve corresponds to the increase of the human population on Earth in billions of inhabitants. On the right, the different color scales and corresponding curves indicate the world land-use transformation for cropland and grazing land (in green) in billions of hectares, the world deforestation and woodland loss (in orange) as a percentage of its value in the year 1700, the world urbanization rate over the last 500 years (in blue) as a percentage, and the global air travel in terms of people transportation (in purple) in millions of passengers. Important periods in human evolution and history are indicated with vertical bars. Re-interpreted and modified from Balloux and van Dorp (2017). For a color version of this figure, see www.iste.co.uk/loreau/biodiversity.zip*

11.5. The new ecology and evolution of zoonotic and sapronotic establishment in the Anthropocene

Before infecting humans and animals, bacteria of the genus *Bordetella* are frequently found in soil, water, sediment, and plants, and bacterial strains recovered from environmental samples are evolutionarily ancestral to animal-associated species (Soumana *et al.* 2017). Notably, *B. bronchiseptica*, despite being an animal-adapted pathogen, has preserved the ability to grow and proliferate in soil, and soil or plant rhizosphere can be considered as a probable natural habitat of this group of bacteria, including the animal pathogenic lineages. *Bordetella* thus appears to be a bacterium of environmental origin that adapted and became pathogenic via the acquisition of factors mediating specific interactions with animal hosts and humans, an evolutionary explanation which is still in progress concerning *B. pertussis* strains circulating in humans under the effect of vaccine pressures (Bart *et al.* 2014). Medicine and medical doctors may have not yet fully considered the consequences of this important information in their practice (even if they are aware of this), whereas what is happening before our eyes now is a full-scale experience of what has happened in the evolution of human civilization since the Neolithic period.

In this period of the Anthropocene, the different epidemiological transitions of the past come together, with important differences from one region of the world to another, however. So, rather than being successive, they coexist with each other (Guégan *et al.* 2020): international agriculture developments, for example deforestation, impacting on large natural biomes and their biodiversity; big cities developing in many regions of the world, notably in the tropics; and intercontinental transportation of people, livestock, and goods all increased in an exceptional way (Figure 11.3). SARS-CoV-2 has benefited from these pathways that we have traced and been able spread and cause the COVID-19 pandemic.

The recent evolution of human civilization with a human demography that could reach 11 billion inhabitants by 2100 has created new ecological settings and conditions that may promote interpenetration and contact between wildlife, domestic animals, and humans. Ecological and physical barriers that formerly kept the territories separated between species, peri-urban areas where agriculture and livestock are developed adjoining natural ecosystems and their biodiversity and feeding the markets in cities; all these new constructed conditions today constitute new ecological settings favoring the development of (re-)emerging infections and spillovers through inter-species transmission and host-switching to human individuals and communities.

These current conditions, for example, higher contacts between different animal species, between humans and wildlife, natural habitat alteration and new disease hazards, may expose more humans to animal infections, for example zoonosis, but at the same time put more animals into contact with human infections, for example contagious diseases or reverse zoonosis. These ecological settings facilitate disease transfer, making it very difficult to distinguish between the egg, that is, the natural host reservoir, and the hen, that is, the target host, if it is not the opposite pathway, as evidenced by South American malaria (de Thoisy *et al.* 2021). These independent but interconnected malaria lifecycles allow for disease propagation, transmission, and re-emergence from wildlife in the absence of an infected human host, and the same could happen today for many infectious diseases, as evidenced for dengue fever for instance (de Thoisy *et al.* 2021).

While viruses, and in particular RNA viruses, have become a priority for international health security, let us not forget that there are other forms of pathogenic microbes. Notably, people in the world today, especially children, are currently dying from more infections caused by bacteria, such as bacterial diarrhea, than viruses. Also, while a large number of human infectious and parasitic disease agents are of animal origin (i.e. zoonosis), before this stage, for particular bacteria and microscopic fungi, they have an environmental origin (i.e. sapronosis), as already illustrated with the bacterium causing pertussis. Our modern scientific and medical culture seems to have forgotten this striking fact in the endless pursuit of wanting to understand and fight against the last epidemic and the last circulating virus. The resemblance between plant responses to bacterial virulence factors and the responses of mammalian immune cells serve as evidence that bacterial–plant interactions may have paved the way for bacterial adaptation to animals and humans in a "training-ground" manner (Berg *et al.* 2005).

In particular, the plant rhizosphere and phyllosphere, that is, the phytosphere, should be considered as important reservoirs of microbial forms, some being important obligate or opportunistic human pathogens (e.g. *Clostridium tetani,* which causes tetanos; *Mycobacterium tuberculosis*, responsible for tuberculosis; *Escherichia coli,* frequently found as mutualistic or commensal, or as pathogens in plants and causing intestinal and urinary diseases in humans). Species belonging to the genera *Enterobacter, Serratia, Klebsiella, Erwinia, Pantoea, Burkholderia, Mycobacterium, Bordetella*, and *Pectobacterium* are plant-associated bacteria and have adapted to circulate in animals and humans through horizontal gene transfer from different microbial communities present in their natural and newly occupied environments (van Overbeek *et al.* 2014). The persistence in the plant environment and the arable ecosystem of these types of bacteria or microscopic fungi would increase their chances of transmission to humans via consumption of plant-derived food; van Overbeek *et al.* (2014) coined the term "phytonosis" for this category of disease agents transmitted via plants.

Many *Mycobacterium* species, for example *M. ulcerans, M. avium, M. chelonae*, and *M. marinum*, form a group of nontuberculous mycobacteria (NTMs) that are most commonly found in water-distribution systems and water bodies, including lakes, rivers, and streams, and in soil, dust, and the plant rhizosphere (Falkinham 2013). Many of these NTMs can be acquired by animals and humans from repeated exposure to their environments and notably through contact with water and bioaerosols that can be inhaled into the lungs (Honda *et al.* 2018). Although the origin of many NTMs remains a mystery today, they are intrinsically linked to humans' interactions with their environments, both natural or built, for example mechanical devices, like air conditioners. It is plausible that other types of these bacteria, like *M. tuberculosis,* were historically aerosolized from soil with increased fire-making and infected human lungs, or, like *M. leprae*, transmitted through long-term exposition to mud, which was used in the past and still used in some countries in the soil or walls of huts. *M. ulcerans* causing Buruli ulcer skin ulcerations in patients in many tropical regions of the world is an environmentally persistent bacillus in lentic freshwater ecosystems. Older people in remote Cameroon provinces say that Buruli ulcer appeared in these infected areas with the first big bushfires, although no formal demonstration has been made. The first human contagions by *M. tuberculosis* are interpreted to have taken place during ancient peat samples in peatlands, although medicine has completely forgotten this natural origin.

Biomedecine has been interested in recent decades in understanding the serotypes adapted to intra-human circulation, that is to say, the tip of a gigantic iceberg. Many opportunistic bacteria and microscopic fungi exist in soil, water, and the phytosphere, and returning to a better consideration of the biodiversity–disease links and to knowledge about them constitutes an important challenge for modern science and medicine.

11.6. The process of globalization of human infectious diseases

The reported number of outbreaks of zoonotic and vector-borne diseases has increased over recent decades according to the GIDEON database (www.gideononline.com), which contains information on the presence of endemic infectious diseases and the occurrence of epidemics of human infectious diseases in each country. Since the 1980s, both the annual total outbreak number and the annual total number of countries affected by outbreaks have dramatically increased (Smith *et al.* 2014; Vourc'h *et al.* 2021). Moreover, endemic infectious diseases seem to be more prone to developing as epidemics and expanding regionally or globally. Changes in local conditions, such as forest conversion, agriculture extension, poor health care, inadequate socio-economic conditions, or increased urbanization may enhance local disease transmission, while changes in global conditions, such as international trade and travels, may favor the global spread of disease transmission (Figure 11.4).

According to World Bank data, the total global numbers for flight passengers and air freight have increased exponentially (see Figure 11.3). The total number of flight passengers increased from around 332 million in 1970 to more than 4 billion in 2017 (an increase of more than 1,250%), while the total amount of air freight increased from around 15,500 million tons (MT) per km in 1970 to more than 220,000 MT per km in 2017 (an increase of more than 1,400%). The international tourist arrivals worldwide increased from 25 million in 1950 to 1400 million in 2016 (5,600% increase) and are expected to reach 1.8 billion by 2030 (UNWTO 2021). In the same manner, the global GDP per capita increased from 805 US dollars in 1970 to 10,778 US dollars in 2017 (an increase of more than 1,300%). The observed correlation between the increasing number of outbreaks of infectious diseases with increasing global transportation and wealth supports the hypothesis that economic globalization also speeds up the globalization of infectious diseases (Morand and Walther 2020).

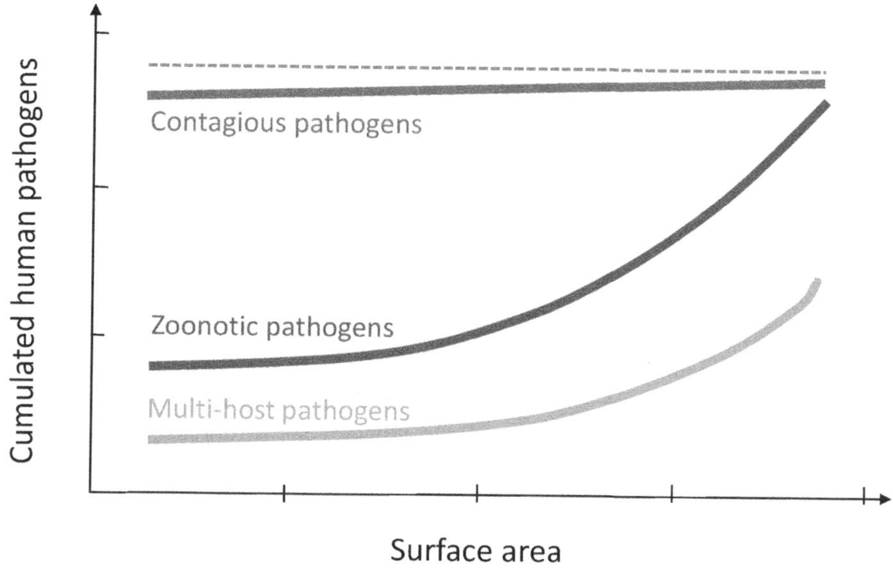

Figure 11.4. *Cumulated pathogen species richness in the human population, at a country level, against surface area, showing the degree of homogenization for three host groups. Human contagious pathogens (in blue) are highly homogenized across spatial scales and they are near everywhere, from local to global. Zoonotic pathogens (in red) show an exponential curve form which indicates that zoonotic pathogen dispersion is conditioned by animal host ecology, and these are followed by multi-host pathogens with humans as potential hosts that may help to spread disease agents worldwide. The gray dashed line indicates pathogen richness saturation over whatever spatial scale. Modified and stylized from Smith and Guégan (2010). For a color version of this figure, see www.iste.co.uk/loreau/biodiversity.zip*

Theoretical models have predicted that increased mobility will lead to faster and more global spread of disease outbreaks and that measures that decrease mobility tend to slow disease agent spread (Hufnagel *et al.* 2004; Hollingsworth *et al.* 2006; Colizza *et al.* 2007). Theoretical models also show that early travel reductions implemented at the onset of epidemics drastically reduce the number of cities around the world affected by major outbreaks (Cooper *et al.* 2006). Even if both empirical and theoretical investigations show that reducing global trade and travel decreases the global risk of pandemics, these restrictions are not sufficient to prevent the occurrence of local disease emergence and transmission. Local conditions prevail in the birth of new emerging infectious diseases and their geographic expansion, and among these conditions, agriculture is of particular interest in the context of recurrent epidemics and pandemics, notably influenza pandemics, for example A(H1N1)*pdm* and H5N1.

11.7. A livestock-dominated planet

The domestication of flora and fauna has significantly transformed the Earth's biosphere, altering the path of human evolution and allowing the rise of agrarian civilizations (Young 2015; Larson and Fuller 2014). The domestication of animals has also strongly affected human health by favoring the spread of zoonotic diseases and the rise of epidemics in agrarian civilizations (Pearce-Duvet 2006). The role played by domestic animals in expending the diversity of human infectious diseases was hypothesized by the environmental historian William McNeill (1976), who first suggested a positive relationship between the pathogens of domestic animals and humans through animal domestication history. This hypothesis was later popularized by Jared Diamond (1997). However, this hypothesis was only recently quantitatively examined by studying a small number of domestic mammals (Morand *et al.* 2014). A significant positive relationship was observed between time since domestication and the number of shared infectious and parasitic diseases amongst several domesticated mammals and humans. Hence, domestication time not only increases the sharing of pathogens with humans but also pathogen-sharing within the whole community of domesticated animals (at least mammals), stressing the importance of cohabitation time (Wolfe *et al.* 2007), that is, the so-called time hypothesis, in pathogen agent acquisition from an animal origin by humans. Interestingly, more recently domesticated animal hosts, for example new pets and new livestock species, like grass-cutters, can act as potential reservoirs of pathogens yet to become zoonotic infections for humans.

Importantly, domestic animals may serve not only as reservoirs but also as bridge hosts allowing pathogen transmission with wildlife. Wells *et al.* (2020) analyzed a database of associations between 1,785 virus species and 725 mammalian host species and showed that domesticated mammals hold the most central positions in networks of known mammal–virus associations. Domestic animals were important in the building of human infectious diseases but they are still important as amplifiers of pathogen spillover from wildlife (Morand 2015).

The dramatic increase in livestock over recent decades is a major cause of the decline in biodiversity and of natural habitat alteration. The number of cattle exceeded 1.6 billion in 2016 when it was under a billion in 1960. Over the same time, the number of pigs has grown from 500 million to 1.5 billion and chickens from 5 billion to 22 billion (not counting laying hens). Nowadays, livestock have a higher biomass than all humans on Earth (Smil 2002), while the number of poultry animals (chickens, hens, ducks) exceeds the number of wild birds. The causes of the growth in livestock depend on several factors, such as the growth of human population, the changes in diet linked to economic growth and Westernization of habits, agricultural industrialization, and integration into the global market trade, but also the cultural values assigned to livestock, which may vary from country to country. The global expansion of livestock, by threatening biodiversity, modifying land use, and homogenizing landscapes, seriously increases epidemiological bridges between wildlife, domestic animals, and humans.

The number of infectious disease outbreaks affecting humans or livestock has been found to be correlated with the continuous increase of livestock (Morand 2020). Livestock and poultry abundance contributes to the increase in the number of outbreaks but also to the increase in the number of threatened wildlife species (Steinfeld *et al.* 2006; Reid *et al.* 2010) and natural and traditionally managed habitats (Alkemade *et al.* 2013), affecting in turn human health globally.

11.8. Conclusion

In 2020, the COVID-19 pandemic crisis unfortunately reminded us as humans of our vulnerability as an animal species with an infectious disease history shared with other animal species and the environment. As the technical progress of medicine continues, which we do not dispute and which has led to very clear improvements in health and living conditions for many populations of the world, it has isolated itself from these founding naturalist bases and the corresponding knowledge.

Recent ecological findings have demonstrated the essential role of parasites and microbes in the organization and functioning of natural ecosystems. Without parasites and microbes, an ecosystem is obviously less rich in species number, but it also becomes less structured in terms of species interactions and less resilient to anthropogenic pressures and natural disasters. This result must be compared to several important components of biomedicine in the 20th century.

The subject of eradication constitutes the goal of the war in infectious disease medicine, and although this goal is entirely commendable it is clear that extremely few disease agents have been eradicated up to now: human smallpox in 1980 and poliomyelitis, for which only 22 human cases were known in 2017, and for which high hopes for eradication exist (in animal health, only Rinderpest virus disease is considered to be eradicated). The current COVID-19 pandemic due to SARS-CoV-2 clearly shows all the difficulties that can exist in eradicating a disease agent that is circulating worldwide, even in an international context of very marked improvement in research and vaccine production capacities. Beyond the eradication of one or more microbes, this subject raises the question of the consequences of their removal and the possible cascading impacts that this action can cause. The French medical doctor and philosopher Mirko Dražen Grmek asked himself these same questions about the development of new emerging infectious diseases responsible for causing pandemics, such as HIV/AIDS, in a context of increasingly effective control against older infections (Grmek 1989).

Very interestingly, current research on the human microbiome (e.g. cutaneous, intestinal, oral, vaginal) raises the same types of fundamental questions as developed in Lafferty and collaborators' study (2006), but at finer individual scale. Trillions of microbes have evolved with and continue to live on and within human beings, and research into the microbiome, that is, the indigenous microbial communities (microbiota) and the host environment that they inhabit, has strongly changed medical doctors and clinicians' ideas about microbes in human health and disease (Fan and Pedersen 2021), thus leaving the era of asepsis where a healthy world was a world without microbes. One of the most radical changes that can be observed concerns the realization by biomedicine that most of the microbes hosted by humans supply crucial ecosystem services that benefit the entire host–microbe system; for example, production of important resources, bioconversion of nutrients, and protection against pathogenic microbial forms. An understanding of the dynamics and function of microbiota has altered the medical view of microbes in maintaining homeostasis and causing disease, thus bringing even more biomedicine to community ecology. With the current and future explosion of studies relating the microbiome to health and disease, we should therefore attend to the same questions about the role of the microbiome at different levels of life organization (Ezenwa *et al.* 2015), thus positioning the ecology of host–parasite interactions as a major science of the 21st century.

All these above research facts mark an important paradigm shift in current international public and planetary health strategies. While the fight against and control over infectious diseases developed in the recent past using a vertical approach – research and public action on a single infection was advocated by the Millennium Development Goals (MDGs), focusing on three main human infections only, that is HIV/AIDS, malaria, and tuberculosis – they are now targeted in a more horizontal or transversal approach by the Sustainable Development Goals (SDGs) and particularly Goal 3 on human health and well-being, taking into account all diseases occurring in local human communities (Guégan *et al.* 2018). This constitutes an essential rectification in the way of fighting against these human infections and better corresponds to the reality of what it ecologically exists in the field. From a medical vision based on a simple classification – a disease, a causative agent that is responsible for it, and one vertical action to fight against – a new, more laudable plan is developing today, considering all the infections circulating in a given place (Roche *et al.* 2018). This entirely matches what was called a parasitocoenosis, a term that disappeared at the end of the 1980s, which refers to a local community of coexisting parasites and pathogen species affecting people in, say, a village at a given time. This conception, which is closer to the nature of what really exists on the ground, is another demonstration of the importance of ecological literacy in recent international public health development and strategies. Current research on the ecological interference of different disease agents, for example measles virus and whooping-cough bacteria in humans, demonstrates strong impacts on their respective means of disease transmission, spatio-temporal dynamics, and vaccine consequences (Rohani *et al.* 1999, 2002). Moreover, other findings have demonstrated that the outcome of interactions between two gut parasites in humans, *Ascaris lumbricoides* and *Trichuris trichiura*, was associated with a direct reduction in cerebral malaria risk in *Plasmodium falciparum*-infected patients in tropical regions, but the benefit of the latter worm species was halved in the presence of the former hookworm. This study reveals that interactions among helminth species can jeopardize the beneficial role of individual helminth species, thus emphasizing the critical role that parasite community interactions play in shaping infection outcomes (Abbate *et al.* 2018).

Finally, the third demonstration of the interest of adopting an ecological perspective on health concerns dengue fever in humans, which is due to four different viral serotypes (i.e. DEN-1, DEN-2, DEN-3, and DEN-4). There is a generally well-accepted concept called antigen displacement enhancement (ADE) that helps to explain why in certain conditions infected patients can develop severe and fatal dengue hemorrhagic fever (DHF) and dengue shock syndrome. ADE may occur when cross-reactive antibodies stimulated by a prior infection by a dengue serotype (in general DEN-2) wane to levels that no longer neutralize the heterotypic dengue virus, i.e. from another virus serotype. Instead of preventing infection, the second episode of infection may lead to a process known as ADE, with enhanced viral replication by increased infection of cells bearing the IgG receptor. ADE has faced so many critics that it now seems, at best, inconclusive and empirical studies do not support the theory and even support adverse evidence. Studying the dynamical consequences of different assumptions using epidemiological SIR modeling, Wearing and Rohani (2006) demonstrated that assuming almost no cross-immunity and strong ADE totally mismatched with observed spatial epidemiological data, thus questioning from both an empirical and theoretical perspective the veracity of the ADE hypothesis that has structured research on dengue fever for three to four decades (Wearing and Rohani 2006). All these results converge by demonstrating both the complex nature of infectious disease systems and the scientific formalism adopted by ecology to answer these important health questions.

To the question regarding whether we can use biodiversity as a functional barrier to stop epidemics and pandemics, the answer appears to be much more complicated than it initially seems. First of all, a restoration of biodiversity will not be efficient against a pandemic like COVID-19, since the disease agent is now circulating from person-to-person, even if reverse zoonosis has been observed recently in mink breeding. Biodiversity and its conservation are important to prevent pathogen emergence and spillovers, that is, at the very birth of the process. In addition, the international scientific literature is currently accumulating evidence that tends to show that both dilution and amplification effects can coexist spatially and temporally, thus calling into question the very use of one of the main and today popular ecosystem services provided by biodiversity against infectious disease. However, infectious disease ecology specialists agree that it is necessary to better understand upstream what modifies and disrupts natural cycles of microbes embedded into pristine ecosystems. Positioning scientific and biomedical research upstream of the disease emergence phenomena has now become crucial to prevent new pandemics, and as a condition of this a return of research and researchers to the field is paramount. Within this, the question of human demography and the current agricultural strategies required to meet growing food needs internationally must be understood and analyzed as one of the most preponderant factors in the co-construction of our ills. Understanding how to modulate new emerging disease risks by learning on the ground rather than sticking to an academic understanding only in the lab must constitute one of the new fields of sustainability science in the 21st century. Our ability to avoid future pandemics will depend on our repositioning of our research system to understand the upstream causes that generate these new disease threats for humanity.

The relationships between biodiversity and health are very numerous but also diffuse, because most often indirect, and research does not stop discovering new biodiversity–health relationships every month. These relationships are not just about the links that biodiversity can have with the origin, birth, and transmission of human pathogens. The most recent discoveries on the links between biodiversity and health ultimately reflect the forgetting or even amnesia of biomedicine over 30 to 40 years, which did not look into them as much as the major progress in the field should have warranted. The most recent scientific work specifies the essential role of microbial diversity, that is, the microbiome, in the physiological and immune systems of human species but also the disorders that an alteration of this can cause, such as the development of chronic diseases over the course of life (Hand *et al.* 2016 ; Zheng *et al.* 2020). Many chronic inflammatory diseases are associated with significant shifts in the microbiota towards inflammatory symptoms, which can affect both the human host and alter the production of microbiota-derived metabolites. Further, the evolutionary organization of the gut microbiota in mammals, including humans, has played an important role in the development of brain capacity (Allen *et al.* 2017). Several investigations seem to indicate that gut microbes majorly impact cognitive functions and fundamental behavior patterns, such as social interactions and stress-related behaviors, including anxiety and depression (Dinan *et al.* 2015). More generally, nature and biodiversity may affect human well-being in multiple ways. The more recent findings suggest that the species diversity of several taxonomic groups in the surroundings where people live in Western Europe, and notably bird species richness and bird song, is positively correlated with life satisfaction (Methorst *et al.* 2021). Interactions with biodiversity may truly increase quality of life (Hedblom *et al.* 2017), thus indicating that this value of nature should be considered both by medicine and public health, and by public authorities for the planning of our habitats. To conclude this chapter, the various disciplines dealing with biomedicine and public health, for example, immunology, infectiology, psychology, and sociology, need to be aware of these biodiversity–health interactions, which can help to inform the future research and medical agenda in these new emerging disciplines. We hope that the reader will be convinced of the importance of placing biodiversity at the center of health and well-being in the 21st century.

11.9. Acknowledgements

This work benefited from an Investissement d'Avenir grant managed by the Agence Nationale de la Recherche (CEBA: ANR-10-LABX-25-01). The National Institutes of Health – National Science Foundation Ecology of Infectious Disease Program Grant NSF-1911457 provided support to J.-F.G. J.-F.G. was also supported by the IRD, INRAE, the French School of Public Health (EHESP), and the University of Montpellier. S.M. was supported by the French ANR FutureHealthSEA (ANR-17-CE35-0003-01) and the Thailand International Cooperation Agency (TICA) project "Animal Innovative Health".

11.10. References

Abbate, J.L., Ezenwa, V.O., Guégan, J.-F., Choisy, M., Nacher, M., Roche, B. (2018). Disentangling complex parasite interactions: Protection against cerebral malaria by one helminth species is jeopardized by co-infection with another. *PLoS Neglected Tropical Diseases*, 12, e0006483 [Online]. Available at: https://doi.org/10.1371/journal.pntd.0006483.

Achtman, M. and Wagner, M. (2008). Microbial diversity and the genetic nature of microbial species. *Nature*, 6, 431–440.

Alkemade, R., Reid, R.S., van den Berg, M., de Leeuw, J., Jeuken, M. (2013). Assessing the impacts of livestock production on biodiversity in rangeland ecosystems. *Proceedings of the National Academy of Sciences USA*, 110, 20900–20905.

Allen, A.P., Dinan, T.G., Clarke, G., Cryan, J.F. (2017). A psychology of the human brain–gut–microbiome axis. *Social and Personality Psychology Compass*, 11, e12309 [Online]. Available at: https://doi.org/10.1111/spc3.12309.

Balloux, F. and van Dorp, L. (2017). Q&A: What are pathogens, and what have they done to and for us? *BMC Biology*, 15, 91.

Bart, M.J., Harris, S.R., Advani, A. *et al. (2014).* Global population structure and evolution of *Bordetella pertussis* and their relationship with vaccination. *mBio*, 5, e01074.

Berg, G., Eberl, L., Hartmann, A. (2005). The rhizosphere as a reservoir for opportunistic human pathogenic bacteria. *Environmental Microbiology*, 7, 1673–1685.

Bjørnstad, O.N. and Harvill, E.T. (2005). Evolution and emergence of *Bordetella* in humans. *Trends in Microbiology,* 13, 355–359.

Cleaveland, S., Laurenson, M.K., Taylor. L.H. (2001). Diseases of humans and their domestic mammals: Pathogen characteristics, host range and the risk of emergence. *Philosophical Transactions of the Royal Society B: Biological Sciences*, 356, 991–999.

Colizza, V., Barrat, A., Barthélémy, M., Vespignani, A. (2007). Predictability and epidemic pathways in global outbreaks of infectious diseases: The SARS case study. *BMC Medicine*, 5, 34.

Combes, C. (2005). *The Art of Being a Parasite*. University of Chicago Press, Chicago.

Convention on Biological Diversity (2021). Health and Biodiversity [Online]. Available at: https://www.cbd.int/health [Accessed on 1 May 2021].

Cooper, B.S., Pitman, R.J.,, Edmunds, W.J., Gay, N.J. (2006). Delaying the international spread of pandemic influenza. *PLoS Medicine,* 3, e212 [Online]. Available at: https://doi.org/10.1371/journal.pmed.0030212.

Desenclos, J.-C. and de Valk, H. (2005). Emergent infectious diseases: Importance for public health, epidemiology, promoting factors, and prevention. *Médecine et Maladies Infectieuses*, 35, 49–61.

Diamond J. (1997). *Guns, Germs and Steel: A Short History of Everybody for the Last 13,000 Years*. W. W. Norton, New York.

Dinan, T.G., Stilling, R.M., Stanton, C., Cryan, J.F. (2015). Collective unconscious: How gut microbes shape human behaviour. *Journal of Psychiatric Research*, 63, 1–9 [Online]. Available at: https://doi.org/10.1016/j.jpsychires.2015.02.021.

Ezenwa, V.O., Godsey, M.S., King, R.J, Guptill, S.C. (2006). Avian diversity and West Nile virus: Testing associations between biodiversity and infectious disease risk. *Proceedings of the Royal Society B: Biological Sciences*, 273, 109–117.

Ezenwa, V.O., Prieur-Richard, A.-H., Roche, B. *et al.* (2015). Interdisciplinarity and infectious diseases: An Ebola case study. *PLoS Pathogens* 11, e1004992.

Falkinham III, J.O. (2013). Ecology of Nontuberculous Mycobacteria – Where do human infections come from? *Seminars in Respiratory Critical Care Medicine*, 34, 95–102.

Fan, Y. and Pedersen, O. (2021). Gut microbiota in human metabolic health and disease. *Nature Reviews Microbiology*, 19, 55–71.

Grmek, M.D. (1989). *Histoire du sida. Début et origine d'une pandémie actuelle.* Payot et Rivages, Paris.

Furuse, Y., Suzuki, A., Oshitani, H. (2010). Origin of measles virus: Divergence from rinderpest virus between the 11th and 12th centuries. *Virology Journal*, 7, 52 [Online]. Available at : http://www.virologyj.com/content/7/1/52.

García-Peña, GE., Garchitorena, A., Carolan, K. *et al.* (2016). Niche-based host extinction increases prevalence of an environmentally acquired pathogen. *Oikos*, 125, 1508–1515.

Guégan, J.-F., Prugnolle, F., Thomas, F. (2007). Global spatial patterns of infectious diseases and human evolution. In *Evolution in Health and Disease,* Stearns, S.C. and Koella, J.C. (eds). Oxford University Press, Oxford, 19–29.

Guégan, J.-F., Suzán, G.A., Kati-Coulibaly, S., Bonpamgue, D.N., Moatti, J.-P. (2018). The United Nations, health and sustainability. An analysis of Sustainable Development Goal #3 "Health and well-being". *Veterinaria Mexico AO,* 5.

Guégan J.-F., Ayouba A., Cappelle J., de Thoisy B. (2020). Emerging infectious diseases and tropical forests: Unleashing the beast within. *Environmental Research Letters,* 15, 083007 [Online]. Available at: https://iopscience.iop.org/article/10.1088/1748-9326/ab8dd7/pdf.

Halliday, F.W., Rohr, J.R., Laine, A.-L. (2020). Biodiversity loss underlies the dilution effect of biodiversity. *Ecology Letters*, 23, 1611–1622 [Online]. Available at: https://doi.org/10.1111/ele.13590.

Hand, T.W., Vujkovic-Cvijin, I., Ridaura, V.K., Belkaid, Y. (2016). Linking the microbiota, chronic disease and the immune system. *Trends in Endocrinology Metabolisms*, 27, 831–843.

Hay, S.I., Battle, K.E., Pigott, D.M. *et al.* (2013). Global mapping of infectious disease. *Philosophical Transactions of the Royal Society B: Biological Sciences*, 368, 20120250.

Hedblom, M., Knez, I., Gunnarsson, B. (2017). Bird diversity improves the well-being of city residents. In *Ecology and Conservation of Birds in Urban Environments,* Murgui, E. and Hedblom, M. (eds). Springer, Berlin, 287–306.

Hollingsworth, T.D., Ferguson, N.M., Anderson, R.M. (2006). Will travel restrictions control the International spread of pandemic influenza? *Nature Medicine*, 12, 497–499.

Honda, J.R., Virdi, R., Chan, E.D. (2018). Global environmental nontuberculous mycobacteria and their contemporaneous man-made and natural niches. *Frontiers in Microbiology*, 9, 2029.

Hosseini, P. *et al.* (2017). Does the impact of biodiversity differ between emerging and endemic pathogens? The need to separate the concepts of hazard and risk. *Philosophical Transactions of the Royal Society B: Biological Sciences*, 372, 20160129 [Online]. Available at: http://dx.doi. org/10.1098/rstb.2016.0129.

Hufnagel, L., Brockmann, D., Geisel, T. (2004). Forecast and control of epidemics in a globalized world. *Proceedings of the National Academy of Sciences USA*, 101, 15124–15129 [Online]. Available at: https://doi.org/10.1073/pnas.0308344101.

Kilpatrick, A.M., Salkeld, D.J., Titcomb, G., Hahn, M.B. (2017). Conservation of biodiversity as a strategy for improving human health and well-being. *Philosophical Transactions of the Royal Society B: Biological Sciences*. [Online]. Available at: https://doi.org/10.1098/rstb.2016.0131.

Lafferty, K.D. (2014). Effects of disease on community interactions and food web structure. In *Infectious Disease Ecology*, Ostfeld, R.S., Keesing, F., Eviner, V.R. (eds). Princeton University Press, Princeton, 205–222.

Lafferty, K.D., Dobson, A.P., Kuris, A.M. (2006). Parasites dominate food web links. *Proceedings of the National Academy of Sciences USA*, 103, 11211–11216.

Larson, G. and Fuller, D.Q. (2014). The evolution of animal domestication. *Annual Review of Ecology, Evolution and Systematics*, 45, 115–136.

Lederberg, J., Shope, R.E., Oaks Jr., S.C. (eds) (1992). *Emerging Infections: Microbial Threats to Health in the United States*. Institute of Medicine (US) Committee on Emerging Microbial Threats to Health, National Academies Press, Washington DC.

Levine, R.S., Hedeen, D.L., Hedeen, M.W., Hamer, G.L., Mead, D.G., Kitron, U.D. (2017). Avian species diversity and transmission of West Nile virus in Atlanta, Georgia. *Parasites and Vectors*, 10, 62.

Locey, K.J. and Lennon, J.T. (2016). Scaling laws predict global microbial diversity. *Proceedings of the National Academy of Sciences USA*, 113, 5970–5975.

Luis, A.D., Kuenzi, A.J., Mills, J.N. (2018). Species diversity concurrently dilutes and amplifies transmission in a zoonotic host-pathogen system through competing mechanisms. *Proceedings of the National Academy of Sciences USA*, 115, 7979–7984.

McNeill, W.H. (1976). *Plagues and People*. Anchor Press, New York.

Methorst, J., Rehdanz, K., Mueller, T., Hansjürgens, B., Bonn, A., Böhning-Gaese, K. (2021). The importance of species diversity for human well-being in Europe. *Ecological Economics*, 181, 106917 [Online]. Available at: https://doi.org/10.1016/j.ecolecon.2020.106917.

Morand, S. (2015). Diversity and origins of human infectious diseases. In *Basics in Human Evolution*, Muehlenbein, M.P (ed.). Elsevier, New York, 405–414.

Morand S. (2020). Emerging diseases, livestock expansion and biodiversity loss are positively related at global scale. *Biological Conservation*, 248, 108707.

Morand, S. and Walther, B. (2020). The accelerated infectious disease risk in the Anthropocene: More outbreaks and wider global spread. *bioRxiv* [Online]. Available at: https://doi.org/10.1101/2020.04.20.049866.

Morand, S., McIntyre, K.M., Baylis, M. (2014). Domesticated animals and human infectious diseases of zoonotic origins: Domestication time matters. *Infections, Genetics and Evolution*, 24, 76–87.

Murray, K.A., Olivero, J., Roche, B., Tiedt, S., Guégan, J.-F. (2018). Pathogeography: Leveraging the biogeography of human infectious diseases for global health management. *Ecography*, 41, 1411–1427.

Nicolle, C. (1933). *Destin des maladies infectieuses*. PUF, Paris.

Ostfeld, R. S. and Keesing, F. (2012). Effects of host diversity on infectious disease. *Annual Review of Ecology, Evolution and Systematics*, 43, 157–182.

van Overbeek, L.S., van Doorn, J., Wichers, J.H., van Amerongen, A., van Roermund, H.J.W., Willemsen, P.T.J. (2014). The arable ecosystem as battleground for emergence of new human pathogens. *Frontiers in Microbiology*, 5, 104.

Pearce-Duvet, J.M.C. (2006). The origin of human pathogens: Evaluating the role of agriculture and domestic animals in the evolution of human disease. *Biological Reviews*, 81, 369–82.

Pike, L.J., Viciani, E., Kumar, N. (2018). Microbial diversity knows no borders. *Nature*, 66.

Randolph, S.E. and Dobson, D.M. (2012). Pangloss revisited: A critique of the dilution effect and the biodiversity-buffers-disease paradigm. *Parasitology*, 139, 847–863.

Reid, R.S., Bedelian, C., Said, M.Y. *et al.* (2010). Global livestock impacts on biodiversity. In *Livestock in a Changing Landscape. Drivers, Consequences, and Responses, Vol. 1*, Steinfeld, H., Mooney, H.A., Schneider F., Neville, L.E. (eds). Island Press, Washington, DC, 111–138.

Reingold, A.J. (2000). Infectious disease epidemiology in the 21st century: Will it be eradicated or will it reemerge? *Epidemiologic Reviews*, 22, 57–63.

Roche, B., Broutin, H., Simard, F. (eds) (2018). *Ecology and Evolution of Infectious Diseases. Pathogen Control and Public Health Management in Low-Income Countries.* Oxford University Press, Oxford.

Rohani, P., Earn, D.J.D., Grenfell, B.T. (1999). Opposite patterns of synchrony in sympatric diseases metapopulations. *Science*, 286, 968–971.

Rohani, P., Keeling, M.J., Grenfell, B.T. (2002). The interplay between determinism and stochasticity in childhood diseases. *The American Naturalist*, 159, 569–581.

Sala, O., Meyerson, L.A., Parmesan, C. (eds) (2008). *Biodiversity Change and Human Health: From Ecosystem Services to Spread of Disease*. Island Press, SCOPE, DIVERSITAS, New York.

Salked, D., Padgett, K., Jones, J. (2013). A meta-analysis suggesting that the relationship between biodiversity and risk of zoonotic pathogen transmission is idiosyncratic. *Ecology Letters*, 6, 679–686.

Smil, V. (2002). *The Earth's Biosphere. Evolution, Dynamics, and Change*. Massachusetts Institute of Technology, Cambridge.

Smith, K.F. and Guégan, J.-F. (2010). Changing geographic distributions of human pathogens. *Annual Review of Ecology, Evolution, and Systematics*, 41, 231–250.

Soumana, I.H., Linz, B., Harvill, E.T. (2017). Environmental origin of the genus *Bordetella*. *Frontiers in Microbiology*, 8, 28.

Steinfeld, H., Gerber, P., Wassenaar T.D., Castel, C., Rosales M., de Haan, C. (2006). *Livestock's Long Shadow: Environmental Issues and Options*. Food and Agriculture Organization of the United Nations, Rome.

Suzán, G., Marcé, E., Giermakowski, J.T. e*t al.* (2009). Experimental evidence for reduced rodent diversity causing increased hantavirus prevalence. *PLoS ONE,* 4(5), e5461.

Swaddle, J.P. and Calos, S.E. (2008). Increased avian diversity is associated with lower incidence of human West Nile infection: Observation of the dilution effect. *PLoS ONE*, 25, e2488.

de Thoisy, B., Duron, O., Epelboin, L. *et al.* (2021). Ecology, evolution and epidemiology of zoonotic and vector-borne infectious diseases in French Guiana: Transdisciplinarity does matter to tackle new emerging threats. *Infection, Genetics and Evolution*, 93 [Online]. Available at: https://doi.org/10.1016/j.meegid.2021.104916.

United Nations World Tourism Organization (2021). UNTWO World Tourism Barometer [Online]. Available at: https://www.unwto.org/.

Vourc'h, G., Morand, S., Moutou F., Jourdain, E. (2021). *Les zoonoses*. Quae, Versailles.

Wardeh, M., Risley, C., McIntyre, M.K., Setzkorn, C., Baylis, M. (2015). Database of host–pathogen and related species interactions, and their global distribution. *Scientific Data*, 2.

Wearing, H.J. and Rohani, P. (2006). Ecological and immunological determinants of dengue epidemics. *Proceedings of the National Academy of Sciences USA*, 103, 11802–11807.

Wells, K., Morand, S., Wardeh, M., Baylis, M. (2020) Distinct spread of DNA and RNA viruses among mammals amid prominent role of domestic species. *Global Ecology and Biogeography*, 29, 470–481.

Willis E. (2016). Extrapolating abundance curves has no predictive power for estimating microbial biodiversity. *Proceedings of the National Academy of Sciences* USA, 113.

Wolfe, N.D., Dunavan, C.P., Diamond, J. (2007). Origins of major human infectious diseases. *Nature*, 447, 279–283.

Woolhouse, M.E.J. and Gowtage-Sequeria, S. (2005). Host range and emerging and reemerging pathogens. *Emerging Infectious Diseases*, 11, 1842–1847.

Young, K.R. (2015). Biogeography of the Anthropocene: Domestication. *Progress in Physical Geography*, 40, 161–174.

Young, V.B. (2017). The role of the microbiome in human health and disease: An introduction for clinicians. *British Medical Journal*, 356.

Zheng, D., Liwinski, T., Elinav, E. (2020). Interaction between microbiota and immunity in health and disease. *Cell Research*, 30, 492–506.

12

Economic Valuation of Biodiversity and Ecosystem Services

Seth Binder
[1] *St. Olaf College, Northfield, Minnesota, USA*

12.1. Introduction

Economic valuation of biodiversity and ecosystem services raises awareness of nature's contributions to people, assists individuals and policymakers in navigating difficult trade-offs among competing uses of natural resources, helps to hold irresponsible actors accountable for their degradation of the environment, and can even inform assessments of the overall sustainability of our socioeconomic systems. However, valuation is also often maligned, misused, and misunderstood. This chapter seeks to clarify what economic valuation is (and is not) and provide a brief introduction to the methods of valuation, highlighting applications and limitations.

12.2. What valuation is and is not

For most people, the exercise of non-market economic valuation is abstruse. It begets many misconceptions. A common description of environmental non-market valuation is "putting a price tag on nature". To some, this might conjure images of a panel of experts debating just what that price ought to be. To others, the exercise must be nothing less than sorcery, a divination of the true, inherent worth of nature.

The Ecological and Societal Consequences of Biodiversity Loss,
coordinated by Michel Loreau, Andy Hector, and Forest Isbell.

However, such images belie the essence of the economist's work and the nature of economic value itself. Economists do not impose their own subjective assessments of value to determine a "price". Nor, of course, do they invoke some supernatural powers of perception to identify the intrinsic value of nature. In reality, non-market valuation is more akin to mathematical detective work, using economic theory and statistical methods to uncover and objectively quantify people's preferences for different environmental conditions.

Economic value reflects people's subjective preferences and is measured by their budget-constrained *willingness to pay* for things (Figure 12.1). For goods available in markets, prices tell us about the value of having one more of the good in question – that is, prices tell us about the *marginal value* of goods. Why? If the value to you of having one more bag of rice were greater than the price, you would buy it; and you would continue to buy additional bags of rice (with each successive bag probably less valuable to you than the one before) until the value of the last bag was just equal to the price. So, for anyone who buys rice, no matter how many bags of rice they buy, the price is approximately equal to their marginal value. Unfortunately, when it comes to environmental changes – especially those affecting biodiversity and ecosystem services – deducing their value is not so straightforward. Biodiversity and many ecosystem services are examples of public goods: everyone in a geographic area (in some cases, the whole world) can enjoy their benefits and no one's enjoyment of their benefits reduces anyone else's ability to enjoy them. The upshot is that biodiversity and ecosystem services are not directly bought and sold in markets, and market prices typically do not reflect their full value, even indirectly. We require more sophisticated methods – and a lot more data – to discern the value people place on changes in biodiversity and ecosystem services.

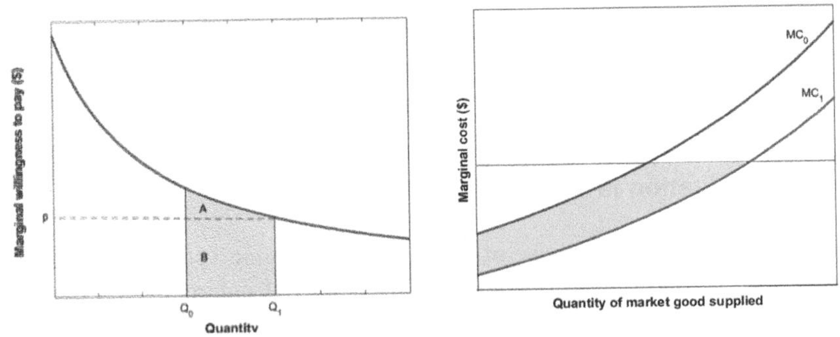

Figure 12.1. *The economic value of changes in ecosystem services is reflected in changes to consumer and/or producer surplus. For a color version of this figure, see www.iste.co.uk/loreau/biodiversity.zip*

NOTE ON FIGURE 12.1.– *The panel on the left illustrates the value of increasing the provision of an ecosystem service from Q_0 to Q_1. At Q_1 the marginal willingness to pay is equal to p. The shaded area (A+B) under the marginal willingness-to-pay curve represents the total willingness to pay for the change. Consumer surplus is the value of people's willingness to pay in excess of what they must spend. If they had to pay p for each unit increase in the ecosystem service, consumer surplus would be the shaded area A. Typically, changes in ecosystem service provision are uncompensated (their price is zero), in which case the change in consumer surplus coincides with willingness to pay for the change. The panel on the right shows the change in producer surplus (revenues minus variable costs) due to a change in biodiversity or ecosystem services that affect producers' marginal costs (i.e. the cost of producing one more unit relative to a given level of production). Shifting costs from MC_0 to MC_1 results in a producer surplus equal to the shaded area between the two curves, which reflects the difference in additional revenues and additional costs.*

The remainder of this chapter provides a non-technical introduction to the primary methods of economic non-market valuation of environmental change. It offers a bit of historical perspective on the development of the methods, as well as examples of their application to biodiversity and related ecosystem services, with attention to their potential, challenges, and limitations.

12.3. Non-market economic valuation methods

Methods of non-market economic valuation have their roots in the mid-20th century (e.g. Ciriacy-Wantrup 1947; Hotelling 1949) and began to grow in their number and use in the 1980s and 1990s. Today they benefit from more sophisticated use of economic theory and econometric techniques, though many methodological challenges remain. For a rigorous introduction to the various methods, the advanced economics student is referred to Freeman *et al.* (2014) and, for a treatment more oriented to biodiversity and ecosystem services, Hanley and Barbier (2009). This chapter presents an overview for a more general audience, not least aspiring natural scientists, and addresses three broad classes of valuation methods: production function approaches, revealed preference approaches, and stated preference approaches.

12.3.1. *Production function methods*

Production function approaches to non-market valuation capture the value of biodiversity and ecosystem functions as inputs to the production of market goods. Changes in diversity or function that affect the costs of production have consequent impacts on producers' decisions and profits. (Figure 12.1, Panel 2, illustrates the economic impact of an environmental change that simply shifts the costs of production.) The net benefits or costs of the environmental change accrue to owners or employees of the affected firms. The corresponding changes in income thus represent the affected individuals' willingness to pay for the environmental change.

A classic example of the production function method comes from attempts to estimate the value of incremental changes in biodiversity for the purposes of "bioprospecting", the search for naturally occurring compounds with potential commercial value. Simpson *et al.* (1996) provide a framework for assessing the marginal value of species preservation for use in pharmaceutical development. Departing from earlier approaches that simply multiplied the estimated probability of discovering a commercially valuable compound from an untested species by the expected commercial value of a novel compound, the authors estimate the pharmaceutical industry's derived demand for species richness, accounting for the potential redundancy in compounds across species and redundancy of different compounds for the same potential pharmaceutical use. Exploring a wide range of values for key parameters, they calculate an upper bound on the bioprospecting value of preserving one more untested plant species globally ($9,431) and they show that the value could fall to fractions of a cent under plausible assumptions.

Like many of the simpler, often early, implementations of the production function approach, Simpson *et al.* do not estimate a fully specified production function that describes how changes in the mix of all key inputs lead to changes in output. In some applications, the connection to an actual production function is even more remote, which can create problems for the validity of the value estimates. Such is the case for the replacement cost method. A prominent example comes from New York City's mid-1990s investment in watershed restoration in the Catskill mountains. As described by Chichilnisky and Heal (1998), the City decided to invest in restoration of the ecosystem's natural filtration capacity to meet national water quality standards, spending $1–1.5 billion to maintain the ecosystem service rather than the $6–8 billion it would have taken to replace that ecosystem service with a human-made filtration plant. The "win-win" outcome of financial savings and watershed protection has inspired numerous similar investments in watershed "natural capital", for example through the growing use of water funds in South America (Goldman-Benner *et al.* 2012). But what exactly is the value of the watershed service? Chichilnisky and Heal take the financial savings as an estimate of the productive value of the watershed restoration. However, the replacement cost "saved" (in New York's case, $6–8 billion) is not equivalent to the benefit of the ecosystem service. The value of the ecosystem service could be higher or lower than the arbitrary cost of its artificial substitute. If a producer rationally would incur the replacement cost, the cost must be less than or (only by coincidence) equal to the benefits of the ecosystem service; it would thus represent only a lower bound on the service's value.

In contrast to replacement costs, avoided damages do accurately reflect the value of biodiversity or ecosystem services for mitigating negative environmental impacts. Assessing the value of these non-market damage-mitigation services involves estimating an ecological production function (Polasky and Segerson 2009) as part of an expected damage function (Barbier 2007). These mathematical functions map ecosystem characteristics to ecosystem function and ecosystem function to economic impacts. This approach has been applied to the valuation of, among others: a mangrove habitat for the mitigation of storm damage to coastal communities in Thailand (Barbier 2007); forest cover for the regulation of baseflow and consequent reduction of diarrheal disease load in Flores, Indonesia (Pattanayak and Wendland, 2007); and wetlands and floodplains for the mitigation of flood damage in a college town in the United States (Watson *et al.* 2016). Though valuation of disease and other health impacts is fraught, the expected damage function approach might ultimately be applicable to the valuation of changes in habitat or biodiversity for the mitigation of zoonotic disease transmission (Pongsiri *et al.* 2009). The method is implicitly used to assess the contributions of biodiversity and ecosystem services to the mitigation of climate change in myriad applications (e.g. Isbell *et al.* 2015) that value the storage or sequestration of carbon using the social cost of carbon (Nordhaus 2017).

Ecological production functions have also been used within the production function framework to estimate the value of various habitats and the species they support as inputs to agricultural, aquacultural, and commercial fishery production. The production function approach has been used to value changes in, for example, a mangrove habitat for coastal fishery production in Thailand (Barbier 2003) and tropical forest cover for drought mitigation benefits to downstream farmers in Indonesia (Pattanayak and Kramer, 2001). Using an experimental approach, Ricketts *et al.* (2004) and Priess *et al.* (2007) assessed the value of pollination services and pollinator habitat for nearby coffee production in Costa Rica and Indonesia, respectively.

Yield experiments have also been useful for assessing the contributions of on-farm biodiversity to agricultural production. Utilizing experimental data from prominent grassland biodiversity experiments, Binder *et al.* (2018) and Schaub *et al.* (2020) estimate the value of planted biodiversity for producing feed from pasture. While Schaub *et al.* find that the risk- and quality-adjusted value of production increases across the full range of observed diversity as measured by the Simpson index, Binder *et al.* account for farmers' optimal choices in the composition and relative abundances of planted species and find an inverse-U relationship, such that the value of adding another species (i.e. the marginal value of species richness) falls – ultimately below zero – as richness rises (Figure 12.2). The latter relationship implies that, while maximally diverse mixtures may still be profitable, mixtures of intermediate richness would be optimal.

While yield experiments can help reveal the effects and value of changing biodiversity or other natural inputs, all else equal, an important limitation of the approach is that the "all else equal" condition is rarely satisfied when agricultural production systems experience a change. Whether the change is endogenous or exogenous, it is likely to be accompanied by or induce other changes in inputs that affect the producer's bottom line. Indeed, where early yield experiments were used to predict the economic effects of climate change on agriculture, they over-estimated agricultural losses because they failed to account for farmers' adaptive responses (Mendelsohn *et al.* 1994), assessing instead a so-called "dumb farmer" scenario. Accounting for endogenous changes in other productive inputs typically requires the specification and estimation of an *economic* production function.

Several studies explicitly estimate economic production functions to assess the private value of agricultural biodiversity. Chavas and Di Falco (2012), using the theoretical framework of Chavas (2009), found that crop diversity boosted productivity by 37% in small farms in the Ethiopian highlands. Di Falco and Chavas (2006) quantified the productivity and risk-mitigating welfare effects of crop genetic diversity for farmers in Sicily. Common to such studies is the assumption that producers themselves know the potential benefit of increasing biodiversity, which is then reflected in their (presumably optimal) production decisions and there to be revealed by econometric detective work. The extent to which this assumption holds true (or true enough) depends on the agroecosystem in question. Whether it is preferable to the yield experiment assumption of "all else equal" depends on the agroecosystem and the purpose of the valuation exercise. In general, it could be said that the economic production function approach provides estimates of the *realized* value of biodiversity, whereas simulations based on yield experiments can provide estimates of the *potential* value of agricultural biodiversity.[1]

Typically, production function approaches to the valuation of on-farm agricultural biodiversity capture its *private* productive value to the farmer. However, species dispersal might create the potential for important spillover effects, such that

1 Even if the potential benefits are high, producers may perceive high costs of switching to a new, biodiverse production system, locking them into less profitable and less diverse systems.

improving on-farm diversity in one location can benefit production in another. If such spillover effects exist, they are likely captured (though not identified distinctly from private benefits) in a notable study by Donfouet *et al.* (2017), who estimate an economic production function aggregated to the level of small agricultural regions in France, taking regional Shannon index values of crop diversity as an input. While an aggregated approach may be helpful for assessing the total productive value of changes in crop diversity, ultimately, the ability to disentangle private from spillover benefits is essential for adequately informing either private or public policy decisions.

Biodiversity and other forms of natural capital determine not only the level of expected ecosystem services but also the variability of those services (Yachi and Loreau 1999; Lehman and Tilman 2000), and the latter must be accounted for in the valuation of environmental change. Let us deal first with the variability of ecosystem service provision for a particular place and set of beneficiaries. An example might be the variability of annual agricultural yields on a given farm, as a function of on-farm diversity and/or near-farm habitat. In this case and others, the value of a reduction in variability depends on the degree of the beneficiary's *risk aversion* and the degree to which the risk can be dissipated through insurance markets (Baumgärtner 2008). If the beneficiary is risk-neutral, only the expected value (mean) of the ecosystem service matters; the spread (variance) of outcomes is irrelevant. However, most people – and even many firms – have some degree of risk aversion. That is, even as the expected value of a distribution of outcomes is held constant, the perceived value of the "lottery" associated with the distribution falls as its variance rises. The difference between the expected value and perceived or *certainty-equivalent* value of a distribution reflects the value of eliminating the risk or variability, either naturally or through financial markets. This "insurance" value can be substantial (Schläpfer *et al.* 2002). Changes in biodiversity affecting an ecosystem service typically affect both its mean and variance. The value of the net effect of both changes is captured in the difference in certainty equivalent values. Certainty-equivalent values are generally not observed; they can be calculated based on the estimation or calibration of the parameters of a specified utility function, as in Di Falco and Chavas (2006), Binder *et al.* (2018), and Schaub *et al.* (2020). The Di Falco and Chavas (2006) work is particularly noteworthy since they additionally account for the skewness of the distribution of outcomes and farmers' differential sensitivity to *downside* risk.

Beyond the local insurance benefits of greater biodiversity, Loreau *et al.* (2003) propose a potential spatial insurance effect on ecosystem function at landscape scales, mediated by species dispersal. This suggests that analysts interested in valuing biodiversity as an input to production might need to account for potential beneficiaries beyond the immediate vicinity of the change, as noted above. The basic valuation methodology, however, remains the same, as individual beneficiaries will perceive the impact in terms of the difference in the certainty-equivalent values of their optimal production decisions before and after the change.

In addition to reducing the variability of ecosystem service flows under a given regime, biodiversity and other forms of natural capital can also reduce the risk of undesirable regime shifts, and this has economic value as well. If the risks of regime shift as a function of biodiversity and the distribution of net benefits under the alternative regime are quantifiable, then the outcome lotteries under each regime can be weighted by the probability of regime shift and collapsed mathematically into a single lottery, evaluated as above for a static or one-time decision context. For dynamic decisions regarding depletable stocks of natural capital (say, a given area of natural habitat), the probabilities and "payoffs" of regime change inform what economists call the *marginal user cost* of incremental resource depletion (Hanley and Barbier 2009). If the probabilities or payoffs of regime change are unknown but can be learned in the future, the dynamic decision context introduces an alternative value associated with preservation, known as "quasi-option value". Note well, though, it is the decision context not the ecosystem that creates this value, which is essentially equivalent to the value of information. See Freeman *et al.* (2014) for an overview of quasi-option value and the related, though largely defunct, concept of option value.

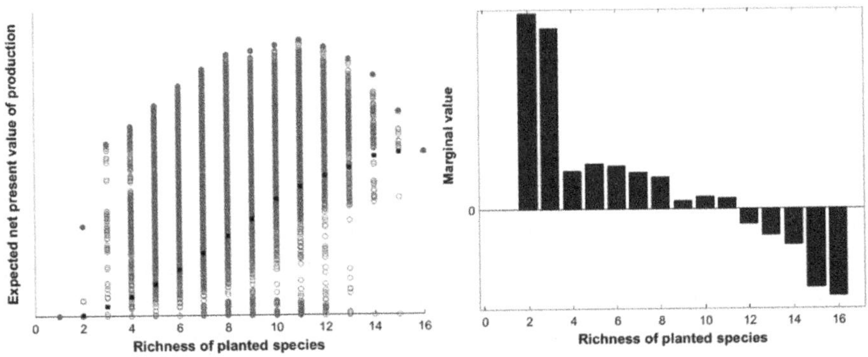

Figure 12.2. *The total and marginal values of species richness*

NOTE ON FIGURE 12.2.– *The panel on the left depicts expected values of production for each of the 65,535 combinations of up to 16 species, represented by the open, gray circles. While yield studies typically focus on the average performance of mixtures at each richness level (solid black squares), only the best-performing combination of a given richness level (solid red circles) is relevant for a producer's choice of richness. When the producer controls not only the number but the identity and proportion of species, the marginal value of richness is the incremental value of moving from the best combination at one richness level to the best combination at the next. These marginal values are depicted in the panel on the right.*

12.3.2. *Revealed preference methods*

Many of the benefits of biodiversity and ecosystem services are not mediated by producers but instead accrue directly to people as "consumers". Revealed preference methods leverage information from markets to estimate consumers' preferences for related non-market goods. By "related" we mean that the market good is a substitute for, or (more typically) complement to, the non-market good. When a market good can substitute (even imperfectly) for some aspect of environmental quality – for example, air purifiers or water filters substituting for unpolluted ambient air or water – economists can use changes in expenditures on the market good as an indicator of the value of changes in environmental quality. Unfortunately, this "averting behavior" method can produce – at best – only lower and upper bound estimates of the value of non-marginal changes in environmental quality (Bartik 1988). Thus, the primary revealed preference methods focus on weak complements to non-market goods. The essential requirement for these methods to produce reliable estimates of value is that participation in a particular market be necessary to obtain the non-market good in question – for example, the purchase of fuel for travel to protected ecosystems or of housing for access to "greener" neighborhood amenities.

Travel cost models exploit the necessity of fuel (and time) expenditures to take advantage of the *in situ* benefits of distant protected areas. First proposed by Hotelling (1949), the method has become a staple of researchers interested in valuing the availability and characteristics of recreational areas (see Lupi *et al.* (2020) for a review of best practices). In the most basic implementation of a travel cost model, economists estimate a demand curve for visits to a site based on associations between the number of annual trips a given user makes and the "price" of their trip, that is, the cost of fuel and value of time required to get to the site.[2] The *consumer surplus* implied by the demand curve represents the total annual value of the availability of the site. A notable early application of the method comes from Tobias and Mendelsohn (1991), who estimated the present value of recreational benefits to visitors to Costa Rica's Monteverde Cloud Forest Biological Reserve and

---

2 Economists traditionally used some arbitrary fraction of an individual's wage as a measure of the cost of their time; improving on this crude approach is an active area of research.

found that, on a per hectare basis, it far exceeded the purchase price of new land for the reserve.

Methodological advances since the 1970s allow analysts to account for the availability of substitute sites and to value particular attributes across a set of recreational areas. Two classes of model, discrete choice models and the "Kuhn–Tucker" model, are used to infer the value of differences across sites that a user might choose. The discrete choice models, based on the pioneering work of McFadden and colleagues (see especially Manski and McFadden (1981)), posits that individuals choose a particular site among alternatives based on the sites' "prices" (fuel and time costs), observable attributes of the site, and the individual, and individual-specific, unobservables accounted for in the random component of the model. The "Kuhn–Tucker" model, developed by Phaneuf *et al.* (2000), also allows the analyst to infer the value of differences across sites. Unlike the discrete choice approach, it does not restrict the analysis to a single choice occasion, so researchers can utilize data on a full season or year's worth of individuals' decisions to visit different sites any number of times. Attribute-based travel cost models have been used to value the landscape diversity of recreational sites in Mallorca, Spain (Bujosa Bestard and Riera Font 2009); the richness of bird species at sites in the U Pacific Northwest (Kolstoe and Cameron 2017) and of marine fish and invertebrates around California's Channel Islands (Viana *et al.* 2017); and the spatially differentiated spillover benefits of restoration for pheasant hunters in Michigan (USA) (Knoche *et al.* 2015).

In addition to estimating the value of environmental attributes among recreational sites, economists have developed methods to assess the value of environmental attributes for homeowners and workers. Hedonic methods, developed by Rosen (1974), decompose the prices of quality-differentiated market goods into the "hedonic prices" of the characteristics associated with the goods. For typical goods, like vehicles, hedonic prices reflect the cost to producers and benefit to consumers of supplying more of a given characteristic – say, horsepower or legroom or trunk space. In some special cases, hedonic prices can also capture benefits inherent to a good that are not purposefully supplied by the seller. For environmental economists, the two most important examples are the environmental characteristics of housing units and the mortality risks associated with different jobs.

Using the hedonic method, economists can estimate the marginal value of environmental amenities (or disamenities) that are capitalized into home prices. The estimated hedonic price of an environmental amenity reflects the average amount homebuyers pay for an incremental change in the amenity, controlling for other housing amenities (see Bishop *et al.* (2020) for a review of best practices). Thanks to the increasing accessibility of data on home sales, including prices and various attributes of the housing units, as well as geospatial information regarding environmental and other neighborhood characteristics, the estimation of hedonic prices for environmental amenities has become quite prevalent. The method has been applied to estimate the value of the proximity and size or extent of various types of habitat or land use/land cover (e.g. Czembrowski and Kronenberg 2016; Gibbons *et al.* 2014; Klaiber *et al.* 2017; Sander and Haight 2012). While, under certain conditions, the estimated hedonic prices correspond to households' marginal willingness to pay for the respective environmental amenities, assessing the value of a hypothetical discrete change in an amenity additionally requires a model of household decision-making to identify a demand curve and to account for endogenous changes in the housing market (Kuminoff *et al.* 2013).

Economists can also use the hedonic method to estimate the value of small changes in mortality risk across different jobs that create "compensating differentials" in the wages paid to workers with similar skills and experience but different exposure to risk. Evans and Taylor (2020) review best practices. In the environmental realm, estimates of the value of changes in risk from hedonic wage models are used primarily in the valuation of the health effects of changes in pollution, but they could also inform the valuation of changes in biodiversity and ecosystem services that imply changes in mortality risk – say, from increased exposure to novel viruses due to human encroachment into previously forested areas. Despite the importance of accounting for the value of changes in mortality risk, use of hedonic wage model estimates has encountered some fierce resistance. One common misapprehension stems from the unfortunate use of these estimates to calculate what is called the "value of statistical life", a practice that Cameron (2010) rightly argues should be "euthanized". There is, of course, no such thing as a statistical life. It is a purely conceptual construct, achieved by multiplying the (change in) probability of death by the affected population to calculate the expected number of lives saved or lost due to a change in mortality risk. The purported economic value of a statistical life is likewise contrived, drawn from observed trade-offs between income and small amounts of risk and extrapolated into meaninglessness for a 100% "risk" of death. Combining this meaningless measure with the ill-considered concept of a "statistical life" gives the impression of putting a price tag on certain deaths – to which many people reasonably object – when what we value economically (and affect through policy) are small changes in mortality risk.

12.3.3. *Stated preference methods*

As production function approaches and revealed preference methods inherently capture only a subset of specific benefits of environmental change, and because they are limited to evaluating environmental changes within observed ranges, economists often rely on stated preference methods to produce estimates of the total economic value of flexibly specified environmental changes. Stated preference methods rely on the analysis of decisions individuals make when confronted with hypothetical situations. Johnston *et al.* (2017) provide a review of best practices for the two main categories of stated preference methods, contingent valuation and choice experiments.

Contingent valuation presents individuals with a "constructed" market for a mooted environmental change and infers the value thereof based on their hypothetical choices (Carson 2011; Carson and Hanemann 2005; Mitchell and Carson 1988). Survey respondents are given a description of the ecosystem and the expected impact of environmental change and then asked about their willingness to pay to avoid (or achieve) the change. Some early contingent valuation studies were open-ended, simply soliciting whatever dollar figure occurred to the respondent. However, many respondents found it challenging to articulate a dollar value for such an unfamiliar type of "purchase" – not only due to the unconventional nature of the good, but also because of the need to "name your price". Other early studies offered an initial price and continued to raise (lower) it incrementally until the respondent rejected (accepted) the offer.[3] Due to the potential for anchoring bias introduced by the arbitrary initial price, most contingent valuation studies now instead present respondents with a single yes/no (referendum) decision to accept or reject a price that is randomly varied across respondents. With a sufficiently large sample, analysts can then estimate average willingness to pay using discrete choice models.

For the assessment of prospective projects that could be adjusted to have more or less environmental impact along various dimensions, conjoint analysis and discrete choice experiments offer the ability to estimate the marginal values of different environmental attributes among which there might be trade-offs. In both methods, researchers present respondents with multiple hypothetical scenarios or bundles of goods, which include the relevant environmental attributes and, typically, some monetary element such as a tax or user fee. In conjoint analysis, which has its origins in the psychology and marketing literatures, respondents are asked to rank or rate the various bundles and the analyst seeks to back out a mathematically

3 An alternative solution is the use of "contingent behavior" methods, in which respondents are asked to describe how their use of the ecosystem would change. Of course, this limits valuation to the specific non-market benefit associated with the particular use or behavior.

consistent preference structure. Alternatively, and more consistent with economic theory (Louviere *et al.* 2010), discrete choice experiments require respondents to choose their most preferred of two or more options and the analyst leverages random utility theory to estimate the marginal value of each attribute (Hanley *et al.* 1998).

From the beginning, stated preference methods have grappled with challenges to their ability to yield reliable value estimates. Despite some continued skepticism (Hausman 2012), there seems to have emerged a consensus among environmental economists that stated preference methods *can* produce reliable estimates (Kling *et al.* 2012). However, the methodological requirements are fairly restrictive. The chief issue is the design of survey instruments that induce participants to respond in ways that are consistent with their true willingness to pay. This requires "consequentiality", the idea that respondents believe there is a real chance that their responses will influence policy to achieve the environmental change described and that they would actually have to pay what they hypothetically committed (Carson and Groves 2007). Stated preference methods would then seem to be restricted to contexts in which: a specific policy or project could plausibly lead to the environmental change of interest; some governmental body has the authority and perceived capacity to pursue such a project; said body has a credible interest in citizen input regarding the project and is likely to receive the survey responses and consider them as legitimate input; said body would plausibly utilize the hypothetical payment vehicle and payment rates presented to the survey respondents; and so on. These criteria no doubt still admit the application of stated preference methods to many important environmental changes, but the methods cannot be considered a panacea for the limitations of production function or revealed preference approaches.[4]

12.3.4. *Benefit transfer methods*

In many instances when estimates of the value of environmental change could usefully inform decisions, the cost of undertaking a new valuation study may not be perceived as likely to justify the benefits of the information obtained, or a decision-making body may simply not be able to afford a new study. In these all-too-common cases, analysts turn to benefit transfer methods as a guide to the value of the environmental changes under consideration. As the name implies, benefit transfer methods involve taking information from existing valuation studies from one context and applying that information to a new context. The most basic method simply takes the value estimate obtained in one study and assumes its adequacy for

4 In fact, some of the most promising uses of stated preference methods appear to be in joint application with revealed preference, especially travel cost, methods (Cameron 1992).

approximating the value of the change under consideration. That method is fundamentally flawed. Even if the environmental changes themselves were identical (and of course they are not), the populations of beneficiaries will be different. They will differ in key determinants of the social value of any environmental change: demographic composition, level and distribution of income, and access to complements or substitutes for the ecosystem services affected.

Economists have developed some creative ways to attempt to overcome the challenges to reliable benefit transfer (Boyle *et al.* 2010; Johnston and Rosenberger 2010). If there exists an adequate number of valuation studies of sufficiently similar environmental changes, analysts can employ meta-analysis to estimate the effects of various observable factors that determine the estimated values; they can then apply the estimated statistical function to data from the focal area and population of beneficiaries to produce an estimate of the value of the environmental change under consideration (Rosenberger and Loomis 2000; Smith and Pattanayak 2002). Instead of estimating a statistical transfer function based on the willingness to pay or consumer surplus values obtained from multiple studies, the analyst would ideally utilize the demand or utility function estimated in a target study (Loomis 1992). Many studies, however, fail to report or even estimate the parameters of a demand or utility function. In such cases, the analyst might instead use preference calibration (Smith *et al.* 2002). Employing this approach, the analyst specifies a utility function and then calibrates the parameters of that function for consistency with willingness to pay or consumer surplus estimates from existing studies, ultimately deriving a willingness to pay function from the calibrated utility function. If existing studies provide sufficient information to identify the parameters of the preference function, this approach ensures that the willingness to pay ultimately estimated is consistent with households' income and the axioms of microeconomic theory more broadly.

12.4. Conclusion

Economics provides a rigorous theoretical framework and associated empirical methods for estimating the economic value of changes in biodiversity and ecosystem services. Non-market valuation methods – including production function, revealed preference, and stated preference approaches – have been widely applied and refined over several decades. While the methods typically require substantial data (and a trained analyst), best practices allow for a reliable quantification of the economic impacts of realized or prospective changes to the natural environment.

Non-market valuation studies demonstrate clear, substantial economic benefits from the preservation or enhancement of biodiversity in many contexts. Changes to biodiversity in agroecosystems have an easily identifiable set of primary beneficiaries (agricultural producers) and well-documented benefits. Such changes might also affect beneficiaries downstream, literally or figuratively. The value of those effects can be estimated just like the effects of changes to restored or relatively intact fragments of natural ecosystems in non-agricultural landscapes. A key challenge is to identify consequent changes to specific ecosystem functions or attributes and the specific groups of beneficiaries to whom those changes matter (Keeler *et al.* 2012). Then analysts must employ the appropriate valuation method for each category of beneficiary or a single, comprehensive stated preference survey. The valuation literature has demonstrated local and global benefits to maintaining the extent and integrity of many natural habitats and the diversity of life therein – from recreational opportunities to climate regulation – though these economic benefits might not always be sufficiently high to cover the opportunity costs of preservation.

Of course, economic value represents only one facet of the value we attribute to nature. Estimated economic values may not always be a sufficient, or even appropriate, input to decision-making processes regarding our right relationships with nature and one another. Indeed, much of the controversy surrounding the economic valuation of biodiversity relates not to its scientific validity but to the normative legitimacy of its *use* in specific contexts.[5] Non-economic values (e.g. religious, spiritual, ideological) may take priority over or even proscribe the use of economic values in societal decisions affecting biodiversity and ecosystem services.

Nevertheless, when used appropriately within cultural and legal bounds, the products of economic non-market valuation can improve decision-making to enhance human well-being. They can help natural resource managers set judicious limits on rates of exploitation. They can reveal when the environmental and financial costs of an otherwise permissible infrastructure project outweigh the expected benefits. They can quantify the damage done by oil spills and thereby help ensure the financial liability of responsible parties and incentivize others to take greater precaution in the future. Though the practice of economic valuation itself cannot keep us from crashing on the rocky shores of ecological destruction or human impoverishment, by acknowledging, quantifying, and ultimately informing the trade-offs we are willing to make among ecosystem services and the many other things we value, it allows us to navigate the murky waters in between with eyes wide open.

5 Estimates of "existence value" are a notable exception, as their ethical legitimacy *and* their scientific validity are strongly contested (Binder 2020).

12.5. References

Barbier, E.B. (2003). Habitat–fishery linkages and mangrove loss in Thailand. *Contemporary Economic Policy*, 21(1), 59–77.

Barbier, E. B. (2007). Valuing ecosystem services as productive inputs. *Economic Policy,* 22(49), 179–229.

Bartik, T.J. (1988). Evaluating the benefits of non-marginal reductions in pollution using information on defensive expenditures. *Journal of Environmental Economics and Management*, 15(1), 111–127.

Baumgärtner, S. (2008). The insurance value of biodiversity in the provision of ecosystem services. *Natural Resource Modeling*, 20(1), 87–127.

Binder, S. (2020). Is existence value appropriate for regulatory benefit–cost analysis? *Journal of Benefit-Cost Analysis,* 11(3), 441–456.

Binder, S., Isbell, F., Polasky, S., Catford, J.A., Tilman, D. (2018). Grassland biodiversity can pay. *Proceedings of the National Academy of Sciences of the United States of America,* 115(15), 3876–3881.

Bishop, K.C., Kuminoff, N.V, Banzhaf, H. S. *et al.* (2020). Best practices for using hedonic property value models to measure willingness to pay for environmental quality. *Review of Environmental Economics and Policy*, 14(2), 260–281.

Boyle, K.J., Kuminoff, N.V., Parmeter, C.F., Pope, J.C. (2010). The benefit-transfer challenges. *Annual Review of Resource Economics*, 2(1), 161–182.

Bujosa Bestard, A. and Riera Font, A. (2009). Environmental diversity in recreational choice modelling. *Ecological Economics*, 68, 2743–2750.

Cameron, T.A. (2010). Euthanizing the value of a statistical life. *Review of Environmental Economics and Policy*, 4(2), 161–178.

Carson, R.T. (2011). *Contingent Valuation: A Comprehensive Bibliography and History.* Edward Elgar Publishing, Cheltenham.

Carson, R.T. and Groves, T. (2007). Incentive and informational properties of preference questions. *Environmental and Resource Economics*, 37(1), 181–210.

Carson, R.T. and Hanemann, W.M. (2005). Contingent Valuation. In *Handbook of Environmental Economics, Vol. 2*, Goran-Maler, K. and Vincent, J.R. (eds). Elsevier, Amsterdam, 821–936.

Chavas, J.-P. (2009). On the productive value of biodiversity. *Environmental and Resource Economics*, 42(1), 109–131.

Chavas, J.-P. and Di Falco, S. (2012). On the productive value of crop biodiversity: Evidence from the highlands of Ethiopia. *Land Economics*, 88(1), 58–74.

Chichilnisky, G. and Heal, G. (1998). Economic returns from the biosphere. *Nature*, 391, 629–630.

Ciriacy-Wantrup, S.V. (1947). Capital returns from soil-conservation practices. *Journal of Farm Economics*, 29(4), 1181–1196.

Czembrowski, P. and Kronenberg, J. (2016). Hedonic pricing and different urban green space types and sizes: Insights into the discussion on valuing ecosystem services. *Landscape and Urban Planning*, 146, 11–19.

Di Falco, S. and Chavas, J.-P. (2006). Crop genetic diversity, farm productivity and the management of environmental risk in rainfed agriculture. *European Review of Agricultural Economics*, 33(3), 289–314.

Donfouet, H.P.P., Barczak, A., Détang-Dessendre, C., Maigné, E. (2017). Crop production and crop diversity in France: A spatial analysis. *Ecological Economics* 134, 29–39.

Evans, M.F. and Taylor, L.O. (2020). Using revealed preference methods to estimate the value of reduced mortality risk: Best practice recommendations for the hedonic wage model. *Review of Environmental Economics and Policy*, 14(2), 282–301.

Freeman, M.A., Herriges, J.A., Kling, C.L. (2014). *The Measurement of Environmental Resource Values: Theory and Methods*. RFF Press, Washington, DC.

Gibbons, S., Mourato, S., Resende, G.M. (2014). The amenity value of English nature: A hedonic price approach. *Environ Resource Econ*, 57, 175–196.

Goldman-Benner, R.L., Benitez, S., Boucher, T. *et al.* (2012). Water funds and payments for ecosystem services: Practice learns from theory and theory can learn from practice. *Oryx,* 46 (1), 55–63.

Hanley, N. and Barbier, E. (2009). *Pricing Nature: Cost-Benefit Analysis and Environmental Policy*. Edward Elgar Publishing, Cheltenham.

Hanley, N., Wright, R.E., Adamowicz, V. (1998). Using choice experiments to value the environment design issues, current experience and future prospects 1. *Environmental and Resource Economics*, 11(4), 413–428.

Hausman, J. (2012). Contingent valuation: From dubious to hopeless. *Journal of Economic Perspectives*, 26(4), 43–56.

Hotelling, H. (1949). Letter quoted by R.E. Prewitt in "Economic Study of the Monetary Evaluation of Recreation in National Parks". Unpublished paper, United States Department of the Interior, Washington, DC.

Isbell, F., Tilman, D., Polasky, S., Loreau, M. (2015). The biodiversity-dependent ecosystem service debt. *Ecology Letters,* 18 (2), 119–134.

Johnston, R.J. and Rosenberger, R.S. (2010). Methods, trends and controversies in contemporary benefit transfer. *Journal of Economic Surveys*, 24(3), 479–510.

Johnston, R.J., Boyle, K.J., Adamowicz, V. *et al.* (2017). Contemporary guidance for stated preference studies. *Journal of the Association of Environmental and Resource Economists*, 4(2), 319–405.

Keeler, B.L., Polasky, S., Brauman, K.A. *et al.* (2012). Linking water quality and well-being for improved assessment and valuation of ecosystem services. *Proceedings of the National Academy of Sciences of the United States of America*, 109(45), 18619–18624.

Klaiber, H.A., Abbott, J.K., Smith, V.K. (2017). Some like it (less) hot: Extracting trade-off measures for physically coupled amenities. *Journal of the Association of Environmental and Resource Economists*, 4(4), 1053–1079.

Kling, C.L., Phaneuf, D.J., Zhao, J. (2012). From Exxon to BP: Has some number become better than no number? *Journal of Economic Perspectives*, 26(4), 3–26.

Knoche, S., Lupi, F., Suiter, A. (2015). Harvesting benefits from habitat restoration: Influence of landscape position on economic benefits to pheasant hunters. *Ecological Economics*, 113, 97–105.

Kolstoe, S. and Cameron, T.A. (2017). The non-market value of birding sites and the marginal value of additional species: Biodiversity in a random utility model of site choice by eBird members. *Ecological Economics*, 137, 1–12.

Kuminoff, N.V., Smith, V.K., Timmins, C. (2013). The new economics of equilibrium sorting and policy evaluation using housing markets. *Journal of Economic Literature*, 51(4), 1007–1062.

Lehman, C.L. and Tilman, D. (2000). Biodiversity, stability, and productivity in competitive communities. *The American Naturalist*, 156(5), 534–552.

Loomis, J.B. (1992). The evolution of a more rigorous approach to benefit transfer: Benefit function transfer. *Water Resources Research*, 28(3), 701–705.

Loreau, M., Mouquet, N., Gonzalez, A. (2003). Biodiversity as spatial insurance in heterogeneous landscapes. *Proceedings of the National Academy of Sciences of the United States of America*, 100(22), 12765–12770.

Louviere, J.J., Flynn, T.N., Carson, R.T. (2010). Discrete choice experiments are not conjoint analysis. *Journal of Choice Modeling*, 3(3), 57–72.

Lupi, F., Phaneuf, D.J., Von Haefen, R.H. (2020). Best practices for implementing recreation demand models. *Review of Environmental Economics and Policy*, 14(2), 302–323.

Manski, C.F. and McFadden, D. (eds). (1981). *Structural Analysis of Discrete Data with Econometric Applications*. MIT Press, Cambridge.

Mendelsohn, R., Nordhaus, W.D., Shaw, D. (1994). The impact of global warming on agriculture: A Ricardian analysis. *American Economic Review*, 84(4), 753–771.

Mitchell, R.C. and Carson, R.T. (1988). *Using Surveys to Value Public Goods: The Contingent Valuation Method*. Resources for the Future, Washington, DC.

Nordhaus, W.D. (2017). Revisiting the social cost of carbon. *Proceedings of the National Academy of Sciences of the United States of America*, 114(7), 1518–1523.

Pattanayak, S.K. and Kramer, R.A. (2001). Worth of watersheds: A producer surplus approach for valuing drought mitigation in Eastern Indonesia. *Environment and Development Development Economics*, 6(1), 123–146.

Pattanayak, S.K. and Wendland, K.J. (2007). Nature's care: Diarrhea, watershed protection, and biodiversity conservation in Flores, Indonesia. *Biodiversity and Conservation*, 16(10), 2801–2819.

Phaneuf, D.J., Kling, C.L., Herriges, J.A. (2000). Estimation and welfare calculations in a generalized corner solution model with an application to recreation demand. *The Review of Economics and Statistics*, 82(February), 83–92.

Plummer, M.L. (2009). Assessing benefit transfer for the valuation of ecosystem services. *Frontiers in Ecology and the Environment*, 7(1), 38–45.

Polasky, S. and Segerson, K. (2009). Integrating ecology and economics in the study of ecosystem services: Some lessons learned. *Annual Review of Resource Economics*, 1(1), 409–434.

Pongsiri, M.J., Roman, J., Ezenwa, V.O. *et al.* (2009). Biodiversity loss affects global disease ecology. *BioScience*, 59(11), 945–954.

Priess, J.A., Mimler, M., Klein, A.-M., Schwarze, S., Tscharntke, T., Steffan-Dewenter, I. (2007). Linking deforestation scenarios to pollination services and economic returns in coffee agroforestry systems. *Ecological Applications*, 17(2), 407–417.

Ricketts, T.H., Daily, G.C., Ehrlich, P.R., Michener, C.D. (2004). Economic value of tropical forest to coffee production. *Proceedings of the National Academy of Sciences of the United States of America*, 101(34), 12579–12582.

Rosen, S. (1974). Hedonic prices and implicit markets: Product differentiation in pure competition. *Journal of Political Economy*, 82(1), 34–55.

Rosenberger, R.S. and Loomis, J.B. (2000). Using meta-analysis for benefit transfer: In-sample convergent validity tests of an outdoor recreation database. *Water Resources Research*, 36(4), 1097–1107.

Sander, H.A. and Haight, R.G. (2012). Estimating the economic value of cultural ecosystem services in an urbanizing area using hedonic pricing. *Journal of Environmental Management*, 113, 194–205.

Schaub, S., Buchmann, N., Lüscher, A., Finger, R. (2020). Economic benefits from plant species diversity in intensively managed grasslands. *Ecological Economics*, 168.

Schläpfer, F., Tucker, M., Seidl, I. (2002). Returns from hay cultivation in fertilized low diversity and non-fertilized high diversity grassland. *Environmental and Resource Economics*, 21(1), 89–100.

Simpson, R., Sedjo, R., Reid, J. (1996). Valuing biodiversity for use in pharmaceutical research. *Journal of Political Economy*, 104(1), 163–185.

Smith, V.K. and Pattanayak, S.K. (2002). Is meta-analysis a Noah's Ark for non-market valuation? *Environmental and Resource Economics*, 22(1–2), 271–296.

Smith, V.K., van Houtven, G., Pattanayak, S.K. (2002). Benefit transfer via preference calibration: "Prudential Algebra" for policy. *Land Economics*, 78(1), 132–152.

Tobias, D. and Mendelsohn, R. (1991). Valuing ecotourism in a tropical rain-forest reserve. *Ambio*, 20(2), 91–93.

Viana, D., Gornik, K., Lin, C.-C. *et al.* (2017). Recreational boaters value biodiversity: The case of the California Channel Islands National Marine Sanctuary. *Marine Policy*, 81, 91–97.

Watson, K.B., Ricketts, T., Galford, G., Polasky, S., O'Niel-Dunne, J. (2016). Quantifying flood mitigation services: The economic value of Otter Creek wetlands and floodplains to Middlebury, VT. *Ecological Economics*, 130, 16–24.

Yachi, S. and Loreau, M. (1999). Biodiversity and ecosystem productivity in a fluctuating environment: The insurance hypothesis. *Proceedings of the National Academy of Sciences*, 96(February), 1463–1468.

PART 5

Zooming Out: Biodiversity in a Changing Planet

13

Feedbacks Between Biodiversity and Climate Change

Akira S. MORI[1], Takehiro SASAKI[1], Maiko KAGAMI[1], Takeshi MIKI[2], and Moriaki YASUHARA[3]

[1] *Yokohama National University, Japan*

[2] *Research Center for Biodiversity Science, Ryukoku University, Otsu, Japan*

[3] *Swire Institute of Marine Science and State Key Laboratory of Marine Pollution, University of Hong Kong, Pokfulam, Hong Kong*

13.1. Introduction

Climate, more specifically temperature, is one of the main drivers of large-scale biodiversity patterns, as warmer tropics harbor more species than colder high latitudes, constituting the latitudinal diversity gradient, the most pervasive ecological pattern on Earth (Worm and Tittensor 2018). While long-term monitoring of biodiversity changes is rare and most of the evidence is based on temperature–diversity correlations from spatial data (space-for-time substitution; Yasuhara and Danovaro 2016), some observational and many paleoecological records consistently indicate a positive relationship between temperature and diversity (Yasuhara *et al.* 2009). For example, deep-sea biodiversity changes are tightly linked to climate changes at 10–10^5 year time scales (Figure 13.1). This positive diversity–temperature relationship has been mainly explained by physiological tolerance or metabolic theory (Hunt *et al.* 2005; Yasuhara and Danovaro 2016). More broadly, the mechanisms behind latitudinal diversity gradients remain under active debate (Colwell Lees 2000) and are probably scale-dependent (Wang *et al.* 2009). Although the latitudinal biodiversity gradient is one of the earliest findings in ecological studies (e.g. van Humboldt 1808; Hawkins

The Ecological and Societal Consequences of Biodiversity Loss,
coordinated by Michel LOREAU, Andy HECTOR, and Forest ISBELL.

and Felizola Diniz-Filho 2004), there is, to date, no consensus about its fundamental mechanism applicable for all ecosystems. What controls biodiversity at large spatial and temporal scales remains a prime question in ecology.

At the same time, however, increasing evidence has suggested that the ongoing anthropogenic warming is negatively affecting biodiversity (Figure 13.2). Such mounting observations and projections of the devastating impacts of climate change on Earth's life are seemingly inconsistent with the positive diversity–temperature relationship, a fundamental ecological rule. A possibility is that rate of temperature change matters and the rapidity of ongoing anthropogenic warming (rather than temperature increase per se) negatively affects biodiversity (Yasuhara and Danovaro 2016). Another possibility for this decoupling, at least partially, is more complex diversity–temperature relationships. Recent studies have suggested that the diversity–temperature relationship may not simply be positive linear, but unimodal with a peak at the intermediate temperature (Yasuhara and Danovaro 2016; Yasuhara *et al.* 2020b). For example, sea-surface temperature >25°C may be too high for pelagic biodiversity (Yasuhara *et al.* 2020b). In a warmer world, temperate regions hold more tropical species, and polar regions gain more temperate species, as they change their distributions to live within their optimum temperature niches, but tropical regions will have no source for such immigrants, resulting in a biodiversity decline in the tropics (Cheung *et al.* 2016; Yasuhara *et al.* 2020b). Indeed, tropical biodiversity decline has been recognized (Chaudhary *et al.* 2021; Yasuhara *et al.* 2020b). In this sense, it is likely that the negative effects of warming on biodiversity will be worse in warmer, low latitude areas, both in the terrestrial and marine realms, which are experiencing unparalleled climatic conditions at present (Burrows *et al.* 2014).

While an understanding of the variable responses of biodiversity to the changing climate is surely needed, there are many challenges. Long-term monitoring with a quantitative measure of biodiversity has rarely been feasible. Even many long-term biological monitoring programs do not have quantitative biodiversity records; instead the majority use only selected species or are based on a coarse taxonomic resolution (Smith *et al.* 2009), although time-series data of biodiversity changes are accumulating. While emerging approaches such as metagenomics may make long-term monitoring of quantitative biodiversity measures reasonably doable (e.g. Smithsonian's MarineGEO; Leray and Knowlton (2015); SoilBON; Guerra *et al.* (2021)), uncertainties remain about biodiversity changes, especially over longer time scales. Different collective efforts are needed to comprehend our understandings of climate–diversity relationships and how species and their assemblages will respond to novel, potentially no-analog climate conditions, which are anticipated to appear and may already be emerging in this era of the Anthropocene.

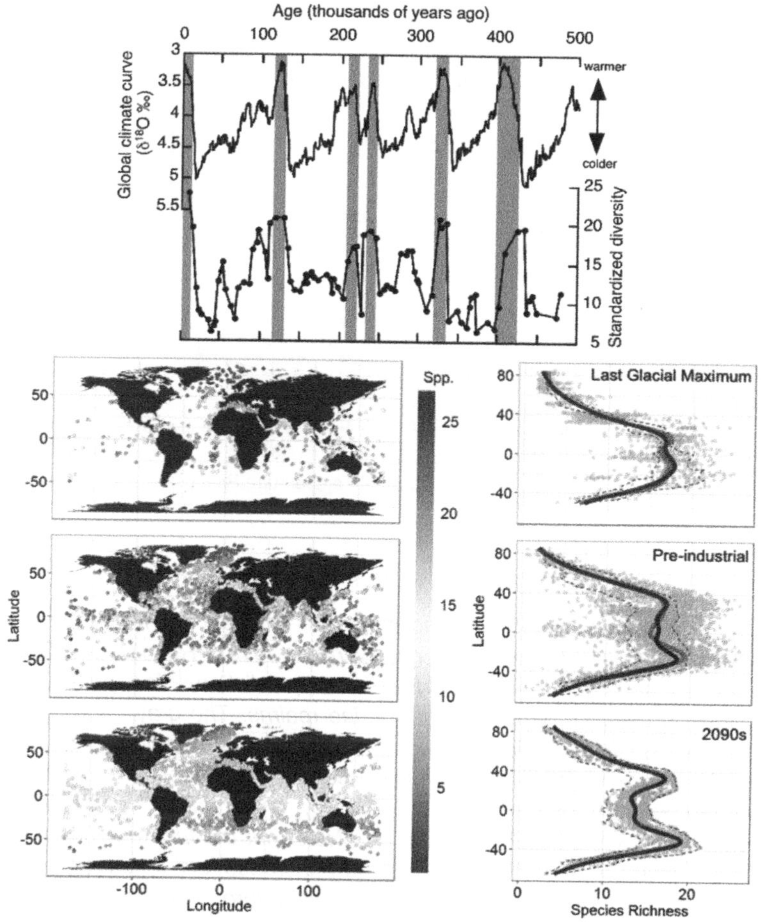

Figure 13.1. *Top panel: Global climate curve (global stack of benthic δ^{18}O records, representing general climatic states; top) vs deep-sea biodiversity (standardized diversity by rarefaction, i.e. number of species when sample size was 50; bottom) (Yasuhara et al. 2009). Orange bars indicate warm peak interglacial periods. Note the clear similarity between them, indicating global climatic impact on deep-sea biodiversity and the positive temperature–diversity relationship. Bottom panel: global distributions and latitudinal gradients of pelagic species richness at different periods (Yasuhara et al. 2020b). Note the tropical diversity depression in the pre-industrial period and 2090s, indicating that tropical climate was already too warm for pelagic biodiversity even in the pre-Anthropocene time. The pelagic diversity–temperature relationship is unimodal across a wide range and the tropical pre-industrial temperature is beyond the ascending part (i.e. the range of the positive temperature–diversity relationship) and reaches the descending part of the unimodal diversity–temperature curve. For a color version of this figure, see www.iste.co.uk/loreau/biodiversity.zip*

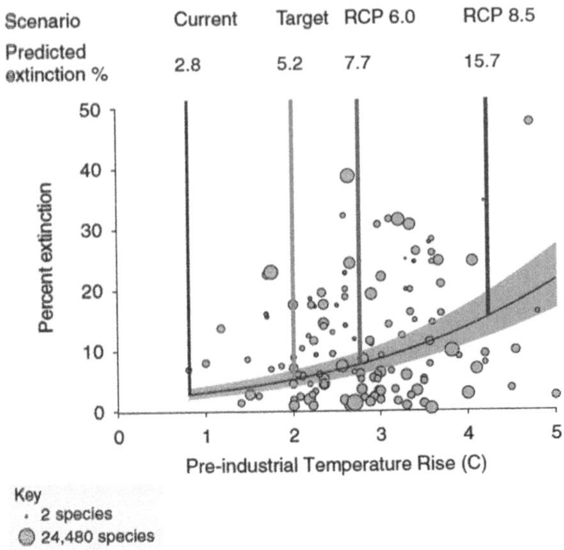

Figure 13.2. *Predicted global extinction risks for species from climate change accelerate with global temperature rise (Urban 2015). Results are based on a meta-analysis, sourced from 131 published predictions for plant and animal assemblages in different terrestrial and marine realms. The gray band indicates 95% credible intervals. Circles indicate means with the area proportional to the sample size (bottom left, key). Extinction risks for four scenarios are provided: the current post-industrial temperature rise of 0.8°C, the policy target of 2°C, and representative concentration pathways (RCPs) 6.0 and 8.5. Note that the future global extinction risk from climate change is predicted to not only increase but to accelerate as global temperatures rise. For a color version of this figure, see www.iste.co.uk/loreau/biodiversity.zip*

As such, the changing climate has been increasingly identified as a substantial and growing threat to life on Earth and many species will likely experience climate-driven redistribution in the near future. Furthermore, the corresponding consequences on ecosystem functioning are largely unknown (Pecl *et al.* 2017). Filling this knowledge gap is urgently required. The rationale of this importance is not only based upon the consideration of pervasive impacts of climate change on biodiversity per se, but also because ecological systems are a vital component as a countermeasure against climate change (natural climate solutions; Griscom *et al.* 2017). Typically, it is well-recognized that forests are massive carbon storehouses, and thus have a critical role in sequestering CO_2 from the atmosphere – one of the most critical, but challenging, issues to address in sustainable development (Tollefson 2019). Many scientific endeavors aim to understand and maximize such benefits from nature. For instance, experimental manipulations of climatic conditions are widely implemented in order to foresee possible future scenarios in terrestrial and aquatic systems. Some of these efforts are globally harmonized: coordinated, distributed experiments are arrayed across different ecosystems and regions, aimed at understanding differential or universal ecosystem sensitivity to chronic and acute changes in climatic conditions (Knapp *et al.* 2017). This and other collective efforts will help understand the potential consequences of climate change on biodiversity and their contributions to the fundamental functionality of ecological systems.

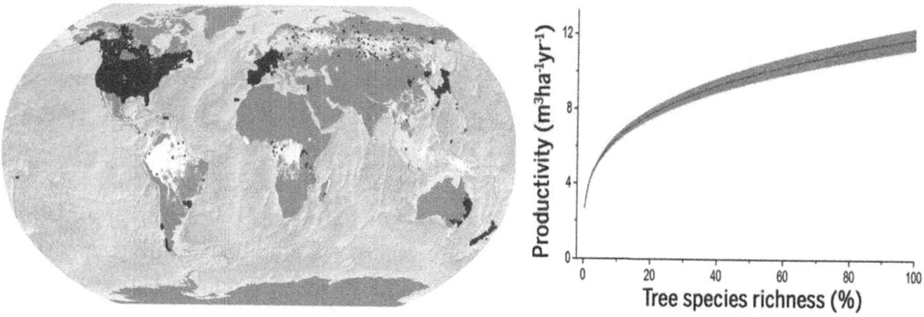

Figure 13.3. *Global effect of tree species diversity on forest productivity (Liang et al. 2016). Ground-sourced data from 777,126 global forest biodiversity permanent-sample plots (dark blue dots, left), which cover a substantial portion of the global forest extent (white), reveal a consistent positive and concave-down biodiversity–productivity relationship across forests worldwide (red line with pink bands representing 95% confidence intervals, right). The result suggests that the ongoing species loss in forests could substantially reduce forest productivity and thereby carbon sequestration, which would in turn compromise the global forest carbon sink. For a color version of this figure, see www.iste.co.uk/loreau/biodiversity.zip*

Biodiversity supports primary productivity. This observation was first made by Charles Darwin in the mid-19th century (Hector and Hooper 2002), at a time when no science existed for anthropogenic global warming. In this 21st century, increasing evidence indicates how and to what extent biodiversity loss could have adverse impacts on primary productivity, seriously threatening carbon storage on the planet. Empirical and theoretical understandings about how the diversity of ecological assemblages, especially those of primary producers in terrestrial systems, are important to sequester carbon and thus stabilize climate (Figure 13.3). However, the existing frameworks of natural climate solutions do not fully acknowledge such substantial roles of biodiversity (Mori 2020). Although biodiversity is predominantly seen as a target for conservation, it is rarely appreciated for the contribution it can make to help mitigate climate change. In sum, while advancing science to realize the environmental twin-wins for the issues of biodiversity and climate change needs further collective efforts at different spatial, temporal, biogeographical, and ecological scales, the time is ripe to synthesize the wealth of knowledge in both disciplines by explicitly considering their interlinkages and feedbacks (Mori *et al.* 2021).

13.2. Vulnerability and responses of biodiversity and ecosystem functioning to the changing climate in different biomes

Terrestrial vegetation such as managed/natural grasslands and forests are essential components of the biosphere that support an enormous number of genotypes and species and have numerous associated ecosystem services, including primary productivity, carbon sequestration, food production, timber supply, provision of foraging, livestock and biofuel production, and cultural benefits, such as tourism, recreation, and landscape aesthetics (Mori *et al.* 2017). The majority of these ecosystem services are threatened by climate change, including warming, changing patterns in precipitation and snow, and increasing atmospheric concentrations of CO_2 (Gibson 2009). Empirical evidence suggests that these benefits gained from nature are fundamentally supported by the diversity of organisms such as plants, animals, protists, and fungi (Cardinale *et al.* 2012). Understanding the impacts of anthropogenic global warming on terrestrial vegetation thus requires a basic understanding of how direct and indirect – mediated by biodiversity – climate change pathways impact ecosystem functioning (Lavorel 2019).

Terrestrial vegetation responds to changes in climatic conditions, such as temperature, precipitation patterns, and CO_2 concentration. These changes are highly likely to reshape the composition and the functionality of terrestrial vegetation systems. For instance, global spatial patterns of plant species richness are largely determined by annual energy input and water supply (Kreft and Jetz 2007), and changing climatic regimes are thus forcing terrestrial systems to a different state. Hirota *et al.* (2011) analyzed data on the distribution of tree cover in Africa, Australia, and South America using remotely sensed images, and found evidence for the existence of three distinct stable states – forest, savanna, and a treeless state – along a gradient of mean annual precipitation. Their results suggest a potential regime shift in vegetation if rainfall patterns are changed due to anthropogenic climate change. In Californian grassland communities, (Harrison *et al.* 2015) reported that an increase in aridity over 15 years had led to directional losses of plant species richness, especially of native annual forbs with low drought tolerance, at both local and landscape scales. In terrestrial systems, the majority of species ranges are predicted to shrink dramatically, even for a rise in the global temperature of less than 2°C (Warren *et al.* 2018). Although empirical evidence generally suggests that climate change will negatively impact biodiversity (Harrison *et al.* 2015), caution is needed because such effects can be highly dependent on changing water balance at a given locality (Sommer *et al.* 2010). For instance, the effects of elevated CO_2 on plant diversity substantially vary among studies, likely due to other resource limitations such as nitrogen and water (Harpole *et al.* 2007). In sum, biodiversity changes are expected to have concomitant effects on ecosystem functioning and the delivery of important provisioning and regulating ecosystem services, which are vital for humanity.

By using 16 years of normalized difference vegetation index data (NDVI, as a proxy for aboveground net primary productivity) across the US, Maurer *et al.* (2020) found potential reductions of primary productivity with decreasing precipitation and increasing aridity, with the most prominent reduction found in the desert and arid grasslands. In forests worldwide, changes to key drivers, such as temperature, CO_2 concentration, and vapor pressure deficit, are likely to force forests towards shorter, younger, and lower-biomass systems, minimizing potential carbon storage (McDowell *et al.* 2020). These studies importantly suggest that the critical carbon storehouse on lands may shrink under the changing climate. Thus, quantifying the climate sensitivity of plant primary productivity at large spatial and temporal scales is essential for predicting how the global carbon cycle will respond to future climatic conditions. Ongoing climate change will also lead to increasing precipitation variability through the increased frequency of extreme events. A recent study reported that precipitation variability has a negative effect on aboveground net primary productivity across global drylands (Gherardi and Sala 2019). Similarly, increasing transient disturbances, including wildfire, drought, windthrow, and biotic attack, are now threatening forest productivity (McDowell *et al.* 2020).

Plant litter decomposition and mineralization are key components of the global carbon cycle and represent more uncertainties in the climate–biodiversity relationship. Prieto *et al.* (2019) reported that leaf litter decomposition can be reduced by warming and rainfall reduction manipulations. This negative impact of climate change can be further exacerbated by reductions in litter quality and bacterial and fungal biomass with warming. Note that while losing biodiversity above- and below-ground can decrease the rate of litter decomposition (Mori *et al.* 2020), suggesting a reduction in the release of CO_2 from the soil, biodiversity decline has opposing effects, such as enhanced nutrient leaching, thus limiting vegetation dynamics (e.g. forest regeneration) and the ability of these systems to store carbon. Net outcomes of biodiversity loss in the carbon budget through the decomposer subcomponent of ecosystems thus feature large uncertainty. In general, warming promotes the degradation of organic material, thus leading to carbon emissions from ecosystems; however, losing plant diversity, material to be decomposed, has opposing effects with the equivalent magnitude of the effect size as climatic changes (Figure 13.4). Synergetic interactions among plants, soil organisms, and abiotic soil conditions – so-called plant–soil feedback – are thus highly context-dependent (Pugnaire *et al.* 2019). Taken together, a substantial knowledge gap exists regarding the relationships and potential feedback between biodiversity changes and carbon dynamics in the biosphere.

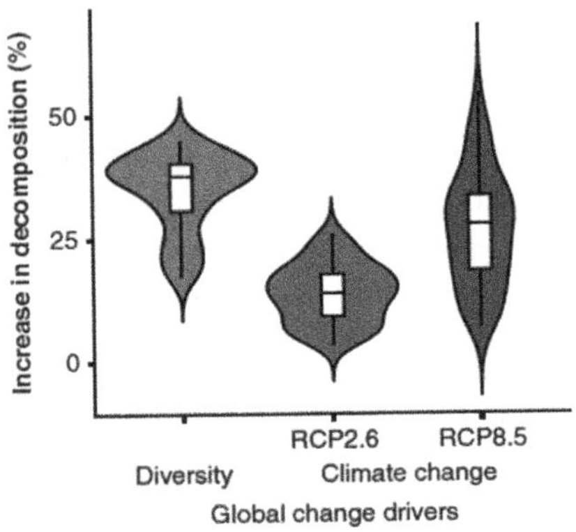

Figure 13.4. *Potential changes in plant litter decomposition rates due to global change drivers (Mori et al. 2020). Violin plots and boxplots showing potential increases (%) in the process of decomposition resulting from diversity change or climatic warming, estimated across 57 study locations in different forest biomes. Diversifying plant litter from mono- to mixed-species could increase the decomposition by 34.7% (mean). Based on future (up to the 2070s) climate change projections in these study locations (increases in annual mean temperature; RCPs 2.6 and 8.5), decomposition could be increased by 13.6% and 26.4% (means), respectively. In sum, the after-life effects of plant diversity are comparable in magnitude to anthropogenic warming as a determinant of decomposition of plant organic matter, and thus its decline through anthropogenic influences, such as creating species-poor plantations, can have substantial and potentially irreversible influences on biogeochemical cycles in the biosphere. For a color version of this figure, see www.iste.co.uk/loreau/biodiversity.zip*

Now, a critical challenge for biodiversity science is to gain further insights about the functional roles of different organism groups beyond primary producers and in different biomes other than well-studied ecosystems, such as grasslands. In aquatic ecosystems, important ecosystem services, such as water regulation and supply, disturbance regulation, nutrient cycling, and waste treatment (Lau 2013), are supported by biodiversity (e.g. Cardinale 2011). Along with other drivers of global change, such as overfishing, eutrophication, deoxygenation, harmful algal blooms, pollution, invasions, emerging diseases, microplastics, land-use change, and sea-floor resource extraction, climate change is now one of the greatest emerging threats to biodiversity in aquatic systems (Breitburg *et al.* 2018; Reid *et al.* 2019). Artificial structures and geomorphological modifications including dams, embankments, riparian works, and water-level manipulations will further amplify these threats, given the increasing frequency of extreme climate events such as heavy rain and storms. Given the importance of aquatic ecosystems, where photosynthetic carbon fixation by phytoplankton accounts for more than half of the global carbon fixation (Falkowski 2012), further efforts are needed to understand the potential impacts of climate change on biodiversity and ecosystem functioning in aquatic ecosystems.

There is a growing consensus that the primal diversity driver in oceans is the temperature (Worm and Tittensor 2018). Time-series data based on fossil records show the strong positive relationship between diversity and climatic temperature for the past tens to hundreds of thousands of years, as mentioned before (Hunt *et al.* 2005; Yasuhara *et al.* 2009). Even in the deeper times of the Cenozoic, over the last 66 million years diversity curves tend to show remarkable similarities, with the global climate curve correlating higher diversity in warmer periods such as the Eocene and middle Miocene (Yasuhara *et al.* 2020b). For example, shark diversity closely tracks the paleoclimatic curve of $\delta^{18}O$ (Condamine *et al.* 2019). Cenozoic diversity trends of Planktic foraminifera are best explained by a combination of climatic change and species' ecology (Ezard *et al.* 2011), as the Planktic foraminifera diversity curve shows partial similarity to the global climatic curve: for instance, distinct Eocene–Oligocene diversity decline and cooling and generally high diversity during the warm Eocene. Now, rapid changes in temperature conditions in the ocean are becoming a major threat to organisms and their interactions in supporting ecosystem functions and services, especially given the wide range of the unimodal diversity–temperature relationship, indicating that overly high temperatures affect biodiversity negatively.

In freshwater ecosystems, climate change potentially threatens biodiversity, and freshwater vertebrates, such as fish, have declined by more than 80% from 1970 to 2018 (reviewed in Reid *et al.* 2019). Rising temperatures can reduce the body sizes of fish and alter species distribution, phenology, and disease outbreaks. Not only warming but also cold shocks can cause mass mortality among fish. Variations in precipitation, storm events, floods, and droughts endanger mollusks, waterbirds, and other species. Climate change affects freshwater ecosystem functioning and increases the risk of regime shift from clear water to turbid states with cyanobacteria blooms (reviewed in Scheffers *et al.* 2016). Synergistic effects of temperature and anthropogenic eutrophication cause bottom-water deoxygenation, which negatively affects biodiversity and ecosystem stability (Breitburg *et al.* 2018, Jane *et al.* 2021).

Note that warming does not necessarily increase primary productivity in aquatic systems. Changes in water column mixing-regime and stronger thermal stratification can decrease the nutrient supply from deep layer to surface layer, thus not stimulating primary productivity (Doney 2006). Also, the temporal stability of ecosystem functions such as productivity is not necessarily maintained by biodiversity itself. A recent global comparison demonstrated that warming can destabilize lake ecosystems by weakening the diversity effects of primary producers (Figure 13.5). Long-term time-series data, however, suggest that climatic impacts on biodiversity and ecosystem functioning may be much stronger than biodiversity impacts on ecosystem functioning at long, decadal–millennial time scales (Yasuhara *et al.* 2016). Warming can alter the metabolic balance of the ecosystems, and differential activation energies of heterotrophic and autotrophic processes could shift the carbon balance of many aquatic ecosystems from being net sinks for atmospheric CO_2 to becoming net sources (Yvon-Durocher *et al.* 2010). Size-structured trophic interaction is a common feature of aquatic ecosystems; that is, larger consumers eat larger resources (Sheldon *et al.* 1972), and thus the size diversity of resources and consumers is one of the key determinants of food-web structure and thus energy transfer (Garcia-Comas *et al.* 2016). Therefore, both taxonomic diversity and food-web structural diversity are key determinants of ecosystem functions, such as primary productivity, and their stability. Warming and rising CO_2 levels alter the producer–consumer coupling, often lowering consumer growth due to phenological mismatch or poor food stoichiometry (Winder and Schindler 2004). Warming changes food-web structures by reducing top predators and herbivores and increasing autotrophs and bacterivores. These effects of warming and rising CO_2 levels could be buffered by high biodiversity (Urabe and Waki 2009).

There is likely feedback between microbe-mediated carbon processes and climate in aquatic systems. Primary productivity, respiration, remineralization of sinking particles, and methane production/respiration are all driven by microbial communities, which are determinants of the carbon budget and greenhouse gas dynamics affecting climate. At the same time, ongoing climate change alters these processes (Cavicchioli *et al.* 2019). The alteration of such carbon processes by climate change is not necessarily caused by the reduction of microbial taxonomic diversity at species or strain levels. The functional redundancy within a functional group of microbes is generally high (Langenheder *et al.* 2006) but this might not be always the case (Miki *et al.* 2014). Rather, the great richness (> 20) of distinct metabolic functional groups (e.g. aerobic chemoheterotrophy, methanotrophy, and fermentation) harbored by a microbial community, which is much greater than in plant and animal communities, drives microbe-mediated ecosystem functioning (Louca *et al.* 2016). Therefore, the impact of climate change on microbial diversity should be evaluated at the level of such distinct metabolic functional groups, rather than fine-scale taxonomic-level diversity. There are growing lines of evidence of microbe–climate feedback, which has been demonstrated by field experiments and long-term observations. For example, the balance between primary productivity and respiration in lakes is altered by summer storms in the temperate region (Giling *et al.* 2017) and tropical and sub-tropical regions (Itoh *et al.* 2017); such storms are projected to change in terms of region, frequency, and magnitude (Cha *et al.* 2020). The methane emission from wetlands increases with warming all over the world because methane production is more strongly enhanced than methane oxidation (Aben *et al.* 2017), which will further enhance warming. Methane production is also affected by more frequent occurrences of summer storms (such as typhoons) and stronger winter stratification in deep lakes under climate changes because these alter water-mixing patterns and thus the vertical distribution of dissolved oxygen (e.g. Itoh *et al.* 2017).

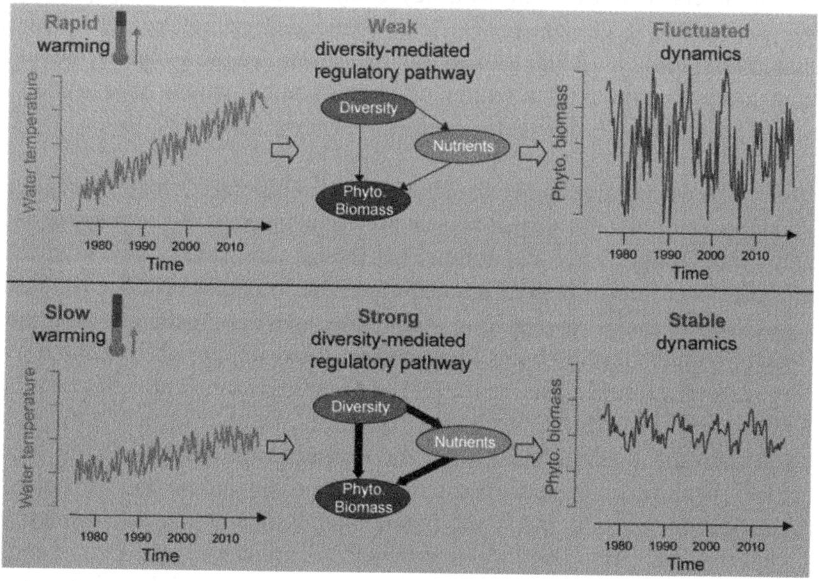

Figure 13.5. *A cross-system study analyzing long-term time series data collected from 10 aquatic ecosystems revealed novel mechanisms explaining how warming can destabilize the dynamics of ecosystem functioning (e.g. phytoplankton biomass) (Chang et al. 2020). The core analysis centered on empirically quantifying causal relationships between the big-picture ecosystem variables: diversity, nutrient cycling, phytoplankton biomass, and others. Systems experiencing stronger long-term warming had weakened diversity-mediated regulatory pathways, and the weakened regulatory pathways made ecosystems less stable. The findings emphasize the importance of a holistic network view of ecosystems, indicating that integrated regulatory pathways, instead of individual variables or interactions, best predict ecosystem stability. For a color version of this figure, see www.iste.co.uk/loreau/biodiversity.zip*

13.3. Societal and political challenges to these twin crises and their interlinkages

Since the 1992 Rio Conventions, there has been growing interest in, and evidence of, how much global environmental changes have impacted nature and people. These Conventions for the environment include the Convention on Biological Diversity (CBD) and the UN Framework Convention on Climate Change (UNFCCC), which face the issues of biodiversity and climate change and have also stimulated many discussions about societal issues, including human health and rights. The UN World Health Organization now emphasizes the importance of a healthy planet, both ecological and climatic systems, to ensure human health. The UN Development Program and Environmental Program consider the efforts to conserve biodiversity and climate as a key component for alleviating poverty and ensuring sustainable development. Societal challenges cannot be separated from environmental issues, as demonstrated by the strong ongoing social movements protesting global warming as a matter of global justice, which have spread over regions and generations (Schiermeier *et al.* 2019; Marris 2019a, 2019b).

Currently, new initiatives, such as nature-based solutions (Cohen-Shacham *et al.* 2016; Seddon *et al.* 2020), which are in principle inspired by and supported by nature to face many socio-environmental issues, are under development. In response, ecosystem restoration as a countermeasure to climate change has been on the rise, as seen in many reforestation (and, sometimes, afforestation) programs on lands that are under development. In the ocean, there is the UN's Second World Ocean Assessment (United Nations 2021). The Intergovernmental Oceanographic Commission of the UN Educational, Scientific and Cultural Organization (UNESCO) and the International Union for Conservation of Nature (IUCN) have been actively working on ocean climatic impact issues, including deoxygenation and ocean acidification. In 2020, the World Economic Forum stressed that business has a strong interest in stopping environmental degradation (World Economic Forum 2020). In light of environmental impacts and human rights, there are now behavioral changes in economic sectors driven by local campaigns for sustainability (Cohen-Shacham *et al.* 2016). These local to global scale movements and initiatives are becoming important in raising public awareness and revising narratives that are influential for policymakers.

Some argue that the environment is finally having its moment. The environmental twins, however, are not gaining equal attention. The Intergovernmental Science-Policy Platform on Biodiversity and Ecosystem Services (IPBES) released its first Global Assessment report in 2019. The report emphasized unprecedented and accelerating rates of species extinction – at least 1 million animal and plant species are threatened with extinction (IPBES 2019). However, despite the surprisingly large number reported, the urgent call for action to maintain biodiversity might have been overshadowed by media coverage of climate strikes (Gardner *et al.* 2020). Indeed, significant work is needed to effectively communicate the importance of biodiversity to the public and in policies compared to climate change (Legagneux *et al.* 2018). Policies exist to formally declare a climate emergency, but no such policies exist for biodiversity (Gardner *et al.* 2020).

Such imbalanced discourses are also notable in economic sectors. Now, many central banks and supervisors are examining how climate risks can be integrated into financial activities. In 2019, the Principles for Responsible Banking were launched by 130 banks from 49 countries (in total, approximately USD 47 trillion in assets). This UN initiative also has a strong focus on the commitment to the decarbonized economy. In contrast, business risks related to biodiversity are still undervalued, although a 2020 report of the World Economic Forum emphasized the role of biodiversity in conserving carbon sinks and thus has called for many sectors, including fishery, construction, energy, fashion, and so on, to avoid the destruction of ecosystems that can amplify climate risks (World Economic Forum 2020). Scholars argue that the issues of climate change are structurally global and primarily result from a global sum of greenhouse gas emissions, while most of the mechanisms involved in biodiversity change are local and only become a global problem by aggregation (Legagneux *et al.* 2018).

Importantly, while the changing climate has been widely identified as a threat for organisms in a series of policy documents and scientific studies, the role of biodiversity in the climate emergency has been a minor focus (Mori 2020). Ecosystems are widely recognized for their ability to stabilize climate, as exemplified by the ability of forests to sequester and store carbon; however, differences between species-rich and -poor forests in this functional context – the substantial potential of the former over the latter – has been underexplored and thus undermined by stakeholders, including scientists. In response to increasing expectations and the need for forests to store carbon, the World Economic Forum launched a platform to grow, restore, and conserve one trillion trees within this decade. The Bonn Challenge, aimed at restoring 350 Mha of forests by 2030, is another global pledge. However, a substantial part of reforestation effects concerns monocultures, which likely limit the potential for climate change mitigation (Lewis *et al.* 2019). Interestingly, Cook-Patton *et al.* (2020) quantified terrestrial global carbon accumulation potential through natural regeneration (so-called passive restoration), instead of tree planting (i.e. active restoration), and found that default rates from the IPCC may underestimate aboveground carbon accumulation rates by 32% on average (Figure 13.6).

Again, it is now increasingly recognized that conserving and restoring the diversity of primary producers have the potential to mitigate the impacts of climate change (Mori *et al.* 2021; Mori, 2020). The importance of biodiversity can be also extended to animal assemblages (Gardner *et al.* 2019). Identifying priority areas based on co-benefits of carbon and biodiversity storage for ecosystem restoration and conservation has been actively debated (e.g. Brancalion *et al.* 2019; Dinerstein *et al.* 2020). Strassburg *et al.* (2020) discuss the cost of realizing such co-benefits, adding further considerations to the debate. However, although well-considered, these assessments for climate change mitigation and adaptation omit biodiversity from the equation. The next critical task for biodiversity management/maintenance, including IPBES, will be to explicitly add biodiversity to calculations for nature-based climate solutions that aim to limit temperature increases by 2.0°C and well below the pathway of the climate pledge of the Paris Agreement. Given the growing climate emergency awareness, this perspective – biodiversity as a solution – which can reinforce initiatives of the nature-based climate solution, deserves further attention from all stakeholders, from scientists to practitioners (Mori 2020; Mori *et al.* 2021).

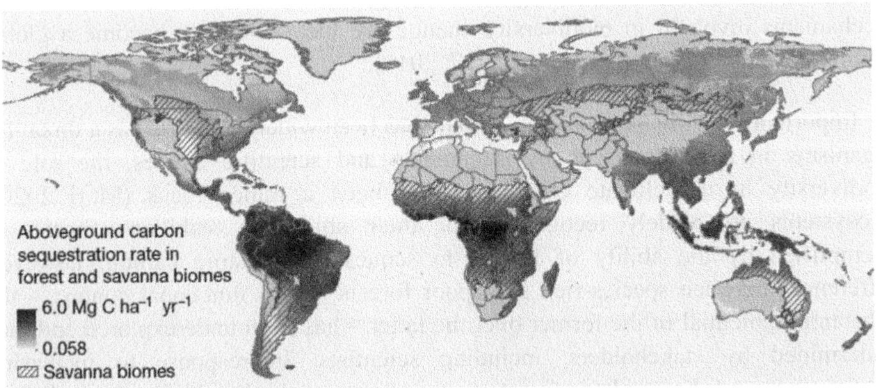

Figure 13.6. *Predicted aboveground carbon accumulation rates in naturally regrowing forests in forest (solid colors) and savanna biomes (hatched colors) (Cook-Patton et al. 2020). The map only predicts accumulation rates where natural forests have been growing for 30 years or less. For a color version of this figure, see www.iste.co.uk/loreau/biodiversity.zip*

13.4. The potential of biodiversity to cope with the changing climate

The UN General Assembly proclaimed the decade of 2021–2030 as the UN Decade on Ecosystem Restoration, which aims to halt ecosystem degradation and restore ecosystems for the benefit of people and nature. Healthy functioning ecosystems are considered to be able to enhance the livelihoods of people, counteract climate change, and halt biodiversity declines. There is now greater enthusiasm for nature-based approaches. In 2021, at the One Planet Summit, over 50 countries pledged to protect at least 30% of the world's land and ocean by 2030 (the High Ambition Coalition for Nature and People), reflecting the interests and demands from global societies to conserve both carbon and biodiversity on lands and in the ocean. However, the win-win narrative, still largely lacks explicit focus on the contributions of biodiversity to climate change mitigation and adaptation.

Now, to tackle the changing climate with nature-based approaches, there is a strong need to focus on possible feedback at multiple scales. In terrestrial ecosystems, strong positive feedback between temperature and CO_2 emission resulting from elevated rates of soil respiration (Arnone *et al.* 2008) is very likely to amplify further climatic warming, causing a further positive feedback between the biosphere and the climate. Similarly, the roles of the ocean as a carbon storehouse are also expected to shrink (Lopez-Urrutia *et al.* 2006). At the global scale, there is thus a consensus that greenhouse gas warming can increase the potential for strong long-term positive feedback to further global warming. However, processes operating at local scales still involve large uncertainties. For instance, positive and negative plant–soil feedback is likely to occur in response to the changes in temperature, precipitation, and CO_2 concentration (Pugnaire *et al.* 2019), which makes it challenging to predict the impact on regional and global carbon dynamics. In particular, knowledge is still limited on the roles of biodiversity in determining the direction of local plant–soil feedback, although its magnitude can be far from negligible when considering carbon cycles.

As such, given the essential and often irreplaceable roles of biodiversity in supporting ecosystem functions such as primary productivity, biodiversity loss due to anthropogenic global warming and other drivers will enhance the positive feedback between climate change and ecosystem function decay. Hence, biodiversity conservation is critical in the prevention, or weakening, of the magnitude of the positive feedback that otherwise leads to a hothouse earth. More ideally, the next target should be to realize a negative stabilizing feedback between biodiversity conservation and climate change mitigation (Figure 13.7). That is, restoring and conserving biodiversity is an under-appreciated but critical way to slow down climate change, further alleviating the biodiversity loss that will lead to climate change (Mori *et al.* 2021). Realizing this will be challenging but science and society need to harmonize to finally reach this ambitious goal (Mori 2020).

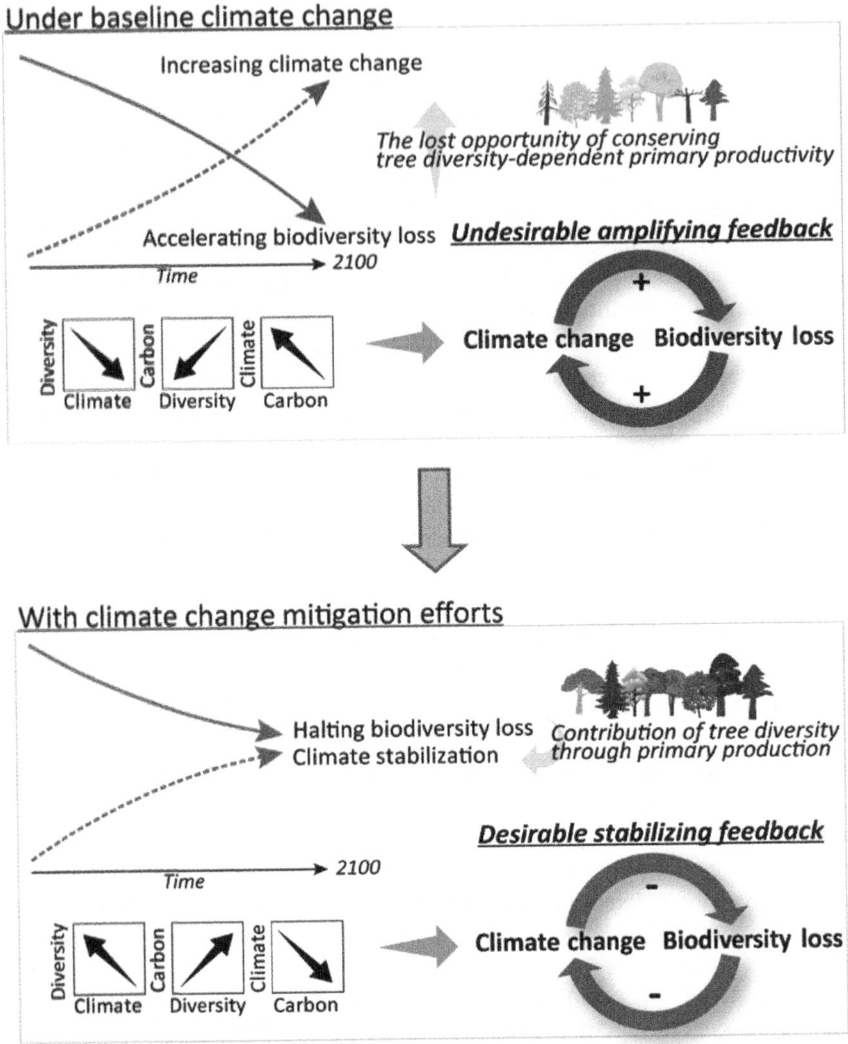

Figure 13.7. *Interdependence between climate change and biodiversity (Mori et al. 2021). There is much emphasis on the undesirable feedback where climate change drives biodiversity loss (magenta arrow feedback). Now, more emphasis should be given to the contribution of underutilized positive feedback in which biodiversity-dependent productivity could contribute to climate change mitigation (green arrow feedback). Mori et al. (2021) showed that the conservation and restoration of tree diversity could enhance this feedback and promote the desirable pathway whereby forest biodiversity contributes to climate stabilization. For a color version of this figure, see www.iste.co.uk/loreau/biodiversity.zip*

13.5. Conclusion

Biodiversity plays a crucial role in the provisioning of ecosystem services. Maintaining and restoring the vital functionality of ecosystems is urgently required in this era of global change. Climate change is expected to have one of the most significant and potentially irreversible impacts on ecosystems. To date, however, the potential roles of biodiversity as a countermeasure against the changing climate have not been acknowledged enough. While many challenges exist, now is the time to reconsider and emphasize the linkages and feedback between biodiversity and climate change – the twin environmental crises – for the sake of nature and people.

13.6. Acknowledgements

Akira S. Mori was supported by the Ichimura New Technology Foundation and by the Environment Research and Technology Development Fund (JPMEERF15S11420) of the Environmental Restoration and Conservation Agency of Japan.

13.7. References

Aben, R.C.H., Barros, N., van Donk, E. *et al.* (2017). Cross continental increase in methane ebullition under climate change. *Nature Communications,* 8(1), 1682.

Arnone III, J.A., Verburg, P.S., Johnson, D.W. *et al.* (2008). Prolonged suppression of ecosystem carbon dioxide uptake after an anomalously warm year. *Nature,* 455(7211), 383–386.

Brancalion, P.H.S., Niamir, A., Broadbent, E. *et al.* (2019). Global restoration opportunities in tropical rainforest landscapes. *Science Advances,* 5(7).

Breitburg, D., Levin, L.A., Oschlies, A. *et al.* (2018). Declining oxygen in the global ocean and coastal waters. *Science,* 359.

Burrows, M.T., Schoeman, D.S., Richardson, A.J. *et al.* (2014). Geographical limits to species-range shifts are suggested by climate velocity. *Nature,* 507(7493), 492–495.

Cardinale, B.J. (2011). Biodiversity improves water quality through niche partitioning. *Nature,* 472(7341), 86–89.

Cardinale, B.J., Duffy, J.E., Gonzalez, A. *et al.* (2012). Biodiversity loss and its impact on humanity. *Nature,* 486(7401), 59–67.

Cavicchioli, R., Ripple, W.J., Timmis, K.N. *et al.* (2019). Scientists' warning to humanity: Microorganisms and climate change. *Nature Reviews Microbiology,* 17(9), 569–586.

Cha, E.J., Knutson, T.R., Lee, T.-C., Ying, M. Nakaegawa, T. (2020). Third assessment on impacts of climate change on tropical cyclones in the Typhoon Committee Region – Part II: Future projections. *Tropical Cyclone Research and Review*, 9(2), 75–86.

Chang, C.W., Ye, H., Miki, T. *et al.* (2020). Long-term warming destabilizes aquatic ecosystems through weakening biodiversity-mediated causal networks. *Global Change Biology*, 26(11), 6413–6423.

Cheung, W.W.L. and Pauly, D. (2016). Impacts and effects of ocean warming on marine fishes. In *Explaining Ocean Warming: Causes, Scale, Effects and Consequences*, Laffoley, D. and Baxter, J.M. (eds). IUCN, Gland, 239–253.

Chaudhary, C., Richardson, A.J., Schoeman, D.S., Costello, M.J. (2021). Global warming is causing a more pronounced dip in marine species richness around the equator. *Proceedings of the National Academy of Sciences*, 118(15), e2015094118.

Cohen-Shacham, E., Walters, G., Janzen, C. Maginnis, S. (2016). Nature-based solutions to address global societal challenges. Report, International Union for Conservation of Nature, Gland, Switzerland.

Colwell, R.K. and Lees, D.C. (2000). The mid-domain effect: Geometric constraints on the geography of species richness. *Trends in Ecology and Evolution*, 15(2), 70–76.

Condamine, F.L., Romieu, J., Guinot, G. (2019). Climate cooling and clade competition likely drove the decline of lamniform sharks. *Proceedings of the National Academy of Sciences of the United States of America*, 116(41), 20584–20590.

Cook-Patton, S.C., Leavitt, S.M., Gibbs, D. *et al.* (2020). Mapping carbon accumulation potential from global natural forest regrowth. *Nature*, 585(7826), 545–550.

Dinerstein, E., Joshi, A.R., Vynne, C. *et al.* (2020). A "Global Safety Net" to reverse biodiversity loss and stabilize Earth's climate. *Science Advances*, 6(36).

Doney, S.C. (2006). Oceanography: Plankton in a warmer world. *Nature*, 444(7120), 695–696.

Evans, K., Chiba, S., Bebianno, M.J. *et al.* (2019). The global integrated world ocean assessment: Linking observations to science and policy across multiple scales. *Frontiers in Marine Science*, 6, 298.

Ezard, T.H., Aze, T., Pearson, P.N., Purvis, A. (2011). Interplay between changing climate and species' ecology drives macroevolutionary dynamics. *Science*, 332(6027), 349–351.

Falkowski, P. (2012). Ocean science: The power of plankton. *Nature*, 483(7387), S17–20.

Garcia-Comas, C., Sastri, A.R., Ye, L. *et al.* (2016). Prey size diversity hinders biomass trophic transfer and predator size diversity promotes it in planktonic communities. *Proceedings of the Royal Society B: Biological Sciences*, 283, 20152129 [Online]. Available at: http://dx.doi.org/10.1098/rspb.2015.2129.

Gardner, C.J., Bicknell, J.E., Baldwin-Cantello, W., Struebig, M.J., Davies, Z.G. (2019). Quantifying the impacts of defaunation on natural forest regeneration in a global meta-analysis. *Nature Communications*, 10(1), 4590.

Gardner, C.J., Struebig, M.J., Davies, Z.G. (2020). Conservation must capitalise on climate's moment. *Nature Communications,* 11(1), 109.

Gherardi, L.A. and Sala, O.E. (2019). Effect of interannual precipitation variability on dryland productivity: A global synthesis. *Global Change Biology,* 25(1), 269–276.

Gibson, D.J. (2009). *Grasses and Grassland Ecology.* Oxford University Press, New York.

Giling, D.P., Nejstgaard, J.C., Berger, S.A. *et al.* (2017). Thermocline deepening boosts ecosystem metabolism: Evidence from a large-scale lake enclosure experiment simulating a summer storm. *Global Change Biology,* 23(4), 1448–1462.

Griscom, B.W., Adams, J., Ellis, P.W. *et al.* (2017). Natural climate solutions. *Proceedings of the National Academy of Sciences of the United States of America,* 114(44), 11645–11650.

Guerra, C.A., Bardgett, R.D., Caon, L. *et al.* (2021). Tracking, targeting, and conserving soil biodiversity. *Science,* 371(6526), 239–241.

Harpole, W.S., Potts, D.L., Suding, K.N. (2007). Ecosystem responses to water and nitrogen amendment in a California grassland. *Global Change Biology,* 13(11), 2341–2348.

Harrison, S.P., Gornish, E.S., Copeland, S. (2015). Climate-driven diversity loss in a grassland community. *Proceedings of the National Academy of Sciences of the United States of America,* 112(28), 8672–8677.

Hawkins, B.A. and Felizola Diniz-Filho, J.A. (2004). "Latitude" and geographic patterns in species richness. *Ecography,* 27(2), 268–272.

Hector, A. and Hooper, R. (2002). Ecology. Darwin and the first ecological experiment. *Science,* 295(5555), 639–640.

Hirota, M., Holmgren, M., van Nes, E.H., Scheffer, M. (2011). Global resilience of tropical forest and savanna to critical transitions. *Science,* 334(6053), 232–235.

Hunt, G., Cronin, T.M., Roy, K. (2005). Species-energy relationship in the deep sea: A test using the Quaternary fossil record. *Ecology Letters,* 8(7), 739–747.

IPBES (2019). Summary for policymakers of the global assessment report on biodiversity and ecosystem services of the Intergovernmental Science-Policy Platform on Biodiversity and Ecosystem Services. Report.

Itoh, M., Kojima, H., Ho, P.-C. *et al.* (2017). Integrating isotopic, microbial, and modeling approaches to understand methane dynamics in a frequently disturbed deep reservoir in Taiwan. *Ecological Research,* 32(6), 861–871.

Knapp, A.K., Avolio, M.L., Beier, C. *et al.* (2017). Pushing precipitation to the extremes in distributed experiments: Recommendations for simulating wet and dry years. *Global Change Biology,* 23(5), 1774–1782.

Kreft, H. and Jetz, W. (2007). Global patterns and determinants of vascular plant diversity. *Proceedings of the National Academy of Sciences of the United States of America,* 104(14), 5925–5930.

Langenheder, S., Lindstrom, E.S., Tranvik, L.J. (2006). Structure and function of bacterial communities emerging from different sources under identical conditions. *Applied Environmental Microbiology,* 72(1), 212–220.

Lau, W.W.Y. (2013). Beyond carbon: Conceptualizing payments for ecosystem services in blue forests on carbon and other marine and coastal ecosystem services. *Ocean and Coastal Management,* 83, 5–14.

Lavorel, S. (2019). Climate change effects on grassland ecosystem services. In *Grasslands and Climate Change*, Gibson, D.J. and Newman, J.A. (eds). Cambridge University Press, Cambridge.

Legagneux, P., Casajus, N., Cazelles, K. *et al.* (2018). Our house is burning: Discrepancy in climate change vs. biodiversity coverage in the media as compared to scientific literature. *Frontiers in Ecology and Evolution,* 5, 175.

Leray, M. and Knowlton, N. (2015). DNA barcoding and metabarcoding of standardized samples reveal patterns of marine benthic diversity. *Proceedings of the National Academy of Sciences of the United States of America,* 112(7), 2076–2081.

Lewis, S.L., Wheeler, C.E., Mitchard, E.T.A., Koch, A. (2019). Restoring natural forests is the best way to remove atmospheric carbon. *Nature,* 568(7750), 25–28.

Liang, J., Crowther, T.W., Picard, N., *et al.* (2016). Positive biodiversity–productivity relationship predominant in global forests. *Science,* 354.

Lopez-Urrutia, A., San Martin, E., Harris, R.P. Irigoien, X. (2006). Scaling the metabolic balance of the oceans. *Proceedings of the National Academy of Sciences of the United States of America,* 103(23), 8739–8744.

Louca, S., Parfrey, L.W., Doebeli, M. (2016). Decoupling function and taxonomy in the global ocean microbiome. *Science,* 353(6305), 1272–1277.

Marris, E. (2019a). Act now and avert a climate crisis. *Nature,* 573(7774), 309.

Marris, E. (2019b). Why young climate activists have captured the world's attention. *Nature,* 573(7775), 471–472.

Maurer, G.E., Hallmark, A.J., Brown, R.F., Sala, O.E., Collins, S.L. (2020). Sensitivity of primary production to precipitation across the United States. *Ecology Letters,* 23(3), 527–536.

McDowell, N.G., Allen, C.D., Anderson-Teixeira, K. *et al.* (2020). Pervasive shifts in forest dynamics in a changing world. *Science,* 368, aaz9463.

Miki, T., Yokokawa, T., Matsui, K. (2014). Biodiversity and multifunctionality in a microbial community: A novel theoretical approach to quantify functional redundancy. *Proceedings of the Royal Society B: Biological Sciences,* 281(1776), 20132498.

Mori, A.S. (2020). Advancing nature-based approaches to address the biodiversity and climate emergency. *Ecology Letters,* 23(12), 1729–1732.

Mori, A.S., Lertzman, K.P., Gustafsson, L. (2017). Biodiversity and ecosystem services in forest ecosystems: A research agenda for applied forest ecology. *Journal of Applied Ecology,* 54(1), 12–27.

Mori, A.S., Cornelissen, J.H.C., Fujii, S., Okada, K.I., Isbell, F. (2020). A meta-analysis on decomposition quantifies afterlife effects of plant diversity as a global change driver. *Nature Communications,* 11(1), 4547.

Mori, A.S., Dee, L.E., Gonzalez, A. *et al.* (2021). Biodiversity–productivity relationships are key to nature-based climate solutions. *Nature Climate Change,* 11(6), 543–550.

Pecl, G.T., Araujo, M.B., Bell, J.D. *et al.* (2017). Biodiversity redistribution under climate change: Impacts on ecosystems and human well-being. *Science,* 355.

Prieto, I., Almagro, M., Bastida, F., Querejeta, J.I. (2019). Altered leaf litter quality exacerbates the negative impact of climate change on decomposition. *Journal of Ecology,* 107(5), 2364–2382.

Pugnaire, F.I., Morillo, J.A., Penuelas, J. *et al.* (2019). Climate change effects on plant–soil feedbacks and consequences for biodiversity and functioning of terrestrial ecosystems. *Science Advances,* 5(11).

Reid, A.J., Carlson, A.K., Creed, I.F. *et al.* (2019). Emerging threats and persistent conservation challenges for freshwater biodiversity. *Biological Reviews,* 94(3), 849–873.

Scheffers, B.R., de Meester, L., Bridge, T.C. *et al.* (2016). The broad footprint of climate change from genes to biomes to people. *Science,* 354.

Schiermeier, Q., Atkinson, K., Mega, E.R. *et al.* (2019). Scientists worldwide join strikes for climate change. *Nature,* 573(7775), 472–473.

Seddon, N., Daniels, F., Davis, R. *et al.* (2020). Global recognition of the importance of nature-based solutions to the impacts of climate change. *Global Sustainability,* 3, e15.

Sheldon, R.W., Prakash, A., Sutcliffe, W.H. (1972). The size distribution of particles in the Ocean1. *Limnology and Oceanography,* 17(3), 327–340.

Smith, K.L., Jr., Ruhl, H.A., Bett, B.J., Billett, D.S., Lampitt, R.S., Kaufmann, R.S. (2009). Climate, carbon cycling, and deep-ocean ecosystems. *Proceedings of the National Academy of Sciences of the United States of America,* 106(46), 19211–19218.

Sommer, J.H., Kreft, H., Kier, G., Jetz, W., Mutke, J., Barthlott, W. (2010). Projected impacts of climate change on regional capacities for global plant species richness. *Proceedings of the Royal Society B: Biological Sciences,* 277(1692), 2271–2280.

Strassburg, B.B.N., Iribarrem, A., Beyer, H.L. *et al.* (2020). Global priority areas for ecosystem restoration. *Nature,* 586(7831), 724–729.

Tollefson, J. (2019). The hard truths of climate change – by the numbers. *Nature,* 573(7774), 324–327.

United Nations (2021). *The Second World Ocean Assessment*. United Nations, New York.

Urabe, J. and Waki, N. (2009). Mitigation of adverse effects of rising CO_2on a planktonic herbivore by mixed algal diets. *Global Change Biology,* 15(2), 523–531.

Urban, M.C. (2015). Climate change. Accelerating extinction risk from climate change. *Science,* 348(6234), 571–573.

Wang, Z., Brown, J.H., Tang, Z., Fang, J. (2009). Temperature dependence, spatial scale, and tree species diversity in eastern Asia and North America. *Proceedings of the National Academy of Sciences of the United States of America,* 106(32), 13388–13392.

Warren, R., Price, J., Graham, E., Forstenhaeusler, N., Vanderwal, J. (2018). The projected effect on insects, vertebrates, and plants of limiting global warming to 1.5 degrees C rather than 2 degrees C. *Science,* 360(6390), 791–795.

Winder, M. and Schindler, D.E. (2004). Climatic effects on the phenology of lake processes. *Global Change Biology,* 10(11), 1844–1856.

World Economic Forum (2020). The Global Risks Report 2020. Report, World Economic Forum, Geneva, Switzerland.

Worm, B. and Tittensor, D.P. (2018). *A Theory of Global Biodiversity.* Princeton University Press, Pirnceton.

Yasuhara, M. and Danovaro, R. (2016). Temperature impacts on deep-sea biodiversity. *Biological Reviews,* 91(2), 275–287.

Yasuhara, M., Hunt, G., Cronin, T.M., Okahashi, H. (2009). Temporal latitudinal-gradient dynamics and tropical instability of deep-sea species diversity. *Proceedings of the National Academy of Sciences of the United States of America,* 106(51), 21717–21720.

Yasuhara, M., Doi, H., Wei, C.L., Danovaro, R., Myhre, S.E. (2016). Biodiversity–ecosystem functioning relationships in long-term time series and palaeoecological records: Deep sea as a test bed. *Philosophical Transactions of the Royal Society B: Biological Sciences,* 371, 20150282.

Yasuhara, M., Huang, H.-H.M., Hull, P., Rillo, M.C., Condamine, F.L., Tittensor, D.P., Kučera, M., Costello, M.J., Finnegan, S., O'Dea, A., Hong, Y., Bonebrake, T.C., McKenzie, N.R., Doi, H., Wei, C.-L., Kubota, Y., Saupe, E.E. (2020a). Time machine biology: Cross-timescale integration of ecology, evolution, and oceanography. Oceanography, 33(2), 16–28.

Yasuhara, M., Wei, C.L., Kucera, M. *et al.* (2020b). Past and future decline of tropical pelagic biodiversity. *Proceedings of the National Academy of Sciences of the United States of America,* 117(23), 12891–12896.

Yvon-Durocher, G., Jones, J.I., Trimmer, M., Woodward, G., Montoya, J.M. (2010). Warming alters the metabolic balance of ecosystems. *Philosophical Transactions of the Royal Society B: Biological Sciences,* 365(1549), 2117–2126.

14

Feedbacks Between Biodiversity and Society

Kirsten Henderson
Institut Natura e Teoria en Pirenèus, Surba, France

14.1. Introduction

The relationship between human societies and biodiversity is convoluted: human societies depend on biodiversity, but many activities that sustain societies lead to biodiversity loss (Díaz *et al.* 2006). The organization of socicties has revolved around nature and the diversity of life on Earth for hundreds of thousands of years. Even today's societies are intwined with nature, despite our best efforts to disconnect ourselves from it or dominate nature. This complex relationship between biodiversity and society goes far beyond the ability of ecosystem services and biodiversity to provide society with the goods and services needed for everyday life and future development. Biodiversity and human societies are connected by multilayer feedbacks that alter decision-making, impact agriculture production, influence well-being, alter income groups, and shape ecosystems.

Some interactions between humans and nature are more apparent, such as changes in land cover, urbanization, resource shortages, and fossil fuel consumption, or the way societies benefit from a diverse array of organisms for medicines, food, fibers, and other goods. However, other benefits, such as the importance of biodiversity for ecosystem processes and cultural services and its contribution to the resilience of ecosystems against environmental change (Chapters 3 and 10), are just as important to livelihoods, human well-being, health, and human activities, but are

The Ecological and Societal Consequences of Biodiversity Loss,
coordinated by Michel LOREAU, Andy HECTOR, and Forest ISBELL.

not as obvious. The coupled socio-ecological system includes many feedbacks between biodiversity and society that result in cascading effects, synergies and unpredictable changes within the system (Figure 14.1).

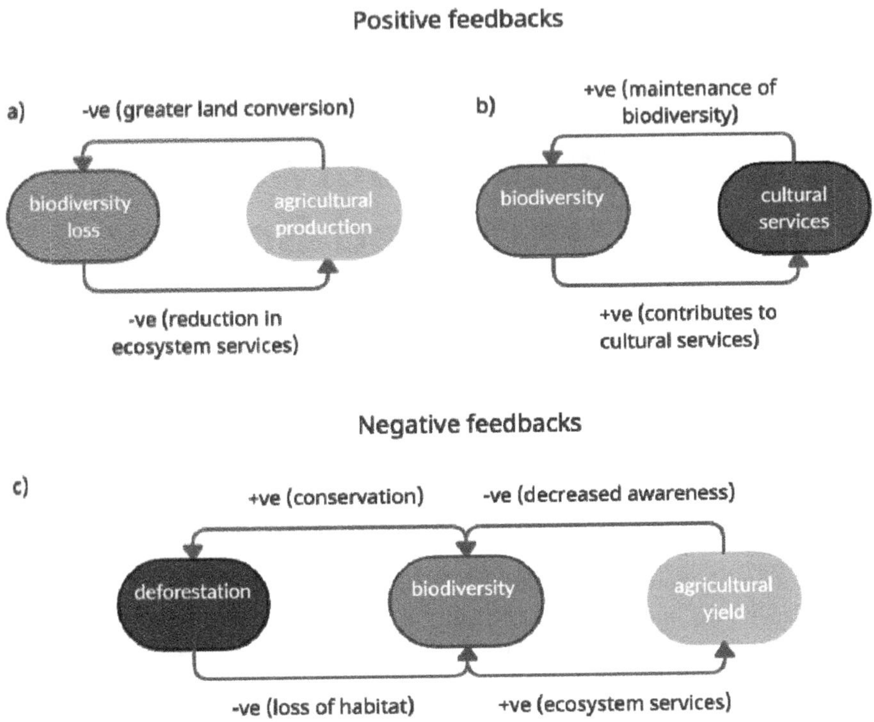

Figure 14.1. *Negative and positive feedbacks*

NOTE ON FIGURE 14.1.– *A positive feedback is an affirming response that leads to maintenance or enhancement of the original action. Positive feedbacks can either be two negative effects or two positive effects. In example (a), biodiversity loss has a negative effect on agricultural production, which further decreases biodiversity loss. In feedback (b), biodiversity promotes cultural services, which in turn maintain biodiversity with its intrinsic value. A negative feedback involves both positive and negative effects. A negative feedback reduces the change or output and has a stabilizing effect. In example (c), both feedbacks relate to scarcity. On the right, biodiversity appears abundant and does not need to be conserved; on the left, deforestation reduces biodiversity, initiating a scarcity response.*

The entangled nature of biodiversity and societal consequences highlights the importance of considering the socio-ecological system as a whole. Coupled human societies can create counter-intuitive and counter-productive effects (Henderson and Loreau 2019). These feedbacks determine the success of policies, in addition to the function of ecological processes and practices related to biodiversity management.

14.2. Society's impact on biodiversity

In Chapter 1, the authors discuss changes in biodiversity throughout history and the main drivers of biodiversity change. It is clear that humans have had an unparalleled impact on biodiversity and, in turn, that biodiversity loss has the potential to dramatically impact human societies. The relationship between biodiversity and society is further complicated by the convoluted feedbacks that make it difficult to discern, or even mask, potential threats and crises. The main pressures that directly impact biodiversity have been identified as habitat change, overexploitation, exotic species, pollution, and climate change (Mazor *et al.* 2019). However, underlying all these pressures are human population growth and consumption demand.

14.2.1. *Agriculture*

Agriculture has defined societies for nearly 12,000 years, so not surprisingly many of the feedbacks between human societies and biodiversity intersect in agricultural lands. In recent decades, biodiversity levels have been declining globally at an unprecedented scale, largely due to agricultural intensification. This trend is likely to continue as the human population increases and the demand for food increases faster still, "requiring" larger areas and/or more agricultural lands (Egli *et al.* 2017). Agriculture often involves replacing complex ecosystems with monocultures and applying heavy doses of agrochemicals, which reduces overall plant diversity and eliminates the many animal species that depend on natural forests. In addition, the conversion of natural land to crops results in fragmented natural land, creating small isolated patches or, worse, resulting in species extinction. Paradoxically, agricultural production and stability is highly dependent on biodiversity and the associated ecosystem services (Renard and Tilman 2019; Seppelt *et al.* 2020). Therefore, the current trends in agriculture management (i.e. intensification) are unsustainable for production and human well-being. However, agriculture does not have to be a major source of biodiversity degradation, it can contribute to the conservation of high-diversity systems that provide important ecosystem services, such as pollination and biological control (Tscharntke *et al.* 2005; Dudley and Alexander 2017). It has been hypothesized that moderate levels of disturbance can promote biodiversity by contributing to ecosystem resilience and increasing adaptive variation, as is the case in the Mediterranean basin (Blondel 2006) and the swidden landscapes in southwest China (Xu *et al.* 2009). The idea that agriculture can contribute to biodiversity maintenance should be seen as a positive opportunity, especially considering that much of the world's biodiversity is located in agricultural lands.

Ironically, regions with high levels of diversity are also highly valued for agricultural production or development projects. The same conditions that promote a significant variety of species also provide good growing environments for crops. Recent biodiversity losses are linked to the conversion of tropical moist forests to palm oil plantations, for which palm oil is the highest producing oil, with large net gains and a wide variety of uses (Vijay *et al.* 2016). The threat to biodiversity-rich areas, mostly from population growth and increased agriculture demand, is widespread and targeted by conservation initiatives (Habel *et al.* 2019). Which feedback loop – positive feedback involving biodiversity loss and agriculture production (Figure 14.1a) or negative feedback involving deforestation and biodiversity (Figure 14.1c) – dominates depends on what society values.

As societies develop, the typical diet shifts towards higher consumption of meat and highly processed food. Furthermore, with greater consumption comes more food waste, all of which requires greater conversion of natural habitats for agricultural purposes (Conrad *et al.* 2018) and biodiversity loss. The problem is not agriculture itself, but the perceptions and values of society, more specifically the perceptions and values of industrialized societies.

14.2.2. *Income*

There is no debating the fact that the wealthiest societies contribute disproportionately to climate change, pollution and habitat destruction through consumption demands and urbanization. Despite the widespread and significant impact of individuals from wealthier societies, much of the focus of biodiversity loss is on developing societies, where much of today's most diverse regions are situated. Although society is more attuned to biodiversity loss today and therefore the focus falls on developing regions, this does not discount biodiversity loss from centuries ago in wealthier societies. Biodiversity loss is not a new phenomenon; throughout history human activities have led to species extinctions: the extinction of mammoths and giant kangaroos upon the arrival of humans in Eurasia, Australia, and the Americas 10,000 to 60,000 years ago, the disappearance of megafauna on islands settled 500 to 5,000 years ago, and the current wave of extinction that began in 1500 (Andermann *et al.* 2020; Turvey and Crees 2019). The relationship between biodiversity and society is complex, and the addition of inequality, income, and development masks many of the feedbacks between biodiversity and society and exaggerates others.

The tropics are biodiversity-rich thanks to the favorable climate but also happen to be home to many of the world's poorest populations (Sachs 2001). This fact alone demonstrates the gross inequality of the world and the social-ecological feedbacks maintaining poverty. The loss of biodiversity in lower income societies is driven in large part by the actions and policies of higher income societies, population growth, and local increases in consumption demands (Henderson and Loreau 2021). Inequality, pressures from commercial agriculture, and externally driven changes transform traditional farming practices and livelihoods. One major concern that arises from rapid development and the exclusion of local practices is the potential to create a viscous cycle (i.e. positive feedback loop) that traps individuals in a state of deteriorating land conditions and poor well-being. For example, as individuals clear forests, convert rangelands, or over-harvest to meet external demands, they simultaneously degrade the land for future use. As the cycle repeats the land becomes severely degraded and so do the ecosystem services required by human societies (Barrett *et al.* 2011). Furthermore, the livelihoods of the rural poor depend disproportionately on natural resources from the local environment and, without a source of alternative resources or institutions and governance structures to secure adaptive management strategies, low-income rural people are vulnerable to over-exploitation and global economic pressures, threatening ecosystem resilience and biodiversity (Fisher and Christopher 2007; Selig *et al.* 2013).

Higher income countries are not immune to biodiversity loss, and many wealthier countries utilized the same development strategies (i.e. the industrial revolution) that resulted in widespread species loss centuries ago in an effort to achieve "progress" in society (Ellis *et al.* 2011). The highest percentages of biodiversity loss compared to intact ecosystems are in parts of the US and Canada, Brazil, Argentina, southern Africa, central Asia, China, Mongolia, and Australia (Figure 14.2; Newbold *et al.* 2016). Although biodiversity loss is presently greater in economically emerging economies, the historical losses and percentages of loss indicate high overall losses across a spectrum of economic groups. At present, there is little biodiversity to lose in higher income regions because it was wiped out centuries ago. The short-term memory and current situation make it easy for wealthier societies to shirk the responsibilities of biodiversity loss and give credence to the idea that once societies become economically, socially, and technologically developed it might seem that the feedbacks from biodiversity disappear. However, as much as people attempt to transform and distort society's dependence on nature, the underpinning of everyday life is still firmly dependent on biodiversity. There is evidence to support the idea that economic growth has actually made humans more dependent upon ecosystem services and biodiversity (Guo *et al.* 2010). Society is asking more and more from natural ecosystems even as we reduce their capacity to meet our needs, but there are also cultural services that impact tourism and well-being (for details see section 14.3.2).

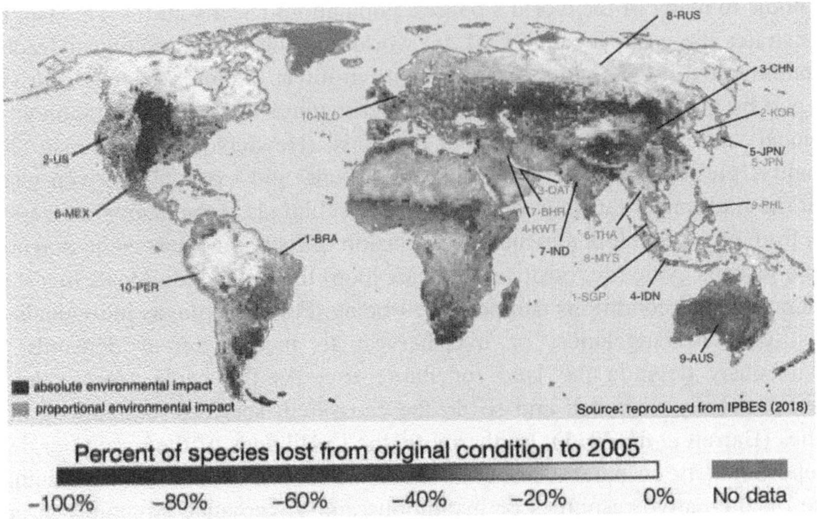

Figure 14.2. *Biodiversity loss and environmental impact. For a color version of this figure, see www.iste.co.uk/loreau/biodiversity.zip*

NOTE ON FIGURE 14.2.– *The map shows the total estimated biodiversity loss compared to an intact state (IPBES 2018). The countries in blue are those that have the highest current absolute environmental impact – the highest combined ranking for natural forest loss, natural habitat conversion, marine capture, fertilizer use, proportion of threatened species, water pollution, and carbon emissions. The countries in orange are those that have the highest proportional impact – the highest combined ranking compared to the resources in the country. The environmental impact rankings from Bradshaw* et al. *(2010) do not take into account whether the impact is for local use or foreign use.*

The difference between affluent societies and the rural poor are the institutions that support resource management. It is not that biodiversity loss does not exist or impact affluent societies, but institutions and power absorb the shock of biodiversity loss. The relationship between economic development and environmental degradation is described by a controversial concept, known as the environmental Kuznets curve. It is hypothesized that environmental degradation initially increases as countries develop and then, when a threshold is reached, there is a minimum level of institutional quality and a willingness to implement conservation measures. Therefore, economic development mitigates biodiversity loss at higher incomes due to a collective demand for improvements in environmental quality and the ability to invest in measures to protect biodiversity. This may be true on a local level for a specific indicator, but not globally or when the indicators (e.g. pollutants) are expanded to include new indicators that replace the outdated ones (Stern 2017). The problem with theories related to development is the assumption that consumption is constant, but rather, as the Jevons paradox shows, when more is available or produced more efficiently people tend to consume more (York and McGee 2016). It is more likely that the problems are shifted to other regions or new pollutants/issues emerge. A true negative feedback between biodiversity loss and development, as Kuznets described, has yet to be observed empirically.

14.3. How societies view biodiversity

14.3.1. *Biodiversity and culture*

There is considerable experimental and observational evidence showing that biodiversity influences provisioning and regulating services (Chapter 10). Furthermore, these ecosystem services are unquestionably essential to human life. However, these two concepts are not evident to all societies. How biodiversity is perceived is dependent on culture and social norms. The very definition of a society refers to a group of people who share a common lifestyle and organization, and how people interact with diversity provides a clear distinction between groups. In general, there are two societies, those that are connected and highly dependent on nature (Fedele *et al.* 2021) and those that are disconnected and technologically minded (Soga and Gaton 2016). This is a gross simplification of global societies, but it categorizes people into those who are aware that they are governed by the natural system and those that attempt to change the natural system to meet their needs, while ignoring the underlying workings of biodiversity.

The number of nature-dependent people is dwindling; it is estimated that 70% of the world's population will live in urban spaces by 2070, which will shift the values and perspectives of society. People in urban environments do not have a direct experience of nature and are less likely to recognize the benefits provided by nature and biodiversity (Soga and Gaston 2016). Over time, individuals are less likely to experience nature, which reduces positive perceptions of nature. The desire to protect it and the value associated with it then diminish. As new generations grow up in urban environments, they become less and less aware of societies' dependence on nature. Urbanization is a leading cause of biodiversity loss, as it requires infrastructure, greater production, dams and reservoirs, and higher air pollution.

However, what is potentially more critical are the feedbacks involved between urbanization, technology, and biodiversity. In theory, technology can spawn temporal mismatches that delay biodiversity and reduce resilience, thus increasing the probability of feedbacks driving the socio-ecological system towards unsustainable trajectories (Lafuite and Loreau 2017). Before biodiversity management and conservation can be implemented, there needs to be a general awareness of the situation, which is at odds with urbanization and development.

Industrialization, technology, and urbanization have created an enormous disconnect between humans and nature (Soga and Gaston 2016). However, technology and industrialization have also contributed immensely to societies' standards of living (e.g. education, medicine, liberties). Technology has managed to make the unimaginable possible, which could be why societies believe it is possible to replace ecosystem services with technology (Fitter 2013). How technology will progress in the future is a big unknown, but it is highly unlikely that technology will be able to reproduce all the benefits of biodiversity or override the feedbacks between society and biodiversity. In most cases, technology has created new problems or has generated positive feedbacks that push the system towards unsustainable outcomes (Henderson and Loreau 2021). Conversely, Selig and colleagues (2013) found that countries with greater knowledge, greater technological development, and more stable governance chose to preserve species and habitats. Where and how these policies are implemented is discussed in the next section.

By contrast, indigenous and nature-dependent societies can often be characterized by their practical knowledge of nature (i.e. fire control, soil management, and watershed protection) that has been passed down through generations. Indigenous and rural populations in lower income regions use biodiversity as a gauge to adjust their demands and practices with respect to land disturbances. For example, the Hani people of southwestern China were dependent on biodiversity for several important ecological and social functions up until 2000, after which the majority of land was converted to tea and rubber plantations (Xu *et al.* 2009). The Hani understood the succession processes that enabled them to carry out long-term land management strategies. They maintained the sacred fig tree (which also happens to be a keystone species in the region), the soil regeneration period was set at 13 years, and certain fields were never cultivated to avoid bad luck. Their reasons may have been spiritual but their practices reflect a keen awareness of society–biodiversity feedbacks. However, recent and rapid development in Asia has been displacing local communities, including the Hani people.

Globally, these rural cultures that respond to changes in biodiversity are being converted to urban areas, which also leads to a conversion in attitudes towards biodiversity. The urbanization of rural farmers represents a positive feedback loop, as urbanization degrades ecosystems, which hinders rural subsistence farmers' livelihoods, which in turn forces them to relocate to urban environments in search of new opportunities, which propagates the cycle.

14.3.2. *Biodiversity and well-being*

Despite biodiversity's numerous benefits to human health and well-being (Naeem *et al.* 2016; Chapter 11), societies typically see development by way of technological advancements and urbanization, both of which contribute to the decoupling of society from nature. There is now evidence of declining well-being in the absence of interactions with nature (Soga and Gaston 2016). Well-being encompasses mental health, physical health, and cultural identity, in addition to basic material needs. The access to material goods has been increasing, but at the expense of other elements, such as cultural connections with biodiversity (Díaz *et al.* 2006). Goodhart's law states that when a measure becomes a target, it ceases to be a good measure. For example, in the pursuit of economic development, other societal issues and biodiversity have been neglected, which are also important for economic development.

14.3.3. *Value of biodiversity*

Biodiversity is not valued by societies for the same reasons nor using the same metrics, which makes the valuation of biodiversity a complex issue that does not translate well to policy or conservation. Biodiversity is often communicated as a quantification of ecological risk in association with biodiversity loss (Spierenburg 2012). Generally speaking, the economic value of biodiversity requires a loss of biodiversity. The supply and demand concept from economics has trickled down into the valuation of biodiversity, such that biodiversity is valued because it is rare, whereas aesthetic appreciation or cultural values place the emphasis on maintaining biodiversity for its intrinsic value.

Assigning a monetary value to biodiversity inevitably involves trade-offs, specifically temporal, spatial, beneficiary, and service trade-offs (Spierenburg 2012). Agriculture and its many feedbacks with biodiversity also has many trade-offs: higher production yield now (i.e. benefit now/here, cost later/elsewhere) versus long-term sustainable agriculture with various species (i.e. cost now, benefit later). Biodiversity conservation comes with trade-offs, for example the regions that are preserved are often not the most biodiversity-rich, as biodiversity-rich regions are often worth more money (i.e. some win, others lose). Likewise, conserving one region can negatively impact another in what is known as a spillover effect (i.e. benefit here, cost there) (Barrett *et al.* 2012).

Many components of biodiversity have cultural value, including appreciation of wildlife and scenic places and spiritual, educational, religious, and recreational values (Díaz *et al.* 2006). Often, individuals value places with a diverse array of species, especially those that include more charismatic animals and plants. All societies value clean water, access to food, and stable environments, which are just some of the benefits provided by biodiversity at the ecosystem level (described in Chapters 6 and 12). Whether we realize the role of biodiversity or not, we still value the end products.

14.4. Biodiversity policy and society

Biodiversity policies need to establish a link between cultural values, emotions, health, livelihoods, and biodiversity. However, the decline in biodiversity and natural areas means that the link between humans and biodiversity is eroding with every generation, which means fewer chances to experience nature and a decreased awareness, or valuation, of biodiversity (Soga and Gaston 2016; Saunders *et al.* 2006). The first step in establishing biodiversity conservation initiatives is for society to recognize there is a problem.

14.4.1. *Awareness and perception*

In nearly all societies, people see some value in maintaining biodiversity, but they tend to overlook the role of human activities in biodiversity loss, blaming instead extreme climatic conditions or soil conditions (Menzel and Bogeholz 2009), both of which list human activities as underlying drivers. Societies that are less economically developed, with minimal social infrastructure, are more likely to make decisions based on less scientific and more utilitarian perspectives of ecosystem services. This bottom-up approach garners greater public participation and awareness of local issues. Societies that are wealthier and include intensively technological social infrastructure implement programs to maintain biodiversity for conservation purposes, known as the top-down approach to decision-making. The top-down approach gives a scientific solution to an administrative understanding of the biodiversity challenges, which often precludes local action and public participation (Rauschmayer *et al.* 2009).

Highly technologically developed and higher income societies are fortunate to have greater access to information, which brings forth an awareness of the benefits of biodiversity, at least on a scientific or administrative level, and the current threats to biodiversity, but that does not mean that people within these societies recognize their impact on biodiversity loss. In a case study of Chilean and German adolescents' awareness of biodiversity loss, it would appear that the proximity to biodiversity hotspots gives the Chilean adolescents a greater awareness of the issue than German adolescents, but neither group was particularly aware of the drivers of biodiversity loss (Menzel and Bogeholz 2009). Individuals in higher income societies recognize the benefits of biodiversity to well-being and can be motivated to conserve biodiversity based on well-being, but primarily in lower income regions.

Establishing biodiversity conservation areas in biodiversity hotspots with local subsistence populations also has its challenges, despite the connectedness to nature. In some cases, local populations believe their good practices have been responsible for maintaining the area and therefore that protecting the area is unnecessary (Muhumuza and Balkwill 2013). In turn, they do not respect the boundaries of the protected areas and may undermine the conservation efforts.

There also exists a phenomenon of seeing problems where they do not exist. Conservation attitudes are influenced by familiarity, positive experiences, recreational activities, and emotions (Kiley *et al.* 2017). What's more, people tend to believe that the landscape they prefer is also the most diverse. Landscape preferences have led to forest bias, where forests are preferentially conserved and viewed as preferable to grasslands, either because forests are green and associated with pro-environmental attitudes or because of the fact that the rapid rate of deforestation globally has become symbolic of ecosystem destruction and forests must be protected above all else. As a result, people shape the environment to fit their ideal image of nature, not necessarily in terms of biodiversity or ecosystem functioning.

How societies perceive problems and communicate them determines the effectiveness of solutions (Saunders *et al.* 2006). Whether it be through agricultural production, recreational activities, goods and services, cultural preservation, or species and habitat preservation, humans and their societies shape the environment through their perceptions. Generally speaking, it is the awareness stage that produces a positive or negative feedback. Based on the literature, there appear to be three main cultural views that impact biodiversity conservation: locally dependent utilitarian societies that are connected to their environment and maintain it for sustainability purposes (Johns *et al.* 2013) – this generally involves a positive feedback between biodiversity and cultural values; higher income industrial societies with a "formed environmental consciousness" (Jacobsen and Hanley 2009) –

more typically a negative feedback that applies scarcity principles; and economically driven societies, which can be high or low income, that prioritize development rather than conservation (Jacobson *et al.* 2009) – generally a positive feedback between degradation and detachment from nature. Each of these cultural groupings approaches biodiversity conservation and maintenance differently, with varying degrees of success.

14.4.2. *Management strategies*

Biodiversity management aims either to set land aside for conservation purposes or maintain biodiversity in agricultural environments. In a meta-analysis of 359 protected areas, Gray and colleagues (2016) found that protected areas are essential to biodiversity conservation. This approach requires strong management and local cooperation. Provisioning services, including agriculture, represent a major threat to biodiversity but are also highly sought after, which leads to socio-ecological conflicts and feedbacks that undermine biodiversity conservation. The silver lining is that agriculture also has the potential to implement policies that are beneficial to biodiversity maintenance (Tscharntke *et al.* 2005). Agricultural biodiversity can bridge the interests of local individuals/small producers and global populations (Johns *et al.* 2013). Rural, locally dependent societies maintain an ancestral knowledge of the environment and the food farmed or wild-harvested on the land, including a knowledge of biodiversity. These traditional food systems establish strong socio-ecological relationships through their cultural link to biodiversity and by doing so favor sustainable agricultural systems and biodiversity. Industrialized agricultural systems are taking cues from traditional farmers, with initiatives to maintain biodiversity in industrialized agricultural systems through ecological corridors, waste reduction, and mosaic crop systems with trees, crops, and grass to mimic natural systems (Scherr and McNeely 2008).

In general, conservation is meant to give non-monetary value to nature, separating nature from economics; however, in reality it transforms nature into a commodity driven by market forces (Büscher *et al.* 2012). Ecosystem services are promoted through payments to guarantee their sustainability. These payment plans make it difficult to assign a value to biodiversity, as the direct links from humans to nature are ecosystem services (biodiversity – ecosystem functions – ecosystem services – human well-being), and societies tend to favor provisioning services for their tangible benefits. As can be seen from the chain, biodiversity is the foundation for ecosystem services, including provisioning services, but these trade-offs between biodiversity and ecosystem services are generally not considered.

14.4.3. *Conflicts in biodiversity management*

Biodiversity conservation is complicated and intricately related to social values and structure, regardless of the income level or cultural aspects. Affluent societies are driving the majority of research and institutional initiatives promoting biodiversity protection (Fazey *et al.* 2005; Jacobsen and Hanley 2009). However, the bulk of conservation initiatives are set in lower income regions. This can lead to cultural mismatches and undesirable feedbacks. Many biodiversity-rich regions have been identified as zones of conflict between conservation and future intensification. Furthermore, conservation faces extrinsic challenges, such as poverty, institutional support, resources, population growth, urbanization, and political instability. Furthermore, the protected areas are often those with lower economic values and higher aesthetic, recreational, or perceived wilderness values, not necessarily the areas with the greatest ecological values. This personal attachment to certain landscapes does not have to be negative; rather, it can be used to promote conservation and bring awareness to conservation (Saunders *et al.* 2013).

Biodiversity management needs to strike a balance between biodiversity promotion and local socio-economic issues and cultural practices (Muhumuza and Balkwill 2013). When biodiversity conservation limits access to food and social development, social-ecological feedbacks result in negative outcomes for both the society and the environment (Cazalis *et al.* 2018; Klein *et al.* 2015; Naughton-Treves 2005). Strictly protected areas prevent locals from accessing resources within the boundaries of the protected area, thereby increasing the degradation of surrounding areas and creating isolated islands of protected areas, which generate problems of equity with local communities and are ineffective at preserving biodiversity. Protected areas present one of many potential conflicts between societies and biodiversity protection. Other impacts on human societies include displacement, fees for accessing the area, lack of involvement and consultation with local populations, inadequate compensation, contested land ownership, and lack of cohesion between communities with respect to economic and cultural differences, all of which hamper biodiversity protection (Muhumuza and Balkwill 2013). In a nutshell, when society loses accountability of their resources, biodiversity is lost.

Affluent and industrialized regions have conflicting ideals and practices (Johnson *et al.* 2017). In higher income regions, conservation initiatives and sustainability are a symbol of societal development, but this contradicts other aspects of development, which favor technology, consumption, and economic growth. Whether the efforts to maintain nature and biodiversity are strong enough to overcome the tangible benefits of development depends on the feedbacks between human well-being, technology, resource accessibility, and the specific initiatives targeted (Henderson and Loreau 2021).

14.4.4. *Successful initiatives*

It is easy to point out the flaws in biodiversity management, but the question remains: how do societies integrate biodiversity policies without triggering undesirable feedbacks? Agriculture is a leading cause of biodiversity loss globally and should be targeted for biodiversity management, but without negatively impacting livelihoods. In a study comparing whether land sparing or sharing – land sparing is high-yield/intensive agriculture over a smaller area of land, which leaves sections of natural land untouched, and land sharing is low-yield farming that maintains biodiversity within the agricultural landscape – would lead to the largest population sizes of bird and tree species in Indian and Ghanian agricultural areas, Phalan and colleagues (2011) found that land sparing, with the goal of increasing agricultural yields while sparing and restoring natural habitats, was the optimal strategy for reconciling food production and biodiversity conservation. This is an optimistic view that assumes that humans have the capacity to adapt their activities to a far greater degree than do other species, which may be true but, in reality, unless human welfare is a priority, biodiversity conservation is not feasible. Others have found that integrated land-use planning over broad spatial scales has the potential to lead to more efficient maintenance of biodiversity levels (Egli *et al.* 2017).

The key to successful biodiversity management in the hardest-hit regions of the world is the inclusion of local knowledge and a component targeting poverty alleviation and sustained production (Naughton-Treves 2005). This involves a bottom-up decision-making approach, where local social norms govern practices with the support of governments and institutions. Many local groups are aware of the scarcity and demand feedback loops that can either promote biodiversity or lead to rapid declines (Figure 14.1). However, they might not have the power, resources, or infrastructure necessary to bring such initiatives to fruition (Adom 2016).

The surface area of protected zones is critical for the success of initiatives, both for social equity reasons (e.g. not so much that it infringes on people's ability to access resources) and ecological function (e.g. connectedness). Researchers suggest that the area of biodiversity designated for conservation depends on what services are desired (Perrings *et al.* 2010). However, the problem remains that the links between biodiversity and ecosystem services are not well-known. The threshold for an ecosystem service, and an area to be conserved, should be relative to the use and abundance of other ecosystem services (Hussain and Tschirhart 2013). Although the exact surface area of conservation needed presents a major unknown, it is clear that surface area is not a fixed point and shifts with societies' demands and decisions.

14.5. Conclusion

Biodiversity may seem to be negligible when there are other more apparent issues in society, such as poverty, standards of well-being, and food shortages, but all of these issues are dependent on sustaining biodiversity. There are synergies and trade-offs to consider, dynamic shifts in development and cultural norms and livelihoods that alter our perceptions. Thus, biodiversity loss is not only decisively influenced by societal processes but can itself have severe impacts on human societies.

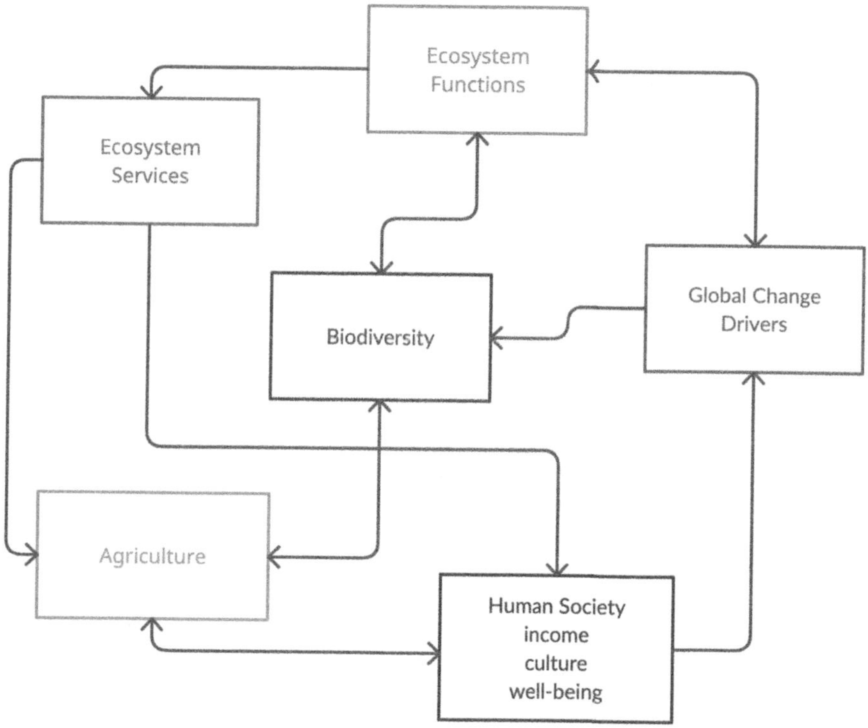

Figure 14.3. *Biodiversity–society interconnectedness. For a color version of this figure, see www.iste.co.uk/loreau/biodiversity.zip*

NOTE ON FIGURE 14.3.– *Biodiversity is the foundation of human society: the ecosystem services we value are dependent on ecosystem functions, which are reliant on biodiversity. Biodiversity is in turn impacted by human societies through global change drivers (climate change, land change, species introductions, biogeochemical cycles). Agriculture technically lies within the ecosystems services category, however, agriculture is at the intersection of many conflicts and solutions between human societies and biodiversity. Agriculture also has direct consequences for both human societies (income, culture, well-being) and biodiversity.*

Given the interdependence between humans and biodiversity (Figure 14.3) and the long-term and global implications of biodiversity loss to both society (i.e. cultural, economic, development, welfare) and ecosystems (i.e. stability, resilience, functioning, productivity), it seems obvious that biodiversity management and conservation require multidimensional policies that incorporate social processes and economic activities and that social demands require biodiversity maintenance. So, why is it that so many human activities degrade biodiversity? The multiple layers and intricate connections between biodiversity and societies, often hidden within positive and negative feedbacks, confuse and mask the potential threats of extensive biodiversity loss. Biodiversity feedbacks are continuously working to maintain ecosystem functioning and ecosystem services, and even more so with economic development and urbanization, even if they appear intangible to society. Agriculture and development feedbacks overwhelm the system, shaping the environment and altering perceptions. Urban development results in positive feedback that promotes urbanization, at the expense of biodiversity, cultural identity, and environmental awareness.

What remains to be known is how much land needs to be conserved, especially considering population growth and increases in consumption. This is further complicated by the masking effect of technology. Technology can improve the efficiency of extracting ecosystem services, but it is unlikely to replace them. Therefore, technological advancements create a time delay in the perception of changes to the quality and abundance of ecosystem services, which is already a delayed response to biodiversity loss. These time delays are particularly important for negative feedbacks that rely on human perception and awareness. The delay may leave enough time for a positive feedback to dominate, which could accelerate biodiversity loss.

Successful integration of biodiversity policy requires targeted approaches for different income groups within society. Higher income societies would benefit from a reintroduction to nature, to increase local awareness of biodiversity issues and the impact of their policies and actions on global biodiversity. Lower income societies are primarily focused on poverty alleviation, for which biodiversity initiatives could be a solution to reducing undernourishment and poverty, so long as local input and livelihoods are taken into consideration. All societies should be educated on the importance of maintaining biodiversity for sustainable development according to unique cultural identities. There are endless factors to consider in the human–biodiversity system, from biogeochemical cycles, species composition, and economic gains to disturbance regimes, social norms, and decision-making, but it is critical to remember that humans are one species among many and that for humans to thrive, others to need to survive.

14.6. Acknowledgements

This chapter was immensely improved by the editors' comments. I would like to thank Forest Isbell and Michel Loreau. I am grateful to Matthieu Barbier and Victor Cazalis for their honest and encouraging feedback.

14.7. References

Barrett, C.B., Travis, A.J., Dasgupta, P. (2011). On biodiversity conservation and poverty traps. *Proceedings of the National Academy of Sciences*, 108(34), 13907–13912.

Blondel, J. (2006). The "design" of the Mediterranean landscapes: A millennial story of humans and ecological systems during the historic period. *Human Ecology*, 34, 713–729.

Bradshaw, C.J.A., Giam, X., Sodhi, N.S. (2010). Evaluating the relative environmental impact of countries. *PLoS ONE*, 5(5), e10444.

Büscher, B., Sullivan, S., Neves, K., Igoe, J., Brockington, D. (2012). Towards a synthesized critique of neoliberal biodiversity conservation. *Capitalism Nature Socialism*, 23(2), 4–30.

Cazalis, V., Loreau, M., Henderson, K. (2018). Do we have to choose between feeding the human population and conserving nature? Modeling the global dependence of people on ecosystem services. *Science of the Total Environment*, 634, 1463–1474.

Conrad, Z., Niles, M.T., Neher, D.A., *et al.* (2018). Relationship between food waste, diet quality, and environmental sustainability. *PLoS ONE*, 13(4), e0195405.

Díaz, S., Fargione, J., Chapin III, F.S., Tilman, D. (2006). Biodiversity loss threatens human well-being. *PLoS Biology*, 4(8), e277.

Dudley, N. and Alexander, S. (2017). Agriculture and biodiversity: A review. *Biodiversity*, 18(2–3), 45–49.

Egli, L., Meyer, C., Scherber, C., Kreft, H., Tscharntke, T. (2018). Winners and losers of national and global efforts to reconcile agricultural intensification and biodiversity conservation. *Global Change Biology*, 24(5), 2212–2228.

Ellis, C.J., Yahr, R., Coppins, B.J. (2011). Archaeobotanical evidence for a massive loss of epiphyte species richness during industrialization in southern England. *Proceedings of the Royal Society B: Biological Sciences,* 278(1724), 3482–3489.

Fedele, G., Donatti, C.I., Bornacelly, I., Hole, D.G. (2021). Nature-dependent people: Mapping human direct use of nature for basic needs across the tropics. *Global Environmental Change*, 102368.

Fazey, I., Fischer, J., Lindenmayer, D.B. (2005). Who does all the research in conservation biology? *Biodiversity and Conservation*, 14(4), 917–934.

Fisher, B. and Christopher, T. (2007). Poverty and biodiversity: Measuring the overlap of human poverty and the biodiversity hotspots. *Ecological Economics*, 62(1), 93–101.

Guo, Z., Zhang, L., Li, Y. (2010). Increased dependence of humans on ecosystem services and biodiversity. *PLoS ONE*, 5(10), e13113.

Gray, C.L., Hill, S.L., Newbold, T. *et al.* (2016). Local biodiversity is higher inside than outside terrestrial protected areas worldwide. *Nature Communications,* 7(1), 1–7.

Habel, J.C., Rache, L., Schneider, U.A. *et al.* (2019). Final countdown for biodiversity hotspots. *Conservation Letters*, 12, e12668.

Henderson, K. and Loreau, M. (2019). An ecological theory of changing human population dynamics. *People and Nature,* 1(1), 31–43.

Henderson, K. and Loreau, M. (2021). Unequal access to resources undermines global sustainability. *Science of the Total Environment,* 763, 142981.

Hussain, A.T. and Tschirhart, J. (2013). Economic/ecological tradeoffs among ecosystem services and biodiversity conservation. *Ecological Economics*, 93, 116–127.

IPBES (2018). Summary for policymakers of the assessment report on land degradation and restoration of the Intergovernmental Science-Policy Platform on Biodiversity and Ecosystem Services. Bonn, Germany.

Jacobsen, J.B. and Hanley, N. (2009). Are there income effects on global willingness to pay for biodiversity conservation? *Environmental and Resource Economics*, 43(2), 137–160.

Johns, T., Powell, B., Maundu, P., Eyzaguirre, P.B. (2013). Agricultural biodiversity as a link between traditional food systems and contemporary development, social integrity and ecological health. *Journal of the Science of Food and Agriculture*, 93(14), 3433–3442.

Johnson, C.N., Balmford, A., Brook, B.W., Buettel, J.C., Galetti, M., Guangchun, L., Wilmshurst, J.M. (2017). Biodiversity losses and conservation responses in the Anthropocene. *Science*, 356(6335), 270–275.

Kiley, H.M., Ainsworth, G.B., van Dongen, W.F., Weston, M.A. (2017). Variation in public perceptions and attitudes towards terrestrial ecosystems. *Science of the Total Environment*, 590, 440–451.

Klein, C., McKinnon, M.C., Wright, B.T., Possingham, H.P., Halpern, B.S. (2015). Social equity and the probability of success of biodiversity conservation. *Global Environmental Change*, 35, 299–306.

Lafuite, A.S. and Loreau, M. (2017). Time-delayed biodiversity feedbacks and the sustainability of social-ecological systems. *Ecological Modelling*, 351, 96–108.

Mazor, T., Doropoulos, C., Schwarzmueller, F. *et al.* (2018). Global mismatch of policy and research on drivers of biodiversity loss. *Nature Ecology and Evolution*, 2(7), 1071–1074.

Menzel, S. and Bögeholz, S. (2009). The loss of biodiversity as a challenge for sustainable development: How do pupils in Chile and Germany perceive resource dilemmas? *Research in Science Education*, 39(4), 429–447.

Muhumuza, M. and Balkwill, K. (2013). Factors affecting the success of conserving biodiversity in national parks: A review of case studies from Africa. *International Journal of Biodiversity*, 798101.

Naeem, S., Chazdon, R., Duffy, J.E., Prager, C., Worm, B. (2016). Biodiversity and human well-being: An essential link for sustainable development. *Proceedings of the Royal Society B: Biological Sciences*, 283(1844), 20162091.

Naughton-Treves, L., Holland, M.B., Brandon, K. (2005). The role of protected area in conserving biodiversity and sustaining local livelihoods. *Annual Review of Environmental Resources*, 30, 219–252.

Newbold, T., Hudson, L.N., Arnell, A.P. *et al.* (2016). Dataset: Global map of the Biodiversity Intactness Index. *Natural History Museum Data Portal.*

Pereira, H.M., Navarro, L.M., Martins, I.S. (2012). Global biodiversity change: The bad, the good, and the unknown. *Annual Review of Environment and Resources*, 37.

Phalan, B., Onial, M., Balmford, A., Green, R. E. (2011). Reconciling food production and biodiversity conservation: Land sharing and land sparing compared. *Science*, 333(6047), 1289–1291.

Rauschmayer, F., van den Hove, S., Koetz, T. (2009). Participation in EU biodiversity governance: How far beyond rhetoric? *Environment and Planning C: Government and Policy*, 27(1), 42–58.

Renard, D. and Tilman, D. (2019). National food production stabilized by crop diversity. *Nature*, 571, 257–260.

Sachs, J.D. (2001). Tropical Underdevelopment. National Bureau of Economic Research, Cambridge, Working paper 8119.

Saunders, C.D., Brook, A.T., Myers, O.E. (2006). Using psychology to save biodiversity and human well-being. *Conservation Biology*, 20(3), 702–705.

Saunders, F.P. (2013). Seeing and doing conservation differently: A discussion of landscape aesthetics, wilderness, and biodiversity conservation. *The Journal of Environment and Development,* 22(1), 3–24.

Scherr, S.J. and McNeely, J.A. (2008). Biodiversity conservation and agricultural sustainability: Towards a new paradigm of "ecoagriculture" landscapes. *Philosophical Transactions of the Royal Society B: Biological Sciences*, 363(1491), 477–494.

Selig, E.R., Longo, C., Halpern, B.S. *et al.* (2013). Assessing global marine biodiversity status within a coupled socio-ecological perspective. *PLoS ONE*, 8(4), e60284.

Seppelt, R., Arndt, C., Beckmann, M., Martin, E.A., Hertel, T.W. (2020). Deciphering the biodiversity–production mutualism in the global food security debate. *Trends in Ecology and Evolution*, 35(11), 1011–1020.

Soga, M. and Gaston, K.A. (2016). Extinction of experience: The loss of human–nature interactions. *Frontiers in Ecology and the Environment*, 14(2), 94–101.

Stern, D.I. (2017). The environmental Kuznets curve after 25 years. *Journal of Bioeconomics*, 19(1), 7–28.

Tscharntke, T., Klein, A.M., Kruess, A., Steffan-Dewenter, I., Thies, C. (2005). Landscape perspectives on agricultural intensification and biodiversity–ecosystem service management. *Ecology Letters*, 8(8), 857–874.

Vijay, V., Pimm, S.L., Jenkins, C.N., Smith, S.J. (2016). The impacts of oil palm on recent deforestation and biodiversity loss. *PLoS ONE*, 11(7), e0159668.

Xu, J., Lebel, L., Sturgeon, J. (2009). Functional links between biodiversity, livelihoods, and culture in a Hani swidden landscape in southwest China. *Ecology and Society*, 14, 2.

York, R. and McGee, J.A. (2016) Understanding the Jevons paradox. *Environmental Sociology* 2(1), 77–87.

15

Protecting and Restoring Biodiversity and Ecosystem Services

Forest ISBELL
University of Minnesota, St. Paul, USA

15.1. Introduction

We are currently living in a window of opportunity when many species of animals and plants are threatened with extinction, but most of the threatened species have not yet gone extinct. For the best-studied groups of animals, about 25% of species are currently threatened with extinction (Díaz *et al.* 2019). About 39% of plant species are currently threatened with extinction (Nic Lughadha *et al.* 2020). For both animals and plants, in the order of 1% of species in well-studied taxonomic groups are known to have gone extinct in recent centuries, mostly in recent decades (Ceballos *et al.* 2015; Humphreys *et al.* 2019). Expansion of conservation efforts and ambitious new ideas are needed to prevent many more species from becoming threatened or extinct in the coming decades and to avoid the devastating impacts of biodiversity loss on ecosystems and people.

Now is the time to slow and eventually reverse biodiversity loss. Protecting and restoring biodiversity is crucial for the sustainability of ecosystem functioning, stability, and services. In this chapter, I consider two widespread conservation strategies, protected areas and ecological restoration, while briefly mentioning a few additional strategies. I describe how protecting biodiversity can also help protect ecosystem functioning and services. I also describe how increasing the levels of biodiversity in restoration schemes can increase their efficacy.

15.2. Protecting biodiversity and ecosystems

15.2.1. *What are protected areas and what are they intended to protect?*

Protected areas are one of the primary strategies for conserving biodiversity worldwide. Protected areas have been implemented in a variety of ways, and with various levels of protection, but all are places where people seek to conserve nature. According to the International Union for Conservation of Nature (IUCN):

> A *protected area* is a clearly defined geographical space, recognized, dedicated, and managed, through legal or other effective means, to achieve the long-term conservation of nature with associated ecosystem services and cultural values. (Dudley 2008)

Although this widely used definition does not explicitly mention *biodiversity*, the phrase *conservation of nature* encompasses conservation of biodiversity. The *associated ecosystem services and cultural values* include those that are directly dependent on biodiversity and those that happen to be co-located with biodiversity.

For more than a century, conservationists have debated the reasons why nature should be protected and the extent to which it should be protected from or for various uses by people. For example, in the United States, preservationists, including John Muir, argued that nature must be protected from exploitation by people, whereas Gifford Pinchot and others instead argued that nature must be managed to sustainably produce natural resources for people (Callicott 1990). Both of their visions led to the establishment of extensive protected areas throughout the US, respectively through the US National Parks and the US National Forests. Today, many National Parks are surrounded by National Forests and both agencies work together to protect and manage these lands. The mission of the US National Park Service is to preserve unimpaired the natural and cultural resources and values of the National Park System for enjoyment, education, and inspiration of current and future generations. The mission of the Forest Service, which is a part of the US Department of Agriculture, is to sustain the health, diversity, and productivity of the nation's forests and grasslands to meet the needs of present and future generations. Thus, despite early and ongoing debates about why and to what extent lands or waters should be protected, there is now widespread appreciation that nature should be protected for its own sake (i.e., nature's intrinsic values) and for its many benefits to people (i.e. nature's instrumental values), which include both non-material benefits (e.g. enjoyment, education, and inspiration) and material benefits (e.g. productivity of harvested forests and grasslands).

The IUCN recognizes seven types of protected areas (Dudley 2008), organized by levels of protection and the many different reasons for protection (Table 15.1). Many protected areas include lands or waters that may fit in multiple categories. Thus, the assigned category is often based on the primary management objectives, which should apply to at least three-quarters of the protected area. Note the progression from protection of nature from people in Category I to protection of nature for the sustainable use of natural resources by people in Category VI.

Category	Definitions: Reasons for and characteristics of protection
Ia) Strict nature reserve	Strictly protected for biodiversity and also possibly geological/ geomorphological features, with human visitation, use, and impacts controlled and limited to ensure protection of the conservation values
Ib) Wilderness area	Usually large unmodified or slightly modified areas, retaining their natural character and influence, without permanent or significant human habitation, protected and managed to preserve their natural condition
II) National park	Large natural or near-natural areas protecting large-scale ecological processes with characteristic species and ecosystems, which also have environmentally and culturally compatible spiritual, scientific, educational, recreational, and visitor opportunities
III) Natural monument or feature	Areas set aside to protect a specific natural monument, which can be a landform, sea mount, marine cavern, a geological feature such as a cave, or a living feature such as an ancient grove
IV) Habitat or species management area	Areas to protect particular species or habitats, where management reflects this priority. Many will need regular, active interventions to meet the needs of particular species or habitats, but this is not a requirement of the category
V) Protected landscape or seascape	Areas where the interaction of people and nature over time has produced a distinct character with significant ecological, biological, cultural, and scenic value: and where safeguarding the integrity of this interaction is vital to protecting and sustaining the area and its associated nature conservation and other values
VI) Protected area with sustainable use of natural resources	Areas which conserve ecosystems, together with associated cultural values and traditional natural resource management systems. Generally large, mainly in a natural condition, with a proportion under sustainable natural resource management and with low-level, non-industrial natural resource use compatible with nature conservation seen as one of the main aims

Table 15.1. *IUCN categories of protected areas (Dudley 2008)*

In addition to seeing values in protecting nature from and for various uses by people, many of us also view ourselves as a part of nature. Personally, when I am immersed in a wild place, staying long enough to experience it with all my senses, I am reminded that nature is where I belong. Human activities are not just a destructive force that the rest of nature must be protected from. Some levels and types of human activities are natural disturbances that other species have coevolved with for millennia (Kimmerer and Lake 2001: Takeuchi 2010). Growth in human population and per capita consumption, as well as trade and technology, have fundamentally changed and excessively increased many human impacts on nature. It is nevertheless important to remember that we also have important roles to play as a part of nature.

15.2.2. *What global targets have been established for protected areas?*

The global biodiversity conservation community has consistently set targets for protected areas. For example, the Secretariat of the Convention on Biological Diversity (CBD) previously established 20 Aichi Biodiversity Targets for the decade from 2011 to 2020. Recent international reports concluded that none of these 20 targets were met (CBD 2020a: Díaz *et al.* 2019); however, some progress was made towards Target 11, which focused on protected areas:

> By 2020, at least 17% of terrestrial and inland water areas and 10% of coastal and marine areas, especially areas of particular importance for biodiversity and ecosystem services, are conserved through effectively and equitably managed, ecologically representative and well-connected systems of protected areas and other effective area-based conservation measures, and integrated into the wider landscape... (CBD 2020a)

New global goals and targets are currently being set for the coming decade. In January 2021, more than 50 countries announced their commitment to protect 30% of Earth's land and ocean by 2030. The current draft of the CBD's post-2020 global biodiversity framework includes the following target for protected areas:

> By 2030, protect and conserve through well connected and effective system of protected areas and other effective area-based conservation measures at least 30% of the planet with the focus on areas particularly important for biodiversity. (CBD 2020b)

Thus, the success of protected areas, according to both past and future global targets, is determined by their global extent, spatial overlap with biodiversity, connectedness to one another, and effectiveness. The equitability of decision-making and ecosystem services are not included in the future target for protected areas because they have been elevated to their own high-level targets (CBD 2020b). The CBD, IUCN, and others are beginning to account for other effective area-based conservation measures (OECMs) when quantifying the extent of protected areas. OECMs are governed and managed in ways that achieve desirable and sustained long-term outcomes for the conservation of the biodiversity, ecosystem functioning, and ecosystem services they contain (CBD 2020a). Accounting for these additional areas acknowledges that conservation of biodiversity depends not only on the legal protected status of lands and waters, but also the management of lands and waters by other responsible individuals or groups, including landowners, Indigenous people, and businesses. Some areas formally designated as protected areas may be less protected than OECMs that are not formally designated as protected areas and vice versa.

15.2.3. *Where are protected areas and how effective are they?*

The global extent of protected areas has grown substantially in recent decades, helping reduce some of the major drivers of biodiversity loss (e.g. habitat loss and overexploitation) in some of the most diverse places on Earth. In January 2021, the World Database on Protected Areas included 239,412 terrestrial protected areas worldwide, which together covered 15.3% of terrestrial ecosystems, and 18,416 marine protected areas, which together covered 7.7% of marine ecosystems (Figure 15.1). Many of these lands and waters were only recently protected, with the area of lands protected increasing by more than 50% and the area of oceans protected more than doubling between 2000 and 2020 (CBD 2020a).

As noted above, a key measure of success for protected areas is whether they cover the most diverse places on Earth. Some of the largest protected areas cover areas with relatively low (albeit important) biodiversity, such as the ice-covered lands of Greenland or in the Southern Ocean near Antarctica (Figure 15.1). However, protected areas increasingly cover some of the most diverse parts of the planet. Over the past two decades, the proportion of key biodiversity areas that are protected has increased from 29% to 44% (CBD 2020a). Another recent study found that one out of every four terrestrial ecoregions, which are areas that tend to have unique groups of species, already has 30% of the area protected and a total of two out of every three ecoregions could still possibly meet the goal of 30% protection by 2030 (Dinerstein *et al.* 2019). Spreading protected areas across as many ecoregions as possible helps maintain beta diversity and protect many endemic species.

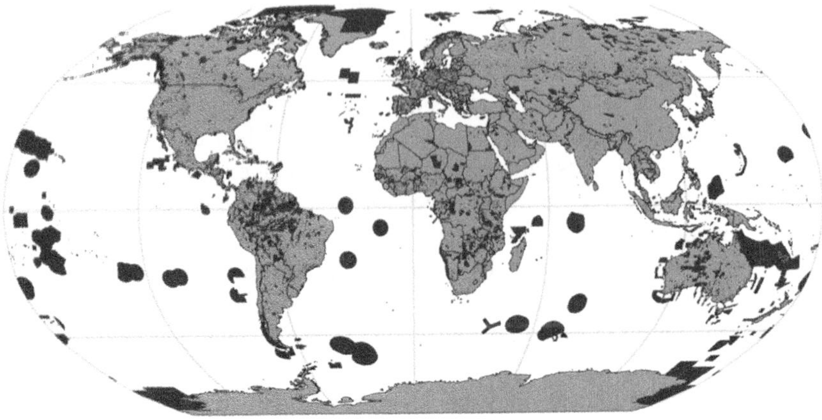

Figure 15.1. *Global protected areas on land and in the oceans (UNEP-WCMC and IUCN 2020)*

Population abundances and local species richness tend to be greater in protected areas than in nearby unprotected areas, but not as high in protected areas as in the most pristine places (Gray *et al.* 2016). Specifically, to compare the effectiveness of many protected areas worldwide on a common scale, consider a measure of effectiveness that would be 0% if sites within protected areas are, on average, as diverse as unprotected sites, and 100% if the biodiversity of protected sites is, on average as high as for pristine sites (i.e. minimally impacted primary vegetation) (Gray *et al.* 2016). Note that this measure of effectiveness is bound neither by 0% nor by 100%. By this measure, the global protected area network is, on average, 41% effective at retaining within-sample species richness and 54% effective at retaining local abundance (Figure 15.2) (Gray *et al.* 2016).

The finding that the abundance and diversity of species are greater in protected than in unprotected areas is encouraging, though perhaps unsurprising. The finding that protected areas fall short of the levels of abundance and diversity observed in the most pristine places draws attention to the fact that not all protected places are pristine and not all pristine places are protected. Some pristine places are outside protected areas, and protected areas often include secondary vegetation that is passively recovering or being actively restored after anthropogenic disturbances.

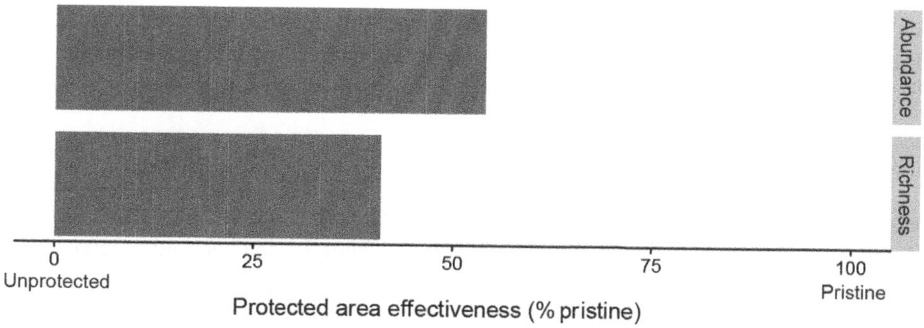

Figure 15.2. *The abundance and diversity of species in protected areas tends to be higher than in unprotected areas, but lower than in pristine areas (Gray et al. 2016)*

15.2.4. *Does protecting biodiversity also protect ecosystem services?*

Protecting biodiversity can help protect ecosystem services that are co-located with biodiversity. Many previous studies have considered the spatial overlap of biodiversity and ecosystem services. To the extent that biodiversity and ecosystem services are found at the same places, and to the extent that the legal protections for biodiversity also protect ecosystem services, protected areas can safeguard places with high levels of both biodiversity and ecosystem services. Although the most diverse places worldwide are not always the places with the highest levels of all ecosystem services (Naidoo *et al.* 2008), there are many win-win locations where high levels of both biodiversity and ecosystem services can be found. For example, many studies have identified locations where current or future protected areas could include both biodiversity and carbon storage (Strassburg *et al.* 2020). Studies such as these often identify tropical forests as areas with many species and considerable carbon stored in plant biomass. These studies may miss other places where there are fewer species, but where conserving biodiversity over time could nevertheless be essential for preventing the erosion of ecosystem services that directly depend on biodiversity.

Protecting biodiversity can also help protect ecosystem services that directly depend on biodiversity. Rather than simply viewing nature as either protected or not, studies are increasingly considering the quality, in addition to the quantity, of protected areas (Beyer *et al.* 2020). Widespread drivers of biodiversity loss, including habitat fragmentation and climate change, are degrading protected areas over time as these habitats become increasingly disconnected and as their environmental conditions become inhospitable to some of the species living within them. Thus far, only a few studies have considered whether the erosion of biodiversity over time, which is projected to occur even within protected areas, could cause the erosion of ecosystem services.

For example, we considered the extent to which nearby habitat loss could lead to extinction debts, the gradual loss of biodiversity, and resultant ecosystem service debts, the gradual loss of biodiversity-dependent benefits that people receive from nature, in remaining fragments of nature (Isbell *et al.* 2015). We estimated that in the order of 2–21 Pg C could be gradually emitted globally from remaining fragments of terrestrial ecosystems because of plant species loss caused by nearby habitat destruction (Isbell *et al.* 2015). Our results indicate that these biodiversity-dependent ecosystem service debts could be globally substantial, even when locally small, if they occur across the many remaining fragments of nature, including in protected areas. We also found that there is tremendous value in conserving biodiversity over time within habitat fragments, including in protected areas. The social cost of carbon emissions resulting from the projected biodiversity loss in remaining habitat fragments amounts to nearly 100 times current annual global conservation expenditures (Isbell *et al.* 2015). This suggests that much greater investments in conservation may be warranted, even though our study considered only one of many drivers of biodiversity loss, land conversion to croplands and urban areas, and only one of many ecosystem services, carbon storage.

15.2.5. *What are the limitations of protected areas?*

Although protected areas will remain an important strategy for global conservation, they cannot, by themselves, adequately conserve biodiversity or ecosystem services (Mora and Sale 2011). Some major limitations of current protected areas, and proposed solutions for addressing them, are shown in Table 15.2.

The establishment of protected areas has often led to undesirable impacts on people. In particular, protected areas have often been established, owned, and managed by wealthy individuals, governmental organizations, and nongovernmental organizations without the prior informed consent of the people that they have displaced or disenfranchised, which include many Indigenous peoples and rural people who work in agricultural, pastoral, or fishing communities (Mora and Sale 2011). The future success of protected areas depends not only on whether they are large enough, located in the best places, and effective at conserving biodiversity, but also whether they are established, owned, and managed by a diverse group of people and interests. The growing recognition and inclusion of other effective area-based conservation measures (CBD 2020a) offers some hope that multiple perspectives will be equitably included in future decisions regarding the protection of biodiversity and ecosystems, which may help reduce leakage (Ewers and Rodrigues 2008).

Some Limitations of Protected Areas	Current Strategies to Address Limitations
Species in protected areas remain threatened because they have reduced abundances and restricted movements	Expand the size and improve the connectivity of protected areas
Protected areas are unable to escape degradation from climate change, nutrient pollution, and invasive species	Address direct drivers outside protected areas, including through policies, and restore degraded protected areas
Climate change may make it infeasible for some species or ecosystems to continue inhabiting protected areas	Proactively manage for the future, such as by assisting range shifts for species or ecosystems that are otherwise trapped
Protection is unable to prevent all illegal deforestation, poaching, and fishing	Increase staffing, resources, and technology for enforcing legal protections
Protection is temporary, leading to downgrading, downsizing, and degazetting	Bolster legal protections and their permanence
Protected areas can be counterproductive when they displace, rather than prevent, human impacts (also known as leakage)	Spatially optimize protections to avoid shifting impacts (e.g. of deforestation, hunting, fishing) to other areas that are even higher priorities for protection

Table 15.2. *Some limitations of protected areas and strategies for addressing them*

15.3. Restoring biodiversity and ecosystems by reversing degradation

15.3.1. *What is restoration and why is it needed?*

Whereas protected areas seek to retain biodiversity in some of the most pristine and least impacted places on Earth, restoration efforts seek to assist the recovery of biodiversity, ecosystem functioning, and ecosystem services in many other places where they have been degraded. Although protection and restoration tend to target places with more or less biodiversity, respectively, the two strategies are not mutually exclusive and often co-occur. Many protected areas include at least some degraded ecosystems (Beyer *et al.* 2020) and thus restoration is often an important component of protected area management. Additionally, restoration of ecosystem services is often needed in unprotected areas (Figure 15.2), such as after degradation of carbon storage or the productivity of livestock forage, wood, crops, or fishes. According to IPBES:

> *Restoration* is any intentional activity that initiates or accelerates the recovery of an ecosystem from a degraded state. *Degraded states* result from the persistent decline or loss of biodiversity, ecosystem functions and services that cannot fully recover unaided within decadal time scales. (IPBES 2018)

Restoration and degradation are relative terms and must be understood with respect to a reference system. Restored towards what state? Degraded below what state? Common references include: (1) past baselines, which were observed prior to all or some human impacts at a place of interest; (2) nearby relatively unimpacted places, which uses a space-for-time substitution; or (3) future targets, which are desirable even if unrelated to the past state (IPBES 2018). Baselines, references and targets are not stati, and can include consideration of historic, natural, or desired ranges of spatial and temporal variability. The information needed for establishing past baselines is rarely available or complete, simply because, in many parts of the world, people have been degrading natural systems for centuries or millennia. Past baselines that were observed in recent decades are likely shifted baselines, observed after some degradation had already occurred. Likewise, nearby relatively unimpacted places are rarely, if ever, completely free from the impacts of people, and thus they may also be shifted baselines that lead to underestimates of degradation. Future targets are often useful for policy and can account for ongoing global changes that may make returning to a former state of nature infeasible or undesirable.

Restoration is needed because the extent of degraded land is enormous and growing. Biodiversity is substantially reduced by many human land uses that cover vast areas of the planet, including pastures and rangelands, croplands, tree plantations, and urban areas (Newbold *et al.* 2015). Managed grazing is the most extensive land use worldwide, covering 3.2 billion ha (Hurtt *et al.* 2020), an area larger than the continent of Africa and about 25% of Earth's ice-free land. Croplands currently cover about half as much area, 1.6 billion ha (Hurtt *et al.* 2020), an area nearly as large as the continent of South America. By comparison, urban areas cover only about 1% of Earth's ice-free land (Hurtt *et al.* 2020). It is important to remember that these agricultural land uses are not static. Instead, this enormous footprint of agriculture moves around to some extent from one year to the next, stamping out biodiversity in new places and relieving some pressures on biodiversity in other places where land is abandoned from agriculture. For the past two decades, more land has been abandoned from agriculture than converted to it worldwide (FAO 2020). Thus, the global extent of agriculture is shrinking but moving, generating lots of degraded land each year. Recovery of secondary vegetation, which includes recovering grasslands and forests, currently covers approximately 3.0 billion ha and is projected to grow to approximately 3.7 to 4.5 billion ha by the end of this century (Hurtt *et al.* 2020). About half of this recovering vegetation is forest that is repeatedly harvested. The other half includes grasslands and drylands where livestock grazing or croplands have been abandoned, in some cases because unsustainable production practices have eroded the profitability of agricultural production.

Restoration is also needed because the unassisted recovery of degraded ecosystems can, in some cases, be very slow. If biodiversity quickly recovers after agricultural abandonment, then the impacts of agriculture could be largely limited to the places where we currently produce food and harvest wood products. However, biodiversity can remain substantially lower in many of the vast and growing areas of secondary vegetation for many decades or centuries after agricultural abandonment (Moreno-Mateos *et al.* 2017). For example, we recently considered changes in local plant diversity and productivity that occurred over nearly four decades in 21 grasslands and savannas with known agricultural land-use histories (Isbell *et al.* 2019). These recovering old fields had been abandoned for agriculture between 1927 and 2015. Plant diversity and productivity were monitored in them from the early 1980s, or whenever they were abandoned if after the early 1980s, until the present day. Thus, these old fields form a chronosequence, allowing consideration of both the changes that occurred over time within fields, as well as comparisons among fields of different ages after agricultural abandonment. We found that, during the nine decades following agricultural abandonment, local plant diversity and productivity recovered slowly and incompletely (Figure 15.3). Specifically, by 91 years after agricultural abandonment, formerly ploughed fields still had only three quarters of the plant diversity (inverse Simpson's index) and half of the plant productivity observed in a nearby remnant ecosystem that had never been ploughed (Figure 15.3).

Thus, active restoration efforts are sometimes needed to enable and accelerate recovery. This may be the case, for example, in places where dispersal limitation, lack of symbionts, or other factors slow or prevent the unassisted recovery of biodiversity. In other places, however, unassisted recovery may be sufficiently rapid, or active restoration may be no better than slow recovery if restoration efforts are unsuccessful, unambitious, or unwise. For example, the unassisted recovery of forests may often be preferred to low-diversity tree plantings (Mori 2020).

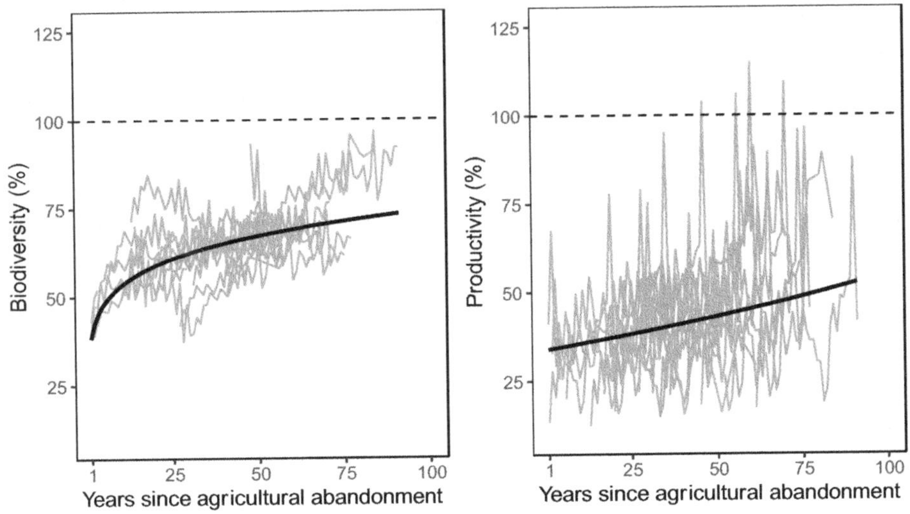

Figure 15.3. *Unassisted recovery can be slow and incomplete (Isbell et al. 2019)*

15.3.2. *What global targets have been established for restoration?*

The global biodiversity conservation community has set goals and targets for restoration. For example, two of the CBD's 20 Aichi Biodiversity Targets for the decade from 2011 to 2020 were related to restoration, including Aichi Target 14:

> By 2020, ecosystems that provide essential services, including services related to water, and contribute to health, livelihoods and well-being, are restored and safeguarded, taking into account the needs of women, indigenous and local communities, and the poor and vulnerable. (CBD 2020a)

and Aichi Target 15:

> By 2020, ecosystem resilience and the contribution of biodiversity to carbon stocks has been enhanced, through conservation and restoration, including restoration of at least 15% of degraded

> ecosystems, thereby contributing to climate change mitigation and adaptation and to combatting desertification. (CBD 2020a)

Recent reports concluded that neither of these targets have been achieved (CBD 2020a, Díaz *et al.* 2019). Goals and targets are currently being set for the coming decade, including the current draft of the CBD's target for restoration:

> By 2030, [50%] of land and sea areas globally are under spatial planning addressing land/sea use change, retaining most of the existing intact and wilderness areas, and allow to restore [X%] of degraded freshwater, marine and terrestrial natural ecosystems and connectivity among them. (CBD 2020b)

Note that the percentages in brackets are still being debated and have yet to be finalized. The success of restoration, according to both past and future global targets, is determined by its global extent, the benefits it provides to people, and the equitability with which those benefits are realized. The equitability of ecosystem services has now been elevated to its own separate high-level target (CBD 2020b).

15.3.3. *How extensive and effective is restoration?*

The global extent of restored areas remains far below the target of 15% of degraded ecosystems, but new ambitious restoration programs are underway (CBD 2020a). Fewer than 10% of the Parties to the Convention on Biodiversity have both established national targets towards Aichi Target 15 and met or exceeded the 15% restoration level (CBD 2020a). Despite this disappointing past progress, there are some signs of hope for the coming decade. For instance, ambitious plans are now underway to implement natural climate solutions, including planting as many as one trillion trees worldwide to help soak up carbon pollution. The diversity of these plantings will strongly influence their effectiveness (Mori 2020) (Aichi Target 15).

Biodiversity and ecosystem services tend to be greater in restored areas than in nearby degraded areas, but not as high in restored areas as in nearby reference ecosystems (Benayas *et al.* 2009). Many studies have tested the effects of various restoration treatments on biodiversity and ecosystem services. For example, restoration efforts may seek to increase water quality, increase the abundance of rare or charismatic species, increase carbon storage to slow down climate change, or increase the production of livestock forage, wood products, crops, or fishes. Progress towards these locally established goals is typically assessed by comparisons with degraded ecosystems where recovery is unassisted and reference ecosystems that have escaped at least some of the degradation that has occurred at the restored site.

Next, I revisit the results of a previous global meta-analysis of restoration studies (Benayas *et al.* 2009) to consider the effectiveness of restoration. To make comparisons at many sites worldwide, I quantified an effectiveness measure that would be 0% if restored sites have, on average, the same levels of biodiversity and ecosystem services as degraded sites and 100% if they are as high as for reference sites. By this measure, restoration is, on average, 78% effective at restoring biodiversity, 48% effective at restoring regulating ecosystem services, and 57% effective at restoring supporting ecosystem services (Figure 15.4).

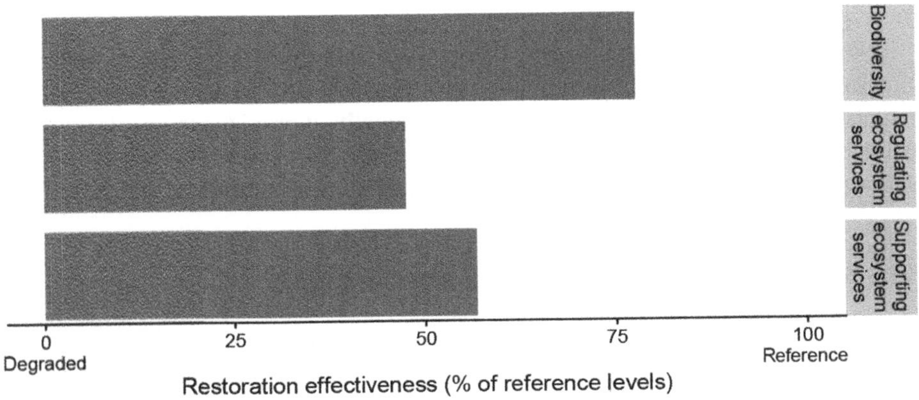

Figure 15.4. *Biodiversity and ecosystem services can be incompletely restored (Benayas et al. 2009)*

15.3.4. *Increasing the diversity of restorations can increase their efficacy*

The results of hundreds of biodiversity studies, which are discussed throughout this book, indicate that increasing biodiversity often causes an increase in ecosystem functioning, stability, and services. Given these findings, it is already clear that increasing the diversity of restorations will often increase their efficacy. Here I briefly highlight a few of the many biodiversity studies that show the importance of using high levels of biodiversity in future restoration efforts.

Restoring high levels of plant diversity can cause an increase in ecosystem services in tropical forests and temperate grasslands. For example, to speed up the recovery of biodiversity and ecosystem functioning in some previously harvested tropical forests, enrichment planting is now used as a restoration technique. The Sabah Biodiversity Experiment is testing the impacts not only of enrichment planting, but also of enriching with low-diversity monocultures of trees or high-diversity mixtures of tree species. Satellite data collected in this experiment have shown that enrichment planting significantly increased plant cover and estimated levels of biomass and that increasing the plant diversity of enrichment caused further increases (Wu *et al.* 2020). Furthermore, in degraded southern English hay meadows, increasing the diversity of the grassland plants sown caused a long-lasting increase in both plant diversity and hay production (Bullock *et al.* 2007). Thus, in both these cases, the restoration treatments were effective, especially when implemented with high plant diversity.

Natural climate solutions, including tree planting, will be most effective if they are implemented with high levels of plant diversity (Mori 2020). For example, a recent experiment found that mixtures of 16 tree species accumulated twice as much carbon over an eight-year period than the average species in monoculture, and as much carbon as two commercial monocultures (Huang *et al.* 2018). Similar results were observed in natural forests (Mori 2018). Indeed, restoring biodiversity enhances carbon storage not only in forests, but also in a wide range of agroecosystems (Yang *et al.* 2020).

Furthermore, different plant species provide different ecosystem services (Hector and Bagchi 2007), and different plant species provide any given service during different years, at different places, and under future environmental conditions (Isbell *et al.* 2011). Thus, restoring highly diverse plant communities can meet not only a single specific objective, such as enhancing carbon storage or hay production, but can also help ensure provision of multiple ecosystem services at multiple times and places in a changing world.

15.3.5. *What are the limitations of restoration?*

The main limitation of restoration is that we cannot fully restore ecosystems, in part because we do not yet fully understand the structure and dynamics of natural systems. Restoration can speed up recovery of biodiversity, and restoration of biodiversity can cause increases in ecosystem functioning and services, but it is not yet possible to fully restore the levels of biodiversity and ecosystem services found in relatively undisturbed ecosystems. Consequently, efforts to mitigate and offset the loss of biodiversity and habitats are inadequate. We cannot simply pave over parts of nature and then recreate them elsewhere. Restoration provides a test of our ecological knowledge. We are not yet passing this test. There are plenty of opportunities for improving our skill in the coming decades, however, because enormous amounts of degraded land are already available, and much more will be generated in the future (Hurtt *et al.* 2020).

15.4. Looking ahead

In the coming decades, what are the prospects for achieving global conservation goals for protected areas and restoration? Conservation efforts have often focused on reducing direct drivers of biodiversity loss, such as land use changes, climate change, and overexploitation. The focus is now shifting to indirect drivers of biodiversity loss, such as human population growth and per capita production and consumption patterns that are upstream from the direct drivers of biodiversity loss. Indirect drivers can be considered the root causes of biodiversity loss and can also be leverage points for identifying solutions to global biodiversity loss (Díaz *et al.* 2019). In this section, I consider past and future trends for human population size and the amount of agricultural land per person. These indirect drivers together determine the global extent of agriculture, which is a major direct driver of biodiversity loss (Newbold *et al.* 2015) and which partly determines the land available for protected areas and restoration. After exploring these temporal trends, I briefly describe two contrasting strategies that seek to meet agricultural demand without compromising biodiversity.

Over the past 60 years, the global human population size more than tripled, increasing by 244%, whereas the global extent of agriculture increased by only 8% (Figure 15.5a,d,e). The overall inefficiency of agricultural land, quantified as the area of croplands and pasture per person, was greatly reduced over this same period of time, dropping from 1.01 ha of pasture and 0.44 ha of cropland per person in 1961 to 0.43 ha of pasture and 0.21 ha of cropland per person in 2017 (Figure 15.5b,c). Thus, rather than the tripling of the global human population leading to a tripling of the global extent of agriculture, from 1961 to 2017, the global extent of croplands increased by only 16% and the global extent of pasture increased by only 5% (Figure 15.5d,e).

Looking ahead to the coming decades, the global extent of agriculture may substantially decrease, even while the human population size continues to grow, creating extensive opportunities for expanding both protected areas and restoration. If the growth in the human population continues to slow, as projected by the United Nations (Figure 15.5a), and if the amount of agricultural land per person continues to decrease (albeit at a slowing rate; dashed lines in Figure 15.5b,c), then the global extent of agriculture may decrease substantially in the coming decades (dashed lines in Figure 15.5d,e). Indeed, the global extent of pasture has already peaked and has been declining for the past two decades (Figure 15.5e). The global extent of croplands continues to expand, but, if the trends over the past 60 years prevail over the next 30 years, then the global extent of croplands may expand very little in the future and may peak within the next decade or two (Figure 15.5d), though land conversion to croplands would continue as long as the footprint of agriculture continues to shift. The future is, of course, far from certain and will depend largely on the choices people make in the coming years.

Minimizing the footprint of agriculture is consistent with a land-sparing strategy, which tolerates the intensification of agriculture to spare biodiversity and ecosystem services from the extensification of agriculture. This strategy seeks to expand protected areas and halt the conversion of nature to agriculture, partly because it is not yet possible to fully restore former agricultural lands. How might we further decrease the amount of agricultural land per person (Figure 15.5b,c) to prevent the further extensification of agriculture? Land-sparing strategies (Foley *et al.* 2011) often seek to minimize the global extent of agriculture by increasing crop yields, shifting to plant-based diets, reducing food waste, and spatially optimizing production and trade. This strategy seeks to reduce the extensification of agriculture enough to leave room for the conservation of biodiversity elsewhere, including in protected areas. In some ways, it favors wealthy countries that previously converted lots of land to agriculture.

In contrast, a land-sharing strategy would tolerate the extensification of agriculture to reduce the undesirable impacts of the intensification of agriculture on biodiversity and ecosystem services. The fertilizers and pesticides typically used to increase crop yields have negative impacts on biodiversity and ecosystem services both near and far from where they are applied. For example, fertilizer contributes to climate change and nutrient pollution of air and water. These negative impacts of intensification must be weighed against those of extensification to determine which poses the greatest threats or benefits in the long term. Furthermore, intensification does not necessarily prevent extensification. For example, as maize yields increased, maize became used in new ways, including as a biofuel, which has led to increased, rather than decreased, land conversion (Hertel *et al.* 2010). Under the land-sharing strategy, many protected areas might include sustainable agriculture and it might be easier to restore lands that were gently used for agriculture. The land-sharing strategy seeks to reduce the intensification of agriculture enough to conserve and restore biodiversity everywhere.

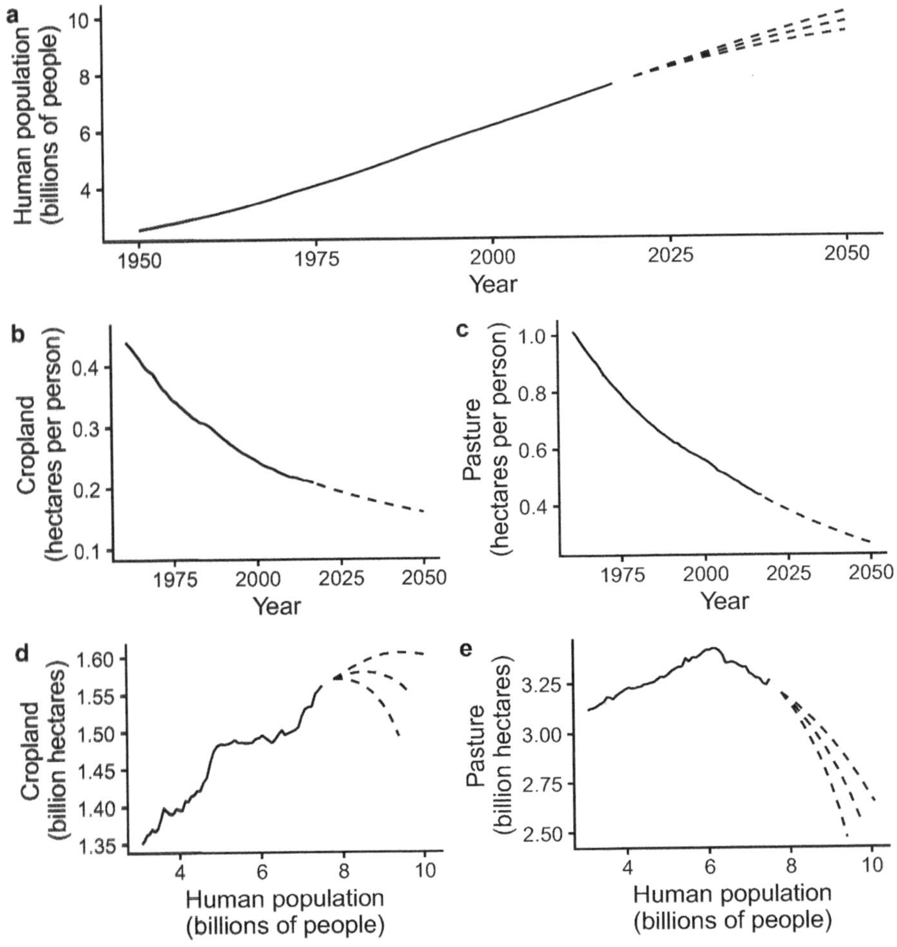

Figure 15.5. *Past and projected changes in the global human population size and the global extent of croplands and pastures*

NOTE ON FIGURE 15.5.– *In all panels, solid lines are observed data for past trends in global human population (United Nations 2019) and global agricultural areas (FAO 2020). Dashed lines in panel (a) are the UN's median projections for the future (middle line) and its 95% lower and upper prediction intervals. Dashed lines in (b) and (c) are extrapolations for the future of nonlinear curves fit to the past data. For dashed lines in (d) and (e), x-axis values match those for dashed lines in (a) and y-axis values are the product of values for dashed lines in (a) and those in (b) or (c), respectively.*

15.5. Conclusion

This chapter considered two of the most widely implemented conservation strategies, protected areas and restoration, and came to several conclusions:

– in the past decade, global goals for protected areas and restoration were not achieved;

– currently, new global goals for the future expansion of these strategies are being established;

– in the coming decades, there will likely be growing opportunities for expanding both protected areas and restoration, if current trends continue and the global footprint of agriculture shrinks;

– protected areas and restoration are only partly effective strategies: it is not yet possible to fully protect or restore biodiversity;

– protecting biodiversity can help protect ecosystem services, especially for the many ecosystem services that directly depend on biodiversity;

– increasing the diversity of restorations increases their efficacy.

Now is the time to expand and, even more importantly, improve the effectiveness of protected areas and restoration. New creative and ambitious ideas are needed to slow and eventually reverse the ongoing loss of biodiversity and ecosystem services. Inspiration will likely come from being in nature. Although biodiversity will face enormous challenges in the coming decades, I hope and trust that we will wisely use our time during this window of opportunity before threatened species vanish forever.

15.6. Acknowledgements

I thank Andy Hector and Akira S. Mori for their constructive feedback that helped to improve this chapter and acknowledge funding support from the National Science Foundation (DEB-1845334 and DEB-1831944).

15.7. References

Benayas, J.M.R., Newton, A.C., Diaz, A., Bullock, J.M. (2009). Enhancement of biodiversity and ecosystem services by ecological restoration: A meta-analysis. *Science,* 325(5944), 1121–1124.

Beyer, H.L., Venter, O., Grantham, H.S., Watson, J.E.M. (2020). Substantial losses in ecoregion intactness highlight urgency of globally coordinated action. *Conservation Letters,* 13(2), e12692.

Bullock, J.M., Pywell, R.F., Walker, K.J. (2007). Long-term enhancement of agricultural production by restoration of biodiversity. *Journal of Applied Ecology,* 44(1), 6–12.

Callicott, J.B. (1990). Whither conservation ethics? *Conservation Biology,* 4(1), 15–20.

CBD (2020a). Global biodiversity outlook 5. Report, Montreal.

CBD (2020b). Update of the zero draft of the Post-2020 Global Biodiversity Framework, Report.

Ceballos, G., Ehrlich, P.R., Barnosky, A.D. *et al.* (2015). Accelerated modern human-induced species losses: Entering the sixth mass extinction. *Science Advances,* 1(5).

Díaz, S., Settele, J., Brondízio, E.S. *et al.* (2019). Pervasive human-driven decline of life on Earth points to the need for transformative change. *Science,* 366(6471).

Dinerstein, E., Vynne, C., Sala, E. *et al.* (2019). A global deal for nature: Guiding principles, milestones, and targets. *Science Advances,* 5(4).

Dudley, N. (2008). *Guidelines for Applying Protected Area Management Categories,* IUCN, Gland.

Ewers, R.M. and Rodrigues, A.S.L. (2008). Estimates of reserve effectiveness are confounded by leakage. *Trends in Ecology and Evolution,* 23(3), 113–116.

FAO (2020). FAOSTAT. Food and Agriculture Organization of the United Nations [Online]. Available at: www.fao.org/faostat [Accessed December 2020].

Foley, J.A., Ramankutty, N., Brauman, K.A. *et al.* (2011). Solutions for a cultivated planet. *Nature,* 478(7369), 337–342.

Gray, C.L., Hill, S.L.L., Newbold, T. *et al.* (2016). Local biodiversity is higher inside than outside terrestrial protected areas worldwide. *Nature Communications,* 7(1), 12306.

Hector, A. and Bagchi, R. (2007). Biodiversity and ecosystem multifunctionality. *Nature,* 448(7150), 188–190.

Hertel, T.W., Golub, A.A., Jones, A.D. *et al.* (2010). Effects of US maize ethanol on global land use and greenhouse gas emissions: Estimating market-mediated responses. *BioScience,* 60(3), 223–231.

Huang, Y., Chen, Y., Castro-Izaguirre, N. *et al.* (2018). Impacts of species richness on productivity in a large-scale subtropical forest experiment. *Science,* 362(6410), 80–83.

Humphreys, A.M., Govaerts, R., Ficinski, S.Z., Nic Lughadha, E., Vorontsova, M.S. (2019). Global dataset shows geography and life form predict modern plant extinction and rediscovery. *Nature Ecology and Evolution,* 3(7), 1043–1047.

Hurtt, G.C., Chini, L., Sahajpal, R. *et al.* (2020). Harmonization of global land use change and management for the period 850–2100 (LUH2) for CMIP6. *Geosci. Model Dev.,* 13(11), 5425–5464.

IPBES (2018). The IPBES assessment report on land degradation and restoration. Report, Secretariat of the Intergovernmental Science-Policy Platform on Biodiversity and Ecosystem Services, Bonn, Germany.

Isbell, F., Calcagno, V., Hector, A. *et al.* (2011). High plant diversity is needed to maintain ecosystem services. *Nature,* 477(7363), 199–202.

Isbell, F., Tilman, D., Polasky, S., Loreau, M. (2015). The biodiversity-dependent ecosystem service debt. *Ecology Letters,* 18(2), 119–134.

Isbell, F., Tilman, D., Reich, P.B., Clark, A.T. (2019). Deficits of biodiversity and productivity linger a century after agricultural abandonment. *Nature Ecology and Evolution,* 3(11), 1533–1538.

Kimmerer, R.W. and Lake, F.K. (2001). The role of indigenous burning in land management. *Journal of Forestry,* 99(11), 36–41.

Mora, C. and Sale, P.F. (2011). Ongoing global biodiversity loss and the need to move beyond protected areas: A review of the technical and practical shortcomings of protected areas on land and sea. *Marine Ecology Progress Series*, 434, 251–266.

Moreno-Mateos, D., Barbier, E.B., Jones, P.C. *et al.* (2017). Anthropogenic ecosystem disturbance and the recovery debt. *Nature Communications,* 8, 14163.

Mori, A.S. (2018). Environmental controls on the causes and functional consequences of tree species diversity. *Journal of Ecology,* 106(1), 113–125.

Mori, A.S. (2020). Advancing nature-based approaches to address the biodiversity and climate emergency. *Ecology Letters,* 23(12), 1729–1732.

Naidoo, R., Balmford, A., Costanza, R. *et al.* (2008). Global mapping of ecosystem services and conservation priorities. *Proceedings of the National Academy of Sciences of the United States of America,* 105(28), 9495–9500.

Newbold, T., Hudson, L.N., Hill, S.L.L. *et al.* (2015). Global effects of land use on local terrestrial biodiversity. *Nature,* 520(7545), 45–50.

Nic Lughadha, E., Bachman, S.P., Leão, T.C.C. *et al.* (2020). Extinction risk and threats to plants and fungi. *Plants, People, Planet,* 2(5), 389–408.

Protected Planet (2020). Protected planet: the world database on protected areas (WDPA) [Online]. Available at: www.protectedplanet.net [Accessed January 2021].

Strassburg, B.B.N., Iribarrem, A., Beyer, H.L. *et al.* (2020). Global priority areas for ecosystem restoration. *Nature,* 586(7831), 724–729.

Takeuchi, K. (2010). Rebuilding the relationship between people and nature: The Satoyama Initiative. *Ecological Research,* 25(5), 891–897.

United Nations (2019). World population prospects 2019. United Nations, Department of Economic and Social Affairs, Population Division [Online]. Available at: https://population.un.org/wpp [Accessed January 2021].

Wu, J., Chen, B., Reynolds, G. *et al.* (2020). Monitoring tropical forest degradation and restoration with satellite remote sensing: A test using Sabah Biodiversity Experiment. In *Advances in Ecological Research*, Dumbrell, A.J., Turner, E.C., Fayle, T.M. (eds). Academic Press, Cambridge.

Yang, Y., Hobbie, S.E., Hernandez, R.R. *et al.* (2020). Restoring abandoned farmland to mitigate climate change on a full Earth. *One Earth,* 3(2), 176–186.

List of Authors

Joey R. BERNHARDT
Yale University
New Haven
USA

Seth BINDER
St. Olaf College
Northfield
Minnesota
USA

Anne CHAO
Institute of Statistics
National Tsing Hua University
Hsin-Chu
Taiwan

Robert K. COLWELL
University of Connecticut
Storrs
and
University of Colorado Museum of Natural History
Boulder
USA
and
Federal University of Goiás
Goiânia
Brazil

Laura E. DEE
University of Colorado
Boulder
USA

Jean-François GUÉGAN
UMR EPIA
INRAE–VetAgroSup
Saint-Genès-Champanelle
France

Yann HAUTIER
Ecology and Biodiversity Group
Utrecht University
The Netherlands

Meghan HAYDEN
University of Colorado
Boulder
USA

Andy HECTOR
University of Oxford
UK

Kirsten HENDERSON
Institut Natura e Teoria en Pirenèus
Surba
France

Forest ISBELL
University of Minnesota
St. Paul
USA

Lin JIANG
School of Biological Sciences
Georgia Institute of Technology
Atlanta
USA

Maiko KAGAMI
Yokohama National University
Japan

Kaitlin KIMMEL
Johns Hopkins University
Baltimore
USA

Michel LOREAU
Theoretical and Experimental Ecology Station
CNRS
Moulis
France

Takeshi MIKI
Research Center for Biodiversity Science
Ryukoku University
Otsu
Japan

Serge MORAND
UMR MIVEGEC
IRD–CNRS–Montpellier University
IRD Montpellier
France

Akira S. MORI
Yokohama National University
Japan

Mary I. O'CONNOR
Biodiversity Research Centre
University of British Columbia
Vancouver
Canada

Sarah K. ORTIZ
University of Texas at Austin
USA

Fons VAN DER PLAS
Plant Ecology and Nature Conservation Group
Wageningen University and Research
The Netherlands

Andy PURVIS
Natural History Museum
London
and
Department of Life Sciences
Imperial College London
Ascot
UK

Chase J. RAKOWSKI
University of Texas at Austin
USA

Benjamin ROCHE
UMR MIVEGEC
IRD–CNRS–Montpellier University
IRD Montpellier
France

Takehiro SASAKI
Yokohama National University
Japan

Bernhard SCHMID
University of Zurich
Switzerland

Christian SCHÖB
Institute of Agricultural Sciences
ETH Zurich
Switzerland

Keila STARK
Biodiversity Research Centre
University of British Columbia
Vancouver
Canada

Jacob USINOWICZ
Biodiversity Research Centre
University of British Columbia
Vancouver
Canada

Shaopeng WANG
Institute of Ecology
Peking University
Beijing
China

Matthew A. WHALEN
Biodiversity Research Centre
University of British Columbia
Vancouver
and
Hakai Institute
Heriot Bay
Canada

Amelia A. WOLF
University of Texas at Austin
USA

Qianna XU
School of Biological Sciences
Georgia Institute of Technology
Atlanta
USA

Moriaki YASUHARA
Swire Institute of Marine Science and
State Key Laboratory of Marine Pollution
University of Hong Kong
Pokfulam
Hong Kong

Index

E

F

G, H

I, M

N ,O

P

R

S

T

Printed and bound by CPI Group (UK) Ltd, Croydon, CR0 4YY
16/09/2026
14963172-0002